Photosynthetic Prokaryotes

BIOTECHNOLOGY HANDBOOKS

Series Editors: Tony Atkinson and Roger F. Sherwood

PHLS Centre for Applied Microbiology and Research
Division of Biotechnology
Salisbury, Wiltshire, England

Photosynthetic Prokaryotes

Edited by
Nicholas H. Mann and
Noel G. Carr

University of Warwick
Coventry, England

Plenum Press • New York and London

Library of Congress Cataloging-in-Publication Data

Photosynthetic prokaryotes / edited by Nicholas H. Mann and Noel G.
Carr.
       p.   cm. -- (Biotechnology handbooks ; v. 6)
   Includes bibliographical references and index.
   ISBN 0-306-43879-8
   1. Photosynthetic bacteria.   I. Mann, Nicholas H.   II. Carr, N.
G.   III. Series.
   QR88.5.P485   1992
   589.9--dc20
                                                     91-42399
                                                        CIP

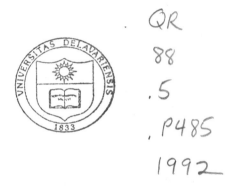

ISBN 0-306-43879-8

© 1992 Plenum Press, New York
A Division of Plenum Publishing Corporation
233 Spring Street, New York, N.Y. 10013

Printed in the United States of America

Contributors

C. Neil Hunter • Krebs Institute for Biomolecular Research, Department of Molecular Biology and Biotechnology, University of Sheffield, Sheffield S10 2TN, United Kingdom

Johannes F. Imhoff • Institut für Mikrobiologie und Biotechnologie, Rheinische Friedrich-Wilhelms-Universität, D-5300 Bonn, Germany

Nigel W. Kerby • Agricultural and Food Research Council Research Group on Cyanobacteria, and Department of Biological Sciences, University of Dundee, Dundee DD1 4HN, United Kingdom

Nicholas H. Mann • Department of Biological Sciences, University of Warwick, Coventry CV4 7AL, United Kingdom

John G. Ormerod • Biology Department, Division of Molecular Cell Biology, University of Oslo, Blindern, 0316 Oslo 3, Norway

Amos Richmond • Microalgal Biotechnology Laboratory, The Jacob Blaustein Institute for Desert Research, Ben-Gurion University at Sede-Boker, Israel 84993

Peter Rowell • Agricultural and Food Research Council Research Group on Cyanobacteria, and Department of Biological Sciences, University of Dundee, Dundee DD1 4HN, United Kingdom

Venetia A. Saunders • School of Natural Sciences, Liverpool Polytechnic, Liverpool L3 3AF, United Kingdom

Hans Utkilen • Department of Environmental Medicine, National Institute of Public Health, Geitmyrsveien 75, 0462 Oslo 4, Norway

Brian A. Whitton • Department of Biological Sciences, University of Durham, Durham DH1 3LE, United Kingdom

Preface

It is necessary to realize that the term "photosynthetic prokaryotes" encompasses the widest and most diverse grouping of bacteria, but in itself has no taxonomic or phylogenetic significance. It means those organisms, other than eukaryotes, which require, obligately or facultatively, light for growth. The recent application 16S rRNA sequencing to microbial phylogeny, associated mainly with the work of Woese, emphasizes the evolutionarily dispersed nature of purple and green photosynthetic bacteria as well as the rather coherent phylogenetic connections of the cyanobacteria. It is not surprising, therefore, that a volume such as this which seeks to give an introduction to this collection of organisms must be highly selective; accordingly, several important features are discussed only superficially, e.g., differentiation, life cycles, and biochemical aspects of nutrition. Rather, we have attempted to provide a description of the essential features which are common and those which are characteristic, e.g., the physiology of photosynthesis and ecological distribution. Bearing in mind the aim of the series, we have asked our authors to emphasize aspects of the importance of these organisms in nature and their industrial applications. As will be seen from the text, these organisms have an ancient history in "biotechnology," having been used as foodstuffs by several cultures, but their exploitation has been limited to their natural patterns and products of growth. Their considerable range of secondary metabolites is now being recognized, although not yet commercially exploited; to do this may require genetic as well as physiological manipulation. As we write this preface, the lay and scientific press are recognizing on the one hand the global importance of some species of photosynthetic prokaryotes in the fixation of carbon dioxide and, in contrast, the local consequences for both humans and animals of toxin-producing species in freshwater environments. We hope that this volume will provide a background to these organisms to a varied audience of biologists as well as to microbiologists and biotechnologists. Finally, we would like to thank the contributors for their excellently prepared chapters.

Nicholas H. Mann
Noel G. Carr

Coventry, England

vii

Contents

Chapter 3

Physiology of the Photosynthetic Prokaryotes 93

John G. Ormerod

Chapter 4

Genetics of the Photosynthetic Prokaryotes 121

Venetia A. Saunders

Chapter 5

Genetic Manipulation of Photosynthetic Prokaryotes

C. Neil Hunter and Nicholas H. Mann

Chapter 6

Mass Culture of Cyanobacteria 181

Amos Richmond

Chapter 7

Cyanobacterial Toxins 211

Hans Utkilen

Chapter 8

**Potential and Commercial Applications
for Photosynthetic Prokaryotes** 233

Nigel W. Kerby and Peter Rowell

Diversity, Ecology, and Taxonomy of the Cyanobacteria

1

BRIAN A. WHITTON

1. INTRODUCTION

The cyanobacteria are photosynthetic prokaryotes possessing the ability to synthesize chlorophyll a and at least one phycobilin pigment; typically water acts as the electron donor during photosynthesis, leading to the release of oxygen. They are by far the largest group of photosynthetic prokaryotes, as judged by their widespread occurrence, frequent abundance, and morphological diversity. Not only are they represented at the present day in most types of illuminated environment, except for those at lower pH values, but they have one of the longest geological records (Schopf and Walter, 1982). Much of the earth's original atmospheric oxygen was probably formed by organisms quite similar to modern cyanobacteria (Knoll, 1985) and they are still responsible for a considerable proportion of photosynthetic oxygen evolution in the oceans.

Their present-day success and presumably also that in earlier geological periods is no doubt a consequence of a number of features widespread in the group. The temperature optimum for many or most cyanobacteria is higher by at least several degrees than for most eukaryotic algae (Castenholz and Waterbury, 1989a). Many terrestrial forms tolerate high levels of ultraviolet irradiation (see Section 4.2), whereas the success of many planktonic forms is favored by their ability to utilize light for photosynthesis efficiently at low photon flux densities (van Liere and Walsby, 1982; Glover, 1986). Tolerance of desiccation and water stress is widespread (Whitton, 1987b) and cyanobacteria are among the most successful organisms in highly saline environments (Bauld, 1981; Borowitzka, 1986). Free sulfide is tolerated by some cyanobacteria at levels much higher than those tolerated by most eukaryotic algae (Padan and Cohen, 1982) and H_2S

BRIAN A. WHITTON • Department of Biological Sciences, University of Durham, Durham DH1 3LE, United Kingdom.

Photosynthetic Prokaryotes, edited by Nicholas H. Mann and Noel G. Carr. Plenum Press, New York, 1992.

is sometimes utilized as the electron donor during photosynthesis (Cohen *et al.*, 1986). Mechanisms exist among cyanobacteria that allow photosynthetic CO_2 reduction to proceed efficiently even at very low levels of inorganic carbon (Pierce and Omata, 1988). The ability of many species to fix atmospheric nitrogen provides a competitive advantage where levels of combined nitrogen are low (Howarth *et al.*, 1988a,b; Paerl, 1990); the ability to form gas vacuoles in some common freshwater plankton species and the marine *Trichodesmium* (van Liere and Walsby, 1982), and hence increase buoyancy, can provide an advantage in waters where the rate of vertical mixing of the water column is relatively low.

Nucleotide base sequence data of 16S and 5S rRNA suggest that the cyanobacteria are a phylogenetically coherent group within the Gram-negative eubacteria (Fox *et al.*, 1980; Woese, 1987), though authors differ as to whether the genera *Prochloron* and *Prochlorothrix* should be included. The term cyanobacterium is used here in the sense of Castenholz and Waterbury (1989a) to exclude those forms which possess chlorophyll *b* and lack phycobilins (Section 2), the Prochlorales (Lewin, 1989); a brief account of them is, however, included in Section 2.5.

In size the cyanobacteria range from subspherical cells less than 1 μm in diameter to trichomes well over 100 μm in diameter. The cells of the latter are always much shorter than wide, but nevertheless cell volume ranges over more than five orders of magnitude. Some species are capable of differentiating several distinct types of cell, and forms which combine these various cell types with various patterns of branching are morphologically the most complex prokaryotes known.

Although apparently all cyanobacteria share certain features such as carboxysomes and, under some environmental conditions, polyphosphate bodies, overall there are considerable ultrastructural differences and a relatively wide range of cell inclusions (Jensen, 1985). *Gloeobacter violaceus* possesses only a cytoplasmic membrane, which includes both photosynthetic and respiratory functions; all other cyanobacteria possess thylakoids internal to the cytoplasmic membrane, but the quantitative and qualitative features of the thylakoids differ among species and according to environmental conditions.

The morphological diversity within the group and the ease with which the larger forms can be viewed under the light microscope have encouraged botanists to describe well over 2000 species. Scepticism concerning this taxonomic approach led to the proposal that the organisms should be classified according to the principles of the Bacteriological Code rather than the Botanical Code (Stanier *et al.*, 1978). Simultaneously several leading bacteriologists advocated that the organisms should be known as blue-green bacteria or cyanobacteria, rather than blue-green algae, and, in spite of eloquent advocacy to the contrary by R. A. Lewin (e.g., Lewin, 1976), cyanobacteria is the name now well established in the literature. This has helped to bring the cyanobacteria to the forefront of modern research, but at the same time has probably led to the neglect of useful older literature.

In view of the morphological diversity of the cyanobacteria, an obvious question is the extent to which this is matched by physiological, biochemical, and molecular diversity. The majority of experimental studies have focused on a limited number of genera and species, and relatively few studies have set out deliberately to investigate diversity at these other organizational levels. Whereas most physiologists and biochemists realize this problem in theory, in practice statements often appear in the literature which fail to take into account the less well-studied organisms.

It is not just with the minutiae that the reader of a review on cyanobacteria needs to be cautious. The history of their research has been one of oversight of important ecological and physiological phenomena. In some cases earlier data were overlooked or their significance not appreciated; sometimes the new insight was sufficiently upsetting to cause a transient period of scepticism. Major steps in understanding have included the realization that the heterocyst is a site of nitrogen fixation (Fay *et al.*, 1968), that some nonheterocystous species also fix nitrogen (Wyatt and Silvey, 1969), that H_2S is an alternative hydrogen donor in some strains (Cohen *et al.*, 1975), that very small-celled (picoplankton) cyanobacteria are a major component of the oceanic phytoplankton (Johnson and Sieburth, 1979; Waterbury *et al.*, 1979), and that they are similarly abundant in many freshwaters (Stockner and Antia, 1986). However, the reader may also need scepticism when reviewing experimental data. For instance, there have been repeated claims—the latest in 1989—that some cyanobacteria growing heterotrophically in the dark lose their chlorophyll, but regain it when reexposed to the light, much as do most angiosperms. Critical study of these claims has led to the general belief that they are untrue, although there is no way the casual reader could be aware of this; further, until someone can show a convincing reason for the inability to lose and resynthesize chlorophyll, there must always remain the doubt that one day a strain may be found where it really does occur.

The aim of this chapter is to give an overview of the morphological and ecological diversity within the cyanobacteria and how this may be explained in terms of their physiology, biochemistry, and molecular biology. Section 5 discusses the successes and limitations of the approaches which have been adopted to classify this diversity. Aspects treated in other chapters, such as photosynthesis and genetics, are mentioned only briefly here.

2. MORPHOLOGICAL DIVERSITY

2.1. Introduction

Castenholz and Waterbury (1989a,b) have reviewed the morphological features of the cyanobacteria and provided a diagnostic key to five "subsections" or orders, based on the more obvious features. These are shown here in Table I as a list of features rather than a key. The groups bear some resemblance to orders given in botanical floras; they also correspond closely to the Groups I–V recognized by Rippka *et al.* (1979), which are now widely used in the physiological and biochemical literature. The orders were recognized for practical convenience, insufficient data being available to produce a more objective classification (Section 5). However a study of Giovannoni *et al.* (1988), in which the sequences of 800–900 consecutive nucleotides of 16S rRNA were compared in over 30 strains, indicates that the strains chosen from Pleurocapsales, Nostocales, and Stigonematales were each phylogenetically coherent, but that the remainder did not cluster according to the groupings into Chroococcales and Oscillatoriales (Castenholz and Waterbury, 1989b).

In view of this recent published review on cyanobacterial morphology, the emphasis here is on aspects of particular relevance to the understanding of cyanobacterial ecology. To what extent is it possible to interpret morphological and physiological features as functional adaptations to particular environments? The account is inevitably highly selective.

2.2. Growth Rate

Various authors have suggested apparent relationships connecting size, shape, and growth rate. However, the situation is complicated by the fact that many cyanobacteria grow in situations differing considerably from those favoring their optimum growth rate. They are successful because they compete effectively with other photosynthetic organisms in environments stressed by limitations in water supply or nutrients like C, N, or P; they encounter conditions permitting optimum growth rates only rarely. As many of the morphological features shown by cyanobacteria appear to be related to their occurrence in environments subject to particular nutrient limitations (see below), it seems likely that the morphologically more complex organisms are associated with the more specialized environments. It is therefore suggested that, in general, the more complex the organism morphologically, the greater is likely to be the difference between its typical growth rate in nature and its growth rate under optimum conditions. Any attempt to relate growth rate with size and shape needs to make clear whether the discussion refers to the rate under optimum conditions, and hence almost always in laboratory studies, or under typical field conditions.

Foy (1980) found a significant positive correlation between growth rate and surface/volume ratio among planktonic *Anabaena*, *Aphanizomenon*, and *Oscillatoria* in laboratory culture, suggesting that cyanobacteria are similar to other phytoplankton in that small cells tend to grow faster than large cells (Paasche, 1960; Fogg, 1975). Part of the explanation almost certainly lies in the fact that strains with a high ratio have higher potential nutrient uptake rates (Gibson and Smith, 1982). Another possibility is that a strain

Table I. List of Features in Orders of Cyanobacteria Recognized by Castenholz and Waterbury (1989b)

Nonfilamentous

Order Chroococcales Wettstein 1924, emend. Rippka *et al.* 1979

   Unicellular or nonfilamentous aggregates of cells held together by outer wall or gel-like matrix; binary division in one, two, or three planes, symmetric or asymmetric; or by budding

Order Pleurocapsales Geitler 1925, emend. Waterbury and Stanier 1978

   Unicellular or nonfilamentous aggregates of cells held together by outer wall or gel-like matrix; reproduction by internal multiple fissions with production of daughter cells smaller than parent; or by a mixture of multiple fission and binary fission

Filamentous

Order Oscillatoriales (see Castenholz, 1989a)

   Binary division in one plane giving rise to 1-seriate trichomes, though sometimes with "false" branches; trichomes do not form heterocysts (or, apparently, akinetes; see Section 2.3)

Order Nostocales (see Castenholz, 1989b)

   Binary division in one plane giving rise to 1-seriate trichomes, though sometimes with "false" branches; one or more cells per trichome differentiate into a heterocyst, at least when concentration of combined nitrogen is low; some also produce akinetes

Order Stigonematales (see Castenholz, 1989c)

   Binary division periodically or commonly in more than one plane, giving rise to multiseriate trichomes or trichomes with true branches or both; apparently always possess ability to form heterocysts

---

with larger cell size (i.e., lower surface/volume ratio) has greater capacity for internal differentiation and is therefore more successful in dealing with a wider range of growth conditions. In a comparison of the growth rate of two *Oscillatoria* strains grown in continuous light or various diel light–dark cycles, Foy and Smith (1980) found that the larger-celled *Oscillatoria agardhii* (diameter 3.54 μm) competed successfully with the narrow-celled *Oscillatoria redekei* (diameter 2.16 μm) when the light period was 6 hr or less. The authors proposed that its larger cell volume gave *O. agardhii* a greater capacity to store carbohydrate during the light period and thus better ability to provide energy for growth during the dark period.

It has also been suggested that unicellular cyanobacteria tend to have more rapid growth rates than filamentous ones (Carr and Wyman, 1986). Based on laboratory studies, the lowest mean generation time reported for a unicellular cyanobacterium is 2.1 hr at 41°C [*Anacystis nidulans** (Kratz

---

Use of strain names. The strain used by Kratz and Myers (1955) was named *Anacystis nidulans* according to the conventions of Drouet (Section 5). It has since become the most widely used cyanobacterial strain for research. Cultures derived from this strain are now held in culture collections under various names: as *Anacystis nidulans* UTEX 625; Tx-20; as *Synechococcus* or *S. leopoliensis* ATCC 27144; CCAP 1405/1; PCC 6301; SAUG or SAG 1402-1. At least one

and Myers, 1955)], whereas the lowest mean generation time reported for a filamentous cyanobacterium is 3.6 hr at 42°C [*Anabaena* sp. in the presence of combined nitrogen (Stacey *et al.*, 1977)]. There are, however, many examples of much longer generation times for both unicellular and filamentous cyanobacteria, even when incubated under optimum growth conditions. The comparison is also complicated by the fact that critical studies have not been made on the larger unicellular forms, so any differences in growth rate between unicellular and filamentous forms may simply reflect differences in cell size (and hence surface/volume ratio) rather than overall morphology.

It is also unclear whether unicellular forms tend to grow faster than filamentous forms in nature. Ideally the comparison should be made for organisms growing in the same microhabitat. If the comparison is applied to environments dominated by unicellular forms (e.g., oligotrophic lakes and the open ocean) versus those dominated by filamentous forms (e.g., eutrophic lakes, rice fields, and some tropical intertidal regions), then factors other than the growth rate may determine why a particular growth form is more successful. The comment about growth rates of unicellular and filamentous forms should at present be regarded as merely a useful hypothesis.

The size (volume) of vegetative cells of a particular strain may vary markedly, sometimes by an order of magnitude or more, but the relationship between size and growth rate within the strain is not clear-cut. In batch culture the cells of many species are smaller during the early growth stages, when the cultures are growing fastest, and become larger in old, slow-growing cultures, as the cells deposit one or more types of storage granule; this occurs, for instance, in most species of *Anabaena* and *Nostoc*. However, if, instead of making the comparison between cells at early and late growth stages, the comparison is made between cells at an early culture stage grown under different conditions, the converse relationship may hold. In young cultures of *Anacystis nidulans* (=*Synechococcus* sp. PCC 6301) (Mann and Carr, 1974) and *Anabaena cylindrica* (C. K. Leach and N. G.

---

other strain, termed *Anacystis nidulans* R2, has been used widely (= *Synechococcus* PCC 7942). In the present account, other strains of *Synechococcus* and *Synechocystis* are also given the culture numbers used by the authors of the papers. Strains belonging to other genera are mostly quoted without the culture number.

A few other points need to be borne in mind when comparing accounts in the literature. The Kratz and Myers strain of *Anacystis nidulans* adapts quite quickly to some environmental changes, so it seems likely that genetic differences occur among strains held in different collections. Authors sometimes quote a culture collection number, even though the culture has been derived from another source and the director of the particular culture collection may not permit this. Until general agreement is reached on an unambiguous way to refer to cyanobacterial strains, authors of papers should be encouraged to give full strain histories. Rippka (1988) gives such details for a number of widely used research strains.

Carr, unpublished; quoted by Carr and Wyman, 1986) grown at different light flux values, the cells with more light grew faster and were larger.

2.3. Examples of Cell Types

In addition to the typical vegetative cell, other types of cell occur in some species (Fig. 1). These include cells where the change is irreversible—heterocysts (usually), necridial cells, and colorless hair cells—and cells which ultimately develop again into one or more vegetative cells—various types of spore (akinete, exospore, endospore or baeocyte, pseudohormocyst), the modified terminal cell of some trichomes, and the cells of hormogonia.

2.3.1. Heterocysts

The heterocyst, a thick-walled cell with a refractive polar nodule at one or both ends, is unique to the group, occurring in two orders (Table I). It is well established that this is a site for nitrogen fixation (Wolk, 1982), and the only site in at least some species with the ability to form heterocysts (Peterson and Wolk, 1978). Heterocysts usually develop only after the concentration of combined nitrogen in the surrounding medium has been lowered (Castenholz and Waterbury, 1989a), though the responses to nitrate and ammonia may differ (Kerby *et al.*, 1989). In *Anabaena* and some other genera heterocysts occur at regular intervals along the trichome, whereas in *Calothrix* there is usually only one heterocyst at the base of the trichome. The frequency of heterocysts in a trichome may reflect not only the nitrogen status of the trichome, but also its iron status, an element important in the functioning of nitrogenase. Deficiency of iron leads to an increased frequency of heterocysts along *Anabaena* trichomes (B. A. Whitton, unpublished observations) and to the replacement of the original heterocyst with a new, and ultrastructurally simpler heterocyst in *Calothrix parietina* (Douglas *et al.*, 1986).

The taxonomic literature reveals that the heterocysts of various species may differ considerably in their morphology, but most studies on ultrastructure, development, and function have been made on a limited number of species of *Anabaena*. Although heterocysts are usually stated to be pale in color (e.g., Castenholz and Waterbury, 1989a), examples frequently occur where they appear more intensely pigmented than adjacent vegetative cells, either bright green or, in some Rivulariaceae, blue-green (Fritsch, 1945; Whitton, 1987a); they are, however, apparently never pink, even in strains forming high levels of phycoerythin in the vegetative cells (B. A. Whitton, unpublished observations). In at least some cases the coloration is associated with the reappearance of phycobiliprotein which had been lost during differentiation [e.g., *Anabaena* sp. (Thomas, 1972)]. It has been

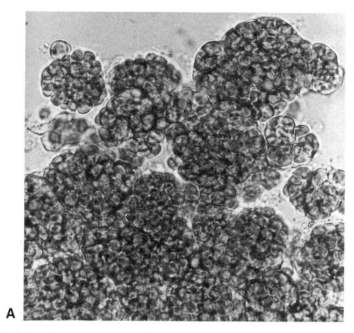

Figure 1. Light micrographs of materials from enrichment cultures of field populations. (A) Aseriate stage of *Nostoc;* (B) *Tolypothrix byssoidea,* showing intercalary heterocyst and false branching; the epiphyte is a narrow form of *Plectonema;* (C) *Calothrix parietina,* showing basal heterocyst, tapering, and part of colorless multicellular hair.

suggested (Whitton, 1987a) that the blue-green heterocysts of rice-field Rivulariaceae may be related to fluctuating levels of combined nitrogen in the environment. [The use of family names such as Rivulariaceae follows Geitler (1932).]

The formation of a mature heterocyst is an irreversible step in most strains, the explanation for which has been shown very clearly at the molecular level for *Anabaena* sp. (Golden *et al.,* 1985; Golden and Wiest, 1988). Nevertheless there are a relatively large number of records for heterocyst germination, indicating that the process may be reversible in some strains. The majority of records are for Rivulariaceae, though it has also been reported for *Anabaena, Brachytrichia, Nostoc,* and *Tolypothrix* (Desikachary, 1946); it has been described in most detail for *Gloeotrichia ghosei* (Singh and Tiwari, 1970). Wolk (1982) questioned whether such heterocysts were fully mature, perhaps having reached only a late proheterocyst stage, but some of the published figures seem to show the germination of a mature heterocyst. The problem will only be resolved unequivocably by showing whether or not examples occur of heterocysts with active nitrogenase which are capable of subsequent dediffentiation.

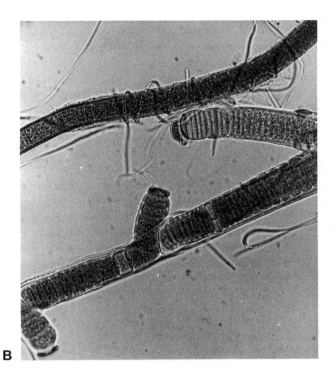

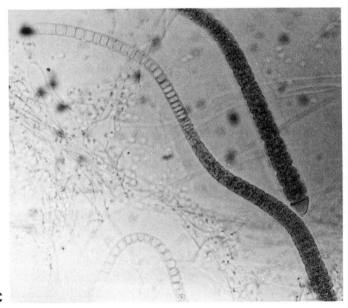

Figure 1. (*Cont.*)

2.3.2. Necridial Cell

A clear example of irreversible change is the breakdown of a particular cell leading to the release of a hormogonium (Drews and Weckesser, 1982). The development of a "sacrificial" necridial cell occurred in all three members of the Oscillatoriaceae studied by Lamont (1969); it was probably the only mechanism of hormogonium production in *Microcoleus vaginatus* and *Oscillatoria chalybea*, but *Schizothrix calcicola* showed both transcellular and intercellular breakage. The release of a hormogonium in many other genera also involves the formation of necridium. Anagnostidis and Komárek (1988) suggest that whether or not a necridium occurs during trichome division is an important taxonomic feature in Oscillatoriales. In several Rivulariaceae the cell of the released hormogonium which was adjacent to the necridium has been shown eventually to differentiate into a basal (one-pored) heterocyst (Whitton, 1987b). A somewhat similar process occurs when the hormogonia of some *Tolypothrix* species are partitioned into several segments (Hoffman, 1988b); in contrast, a necridium forms in the middle of the hormogonium in some other Scytonemataceae, with an intercalary (two-pored) heterocyst developing in the middle of the daughter trichomes.

It is not clear why a necridium should often form as part of the process of hormogonium formation, but not as part of the normal separation process when trichomes divide into two. Several eukaryotic algae (*Microspora, Tribonema*) sacrifice one cell each time a filament divides. Possible explanations for the cyanobacterial necridium are that it is part of a controlled sequence of morphogenetic events leading to the formation of a hormogonium, that it provides a mechanical advantage during the release of the hormogonium, or that the accompanying cell lysis makes it easy to release molecules such as DNA into the environment.

2.3.3. Hair Cells

These provide another example of irreversible change, where the terminal part of a trichome in some or all species of *Calothrix* and other genera in the Rivulariaceae, *Mastigocoleus, Brachytrichia,* and *Kyrtuthrix* metamorphoses into a long, colorless, multicellular hair. The hair is usually a localized site of cell-bound ("surface") phosphatase activity (Whitton, 1987b; Mahasneh *et al.,* 1990), a phenomenon with a parallel in three phyla of eukaryotic algae, the greens, browns, and reds (Whitton, 1988).

2.3.4. Akinetes

Akinetes have been recorded for about one-third of the species forming heterocysts, in two nonheterocystous filamentous genera, *Isocystis* and

Raphidiopsis [which do not fit within the orders recognized by Castenholz and Waterbury (1989b)], and a few examples of Chroococcales within the genera *Gloeocapsa* (especially *G. magma*), *Entophysalis*, and *Siphonema* (Whitton, 1987b). Some strains can still form akinetes when grown on a source of combined nitrogen (Sarma and Khattar, 1986). Unfortunately, no studies have been made to establish whether or not any or all of the non-heterocystous examples are nitrogen-fixers.

Most ecological studies of akinete formation in temperate lakes have shown that the process occurs during the latter part of the growth season (e.g., Cmiech *et al.*, 1984; Watanabe *et al.*, 1985). However, May (1989) found that the maximum period of akinete formation in *Anabaena circinalis* in two New South Wales (Australia) dams occurred soon after its seasonal appearance and then decreased rapidly, in spite of the persistence of the population in the plankton.

Previous reviews have considered the physiological factors leading to akinete formation (Adams and Carr, 1981; Nichols and Adams, 1982; Herdman, 1987), but their interpretation of the factors involved differs from that presented here, in that they largely overlook the role of phosphate. Among 22 examples of akinete formation in heterocystous cyanobacteria described in the literature (Whitton, 1987b), in at least 13 of them the process occurred as a response to increasing phosphorus deficiency. Pandey and Kashyap (1987) suggested that the induction of alkaline phosphatase activity may be one of the important events preceding sporulation in *Anabaena doliolum*, although the former commenced long before the latter. Where phosphorus deficiency does not appear to be closely involved, limitation of energy supply, whether in the form of light or a utilizable carbohydrate source, may be a major factor in the induction of akinete formation (Nichols and Adams, 1982). When this occurs, it seems possible that the transfer of fixed carbon from vegetative cells to the heterocyst may increase in importance. Appreciable nitrogenase activity occurred preceding and during akinete formation in a population of *Aphanizomenon flos-aquae* (Rother and Fay, 1979); in this and *Nodularia spumigena* (Pandey and Talpasayi, 1980), both examples where the akinete develops distant from the heterocyst, there was a reduced C:N ratio at the time of akinete formation. Analyses of the composition of mature akinetes confirm observations from light and electron microscopy: the abundance of cyanophycin granules (Sutherland *et al.*, 1979) and the absence of polyphosphate granules (Simon and Weathers, 1976). Evidence from *Anabaena fertilissima* (Reddy, 1983) indicates that phosphate is probably not stored at high levels in any other way.

There is apparently a relationship between the location of an akinete with respect to the heterocyst and whether or not sporulation occurs as a direct response to phosphorus deficiency. At least 15 of 19 available records (Whitton, 1987b) show that the akinete forms next to a heterocyst if sporulation occurs as a response to phosphorus deficiency, but at the mid-point between two heterocysts (centrifugal versus centripetal development) if it occurs as a response to other factors.

The following hypothesis is proposed to help the reader consider akinete formation in heterocystous species; it is compatible with the majority of observations in the literature. The akinete is a storage cell especially rich in N. In order to synthesize large amounts of cyanophycin, development has to commence at a stage when the trichome can still provide an adequate N supply, either from nitrogenase activity in heterocysts or by mobilizing N stores in vegetative cells. It is suggested that akinete induction usually commences when nitrogenase activity or some other process influencing N availability is stressed by some factor such as lack of phosphate, reduced light energy, or other essential factor, perhaps depending on species. This leads to the diversion of fixed N from growth of vegetative cells to growth of akinetes.

In view of the absence of stored phosphate in akinetes, studies are needed on the phosphorus status of strains where increasing phosphate deficiency is apparently not the immediate factor leading to akinete formation. This absence of stored P in the akinete suggests that germination is likely to occur in microenvironments rich in available phosphate; such conditions occur at the sediment surface of many lakes and ponds, where akinete germination of otherwise planktonic species occurs. The high content of stored N may be needed because of the time required to form a functional heterocyst following akinete germination.

Although the akinetes of *Cylindrospermum* are similar to those of other genera in the abundance of cyanophycin and lack of polyphosphate bodies (Jensen and Clark, 1969), the circumstances leading to akinete formation differ. When an exponentially growing culture of *Cylindrospermum licheniforme* was inoculated into a cell-free supernatant from an akinete-containing culture, sporulation occurred almost immediately (Fisher and Wolk, 1976). The akinete-inducing substance was isolated and its structure partially determined (Hirosawa and Wolk, 1979). Heterocysts, but not active nitrogen fixation, are essential for akinete formation (Van de Water and Simon, 1982). The effect of the akinete-inducing substance is that once sporulation commences in a culture, the process spreads rapidly through the population. A similar phenomenon has been observed during daily observations on colonies of *C. licheniforme* and other *Cylindrospermum* species growing on soils (B. A. Whitton, unpublished observations). *C. licheniforme* forms a characteristic bright-green flat gelatinous thallus several centimeters in diameter, with a sharply delimited outer boundary; akinete formation apparently first develops in trichomes toward the middle of the colony. If extracellular akinete-inducing agents play a similar role in nature, they would have the effect of limiting the size to which a colony develops. The ecological explanation of why this should occur is not obvious.

2.3.5. Hormogonia

These have been defined as short, motile chains of rather uniform cells (Desikachary, 1959). However, the term has been used in quite different ways by various authors, often without a clear definition. Much of the older taxonomic literature treated any motile filament as a hormogonium: in species of *Oscillatoria,* where motility is the normal condition, a hormogonium is thus essentially the same as a young filament. However, in many heterocystous forms the hormogonium is quite distinct, typically with 5–15 cell chains whose diameter is usually less than that of the vegetative trichome, sometimes markedly so. Rippka *et al.* (1979) defined the hormogonium (of heterocystous forms) as being distinguishable from the parent trichome by cell size, cell shape, gas vacuolation, or absence of heterocysts, even when grown without a source of combined nitrogen; they included motile and nonmotile filaments.

Recent authors tend to restrict the term hormogonium to structures whose formation involves a distinct morphogenetic step, suggesting that the activation of a particular gene may be involved. Unlike their parental cells, hormogonial cells contain few copies of the genome (Herdman and Rippka, 1988). However, it is still unclear whether it will prove possible to give an unambiguous definition separating hormogonia from other types of young trichomes.

The presence or absence of hormogonia was used to redefine the generic limits of *Nostoc* and *Anabaena* by Rippka *et al.* (1979); however they regarded the short, mobile trichomes of *Scytonema* to be hormogonia, even if the width is similar to that of a normal trichome. These authors also reported that (laboratory) strains of *Nostoc* spp. with nonmotile hormogonia are not uncommon; this contrasts with the study of Martin and Wyatt (1974), who reported that all 75 strains grown by them in liquid culture were capable of producing motile hormogonia, although strains forming a heavy "slime" were incapable of doing so when grown on agar plates. Lack of mobility also seems rather unlikely in the case of field populations of *Nostoc,* so the topic requires rechecking.

It has been demonstrated (Fattom and Shilo, 1984; Shilo, 1989) for several benthic cyanobacteria that the transition from an attached filament to a hormogonium involves a change from being hydrophobic to hydrophilic. The nonmotile parent trichomes of *Calothrix* PCC 7601 lack fimbriae, whereas the hormogonia are fimbriated and motile (Herdman and Rippka, 1988). In several genera (e.g., *Calothrix, Nostoc, Tolypothrix*) the hormogonia of some species or strains are gas-vacuolate (Canabaeus, 1929:

Walsby, 1972), even though the mature trichomes are not so. The capability for movement and the possession of gas vacuoles have been suggested as alternatives (Herdman and Rippka, 1988), but gas-vacuolate hormogonia can exhibit motility when on a moist agar surface (Whitton, 1987b). In the morphologically more complex cyanobacteria, hormogonia develop in specific regions of the parent thallus, such as at the narrow, tapered end of the trichome or below the hair in Rivulariaceae (Whitton, 1987b) or at the end of lateral branches in some Stigonematales.

The differentiation and liberation of hormogonia appears to be a timed process associated with environmental conditions such as phosphorus repletion (Castenholz and Waterbury, 1989a) or with particular stages of a morphological cycle (Lazaroff, 1973). The addition of phosphate to phosphate-limited material leads to hormogonia formation in many cyanobacteria, including almost all Rivulariaceae (B. A. Whitton, unpublished observations). The hormogonia of *Calothrix* D764 (Islam and Whitton, in press) contain both cyanophycin and polyphosphate granules, in contrast to the akinetes discussed above; it seems likely that this is the typical situation for those hormogonia developing as a response to phosphorus repletion.

Herdman and Rippka (1988) list examples where simple dilution leads to release of hormogonia; in *Nostoc* sp. PCC 7119 an inhibitory compound is formed in late exponential to stationary phase, which suppresses hormogonia formation in a dose-dependent manner. Several different effects of light on hormogonium formation have also been reported, mostly for strains of *Nostoc* spp. (Wyman and Fay, 1987), including stimulation by red light and inhibition by green light (Robinson and Miller, 1970). The effect of light quality is independent of the ability to adapt chromatically (Herdman and Rippka, 1988). Hormogonium formation may not be regulated by light quality *per se*, but may result from a transient metabolic change associated with shift in light quality (Herdman and Rippka, 1988). Transcriptional studies suggest that glutamine synthetase may exert a regulatory function in hormogonium differentiation in two *Calothrix* strains (Herdman and Rippka, 1988).

There are a number of similarities between hormogonia and baeocytes (Herdman and Rippka, 1988), the latter being small spherical cells produced through multiple fission of a vegetative cell and released through the fibrous outer wall of the parent cell (Waterbury, 1989).

2.4. Colonies and Associations

Some planktonic cyanobacteria and probably the majority of non-planktonic forms typically occur in nature as groups of many filaments. This is often the case even with highly motile forms such as *Phormidium*. In some cases individual cells (e.g., *Aphanothece stagnina*) or filaments

(*Gloeotrichia*) are arranged in distinct colonies, the largest of which reach 22 cm diameter in *Nostoc pruniforme* (Dodds and Castenholz, 1987). The colonies may arise from division of a single trichome (probably the normal situation in *Nostoc*) or from aggregation of many trichomes [probably the normal field situation with *Rivularia* (Whitton, 1987a)]. It has proved difficult to recreate typical colony formation in the laboratory. This is perhaps due largely to a lack of understanding of environmental requirements, because culture collections of colony-forming genera often include a few strains which go some way toward forming colonies in the laboratory. Another possibility is that laboratory culture may select for strains lacking the ability to form colonies. An axenic strain of *Nostoc commune* deposited in the former Cambridge Culture Collection of Algae and Protozoa formed typical colonies for about 1 year, but subsequently lost this ability.

Bacteria were reported to be essential for normal colony formation in *Nostoc sphaericum* (Schwabe and Mollenhauer, 1967). However, bacteria appear to be sparse inside healthy colonies of *Nostoc commune* (M. Potts, personal communication), perhaps resulting from inhibitory activities such as reported for isolates of *Nostoc muscorum* (Bloor and England, 1989; Cano *et al.*, 1990). The presence of other cyanobacteria and small eukaryotic algae is frequent in the mucilage of some colonial genera (Pankow, 1986), but not *Nostoc*. However, colonies of *Nostoc parmelioides*, an attached form of streams, have been reported from several countries often to contain a dipteran larva. Upon infection the colonies change shape, developing small lobes. Although the colonies are consumed from the inside out, the ear-shaped colonies infected with *Cricotopus* in an Oregon creek had an increased photosynthetic rate compared with uninfected ones (Ward *et al.*, 1985).

2.5. Prochlorales

The Prochlorales were defined by Lewin (1989) as unicellular or filamentous, branched or unbranched prokarotes resembling cyanobacteria, from which they differ in forming chlorophylls *a* and *b* and lacking accessory red or blue bilin pigments. However, the only examples listed by Lewin are unicellular forms (*Prochloron*), which exist almost entirely as extracellular symbionts of colonial ascidians (Pardy, 1989), and "*Prochlorothrix hollandica,*" an unbranched trichome of indefinite length (Burger-Wiersma *et al.*, 1986; Mathijs *et al.*, 1989). A formal description of *P. hollandica* has now been provided according to the conventions of the Bacteriological Code (Burger-Wiersma *et al.*, 1989). It has trichomes which lack sheaths, motility, or differentiated cells, but possesses gas vacuoles at the ends of the cells. Lewin's (1989) definition of the order to include branched forms seems prepared for organisms yet to be discovered.

Interest in the Prochlorales has increased considerably as a result of the recent account (Chisholm *et al.*, 1988) of a new group of organisms

Figure 2. Natural populations of cyanobacteria. (A) *Nostoc commune* in depression in limestone of Aldabra Atoll, showing partially dried colonies; (B) *Rivularia* colonies in small, highly calcareous stream, Yorkshire, England; the highly calcified colonies are intermittently submerged or (as in photo) out of the water; (C) *Scytonema* mats in intertidal zone of lagoon of Aldabra Atoll.

Figure 2. (*Cont.*)

falling within Lewin's definition of the order. These are free-living cells from deep euphotic oceanic waters, which are retained by a 0.6-μm filter, but mostly pass through a 0.8-μm filter. Like the other prochlorophytes, they possess chlorophyll *b* and lack phycobilins, but differ from them in possessing α-carotene rather than β-carotene and a divinyl chlorophyll *a*-like pigment as the dominant chlorophyll.

3. ECOLOGICAL DISTRIBUTION

Cyanobacteria occur in such a wide variety of habitats that it is at first sight difficult to generalize about them. Wherever they are especially successful, however, it is usually clear that this is due to one or more of the physiological features outlined at the beginning of Section 1, such as tolerance of high temperatures, high UV irradiation, desiccation, and free sulfide, and the abilities to utilize low light flux and CO_2 concentration and to fix N_2. Cyanobacteria play a major role in oceanic waters, mostly as picoplankton (<2 μm), but also as blooms of *Trichodesmium* (Fogg, 1987), which are responsible for about one-quarter of the world's total oceanic nitrogen fixation (Capone and Carpenter, 1982). In freshwater and terrestrial environments they are absent below pH 4.0 and overall their frequency and abundance tend to increase with increasing pH value; they are often conspicuous in calcareous regions (Fig. 2). They are important in a number of environments, which are extreme in the sense that species number (of any group) is very low. They also play an important role in symbiotic associations, examples of which are found with most eukaryotic phyla, both plant and animal.

Picoplankton cyanobacteria appear to be almost ubiquitous in lakes and ponds (Stockner and Antia, 1986; Stockner, 1988; Stockner and Short-reed, 1988; Hawley and Whitton, 1991b), though critical studies are still required on waters at lower pH values and in saline environments. Surveys do not always distinguish between prokaryotic and eukaryotic picophytoplankton, but it seems clear that in general picoplankton cyanobacteria increase in importance in deep, nutrient-poor lakes (Hawley and Whitton, 1991a). However, there are also records of waters with dense populations of *Synechococcus* or similar organisms, though their dimensions are probably often slightly above the picoplankton limit. In lakes whose waters have a high humic content there is a tendency for many of the picoplankton cyanobacteria to be loosely aggregated together by mucilage (G. R. W. Hawley and B. A. Whitton, unpublished data). No comprehensive explanation has been put forward to explain the distribution of small-celled cyanobacteria in freshwaters. However, it seems likely that grazing plays an important role, not just by the small ciliates and similar-sized organisms grazing the picoplankton, but indirect effects due to organisms at higher trophic levels.

It is well known that some gas-vacuolate cyanobacteria, mostly belonging to *Microcystis, Anabaena, Aphanizomenon, Gloeotrichia, Nodularia,* and *Oscillatoria,* form dense populations at the surfaces of lakes and reservoirs, the so-called blooms. The blooms cause serious nuisances because of their visual appearance, the likelihood of deoxygenation, and their frequent formation of toxins (Chapter 7) and unpleasant odorous substances such as goecosmin (Jüttner *et al.*, 1986; Jüttner, 1987; Matsumoto and Tsuchiya, 1988; Naes *et al.*, 1988). The term hyperscum has been introduced (Zohary, 1985, 1989) to describe crusted growths at the surface of a lake, often decimeters thick, which are so densely packed that free water is not evident. Recent reviews about blooms include those by Reynolds (1987), Paerl (1988), and Steinberg and Hartmann (1988) and a series of articles in a volume edited by Vincent (1987). Most recent studies emphasize the role of turbulence. Steinberg and Hartmann suggest that above a threshold of about 10 μg liter^{-1} P, the development of potentially bloom-forming cyanobacteria can be described by physical factors, such as water column stability. The presence or absence of these organisms can be predicted by characterizing different forms of turbulence; for instance, if the mixing depth is much greater than the euphotic depth, the cyanobacteria are outcompeted.

A floating cyanobacterium with a quite different ecological strategy is causing increasing nuisance problems in the southeastern United States (Dyck and Speziale, 1990). Mats of *Lyngbya wollei* persist on the bottom of shallow ponds and lakes during the cooler months, but rise to the surface when the water temperature exceeds about 18°C. The trichomes of this form are very wide (approximately 42 μm) and have thick sheaths (up to 11 μm) resistant to decay. In view of its persistence as a more or less "unialgal" community (L. Dyck, personal communication) and its considerable morphological similarity to the marine *Lyngbya majuscula*, which produces at least two toxins (Moore, 1981), it seems likely that this species will also prove to possess toxins.

Cyanobacteria are important in many terrestrial environments [reviewed by Hoffmann (1988c)], usually ones where there is an intermittent water supply. In deserts they have to withstand not only long periods of matric water stress (Section 4.2), but also extremes of one or more of osmotic stress, high and low temperatures, and high irradiation. The extent to which a visually obvious community develops on soils in hot deserts depends mainly on the extent to which any particular area holds water after the rare periods of rainfall (R. E. Cameron, 1969). The water requirements of most desert algae are, however, probably met by regular dew rather than infrequent rainfall (Friedmann and Galun, 1974). The genera most frequently recorded from desert soils are *Schizothrix*, *Plectonema*, *Microcoleus*, and *Nostoc* (Whitton, 1987b), all of which possess an obvious sheath. Cyanobacteria are often also important in microbial communities growing inside desert rocks, in both hot (Friedmann, 1980) and cold deserts (Friedmann and Ocampo, 1976; Friedmann, 1982). They form a layer on or near the surface of many rocks and buildings in the tropics and sometimes also in temperate regions. Colonial forms of *Nostoc* are sometimes abundant, such as *Nostoc commune* in shallow depressions on limestone.

Cyanobacteria in rice fields have been the subject of many studies (Roger and Kulosooriya, 1980) since their importance as nitrogen-fixers here was first realized (De, 1939). In paddy fields the organisms may grow for extended periods under moist conditions or submerged, but are probably subjected to frequent fluctuations in other environmental factors such as dissolved oxygen. In the deepwater rice fields of Bangladesh, some forms (e.g., *Porphyrosiphon*) occur only on moist soils before the flood, others (e.g., *Gloeotrichia*) only during the flood season, while *Scytonema* has become adapted for growth during both seasons (Whitton *et al.*, 1988b).

In addition to deserts, cyanobacteria are characteristic of several other types of "extreme" environment, of which thermal alkaline streams are probably the best known (Castenholz and Wickstrom, 1975; Brock, 1978; Jørgensen and Nelson, 1988; Ward *et al.*, 1989). *Synechococcus lividus* is the only phototrophic organism growing at temperatures above 70°C. Other examples of "extremes" are highly saline environments (Section 4.2) and some soils and waters rich in heavy metals. Records for metal-enriched sites are mostly for wastes associated with lead–zinc mining (Whitton, 1980), but include some records for soils with elevated copper and other heavy metals

(Rana *et al.*, 1971; Ernst, 1974). The dominants are usually narrow sheathed forms of Oscillatoriaceae.

Cyanobacteria are abundant in many intertidal regions (Whitton and Potts, 1982) and they sometimes form part of complex communities, often grouped together under the term "mat" (Cohen, 1989). Such communities are especially well developed where local conditions lead to hypersalinity. Several aspects of marine mat communities are reviewed in the volume edited by Cohen and Rosenberg (1989). A contrast between intertidal cyanobacterial communities and those of open oceans is the importance of nitrogen-fixers in the former and their rarity in the latter. Because of its importance in determining oceanic productivity, this has been the subject of considerable discussion (Carr and Wyman, 1986; Paerl, 1990). The consensus is that the low availability of iron, and perhaps also molybdenum concentrations, are key factors limiting nitrogen fixation in the open oceans. The case for iron, in particular, is based on its low solubility in surface waters (Huntsman and Sunda, 1980) versus the high requirement for photosynthesis and nitrogen fixation (Rueter, 1988).

It is has long been known that *Trichodesmium*, which is the most important cyanobacterial nitrogen-fixer (Section 4.5) in the oceans, is especially characteristic of certain areas of the oceans. Rueter (1988) and Rueter *et al.* (1990) have made an interesting hypothesis to explain this, which has helped to trigger speculation about the possibility of raising oceanic productivity. Samples of *Trichodesmium* collected off Barbados showed evidence of being iron-limited during subsequent laboratory assays (Rueter, 1988), so the *in situ* population would presumably also have responded positively to added iron. The morphology and buoyancy characteristics of *Trichodesmium* colonies make them very efficient particle interceptors, which may allow *Trichodesmium* to obtain a large portion of the biologically available iron in deposited aeolian dust (Rueter *et al.*, 1990). It was suggested that areas of the ocean to which dust from continental deserts is transported, such as the Arabian Sea and Persian Gulf, are especially suited for *Trichodesmium*. *Phormidium* sp. at a site on the Great Barrier Reef was also found (Entsch *et al.*, 1983) to show signs of iron stress, and based on this and other observations, the authors suggested that iron may be a limiting factor for primary production on coral reefs.

Despite the importance of oceanic cyanobacteria in global primary production, and of heterotrophic bacteria in the consumption of marine organic matter, the causes of cyanobacterial mortality are poorly understood (Procter and Fuhrman, 1990). In the case of the picoplankton, both protozoa (Sherr and Sherr, 1987) and phage (Proctor and Fuhrman, 1990) probably play important roles. Up to 5% of the cyanobacteria sampled from diverse marine locations contained mature phage. Harpacticoid copepods are probably key grazers of *Trichodesmium* (Roman, 1978). Although smaller forms of zooplankton may be abundant in *Trichodesmium* blooms, they apparently graze associated diatoms and dinoflagellates, rather than the cyanobacterium (Nair *et al.*, 1980).

4. INFLUENCE OF ENVIRONMENTAL FACTORS

4.1. Introduction

The extent to which various environmental factors are limiting, adequate, or inhibitory has been shown to have numerous effects on cyanobacterial populations in nature and on thallus and cell morphology, ultrastructure, physiology, biochemistry, and molecular organization. Nevertheless some features appear especially widespread. Others are mentioned here because they are of particular interest and suggest important aspects for further study.

It will be clear from Section 3 that different types of habitat favor the success of particular forms and species. Cyanobacterial species respond in different ways to environmental fluctuations within that habitat. Motile trichomes may move in relation to a light source (Castenholz, 1982; Häder, 1987) and probably other factors. Gas vacuolation usually increases in planktonic species with decrease in light flux (Walsby, 1987). Many open-ocean isolates of *Synechococcus* are capable of swimming motility (Waterbury *et al.*, 1985), apparently with an external rotating organelle resembling a screw propeller (Severina *et al.*, 1989), whereas coastal isolates are non-motile. The motile *Synechococcus* WH8113 responds positively to nitrogenous, but not other compounds (Willey and Waterbury, 1989). The threshold levels of 10^{-9} and 10^{-10} M are lower by several orders of magnitude than those reported for other bacteria and fall within a range that could be ecologically significant. It was suggested that these results support the theory that microzones of nutrient enrichment may play an important role in picoplankton nutrient dynamics.

It seems likely that the morphologically more complex forms are adapted to environments which are especially heterogeneous in space or time. The majority of these forms grow attached to surfaces, such as tropical soils and moist rocks (many Stigonematales), intertidal limestone (*Kyrtuthrix, Brachytrichia*), and streams (*Rivularia, Nostochopsis*). Laminated structures are formed by populations influenced by environmental fluctuations, such as diel changes in light on some stromatolites (Monty, 1965) and a sublittoral population of *Phormidium hendersonii* (Golubic and Focke, 1978), though currents and tidal events probably complicate the pattern in most shallow marine and intertidal structures (Monty, 1979).

Many filamentous cyanobacteria undergo morphological changes when they are subject to limitations, such as desiccation and decreased nitrogen, phosphorus, and iron (see Section 2.3). In the case of nitrogen limitation in heterocystous forms and phosphorus limitation in forms which can produce hairs, the morphogenetic change takes place when the content of the element is still relatively high and in most species with these features the modified form is the typical one in nature (see below). In both filamentous and unicellular forms, many other changes may take place at the subcellular level as a response to nutritional status, including changes in outer membrane proteins (Scanlan *et al.*, 1989), thylakoid membrane proteins (Sherman *et al.*, 1987), pigment content and composition (Carr and Wyman, 1986), and the relative proportions of the glycogen, cyanophycin, and polyphosphate granules and other storage products (Simon, 1987; Carr and Wyman, 1986; Stal *et al.*, 1989).

4.2. Water and Salinity

The ability to tolerate water stress due to desiccation or high salinity is widespread and an important factor in the predominance of cyanobacteria at many sites. Nevertheless the organisms only carry out the major physiological processes under conditions of high matric water potential. For instance, although *Chroococcidiopsis* endolithic in sandstones of the Negev is almost always dry, it fixes CO_2 only when matric water potentials are above -10 MPa, equivalent to a relative humidity of $>93\%$ at 34°C (Potts and Friedmann, 1981). A much higher matric water potential is required for photosynthesis of cryptendolithic cyanobacteria than for lichens with green algae (Palmer and Friedmann, 1990). The great majority of terrestrial cyanobacteria survive periods of desiccation with little obvious morphological change other than that due to loss of water (Whitton, 1987b). Several different types of compound have been found to be induced in desiccated material, including a group of acidic proteins in a natural population of *Nostoc commune*, which can constitute up to 20% of the total protein in colonies subjected to repeated cycles of drying and rehydration (Scherer and Potts, 1989). Vegetative cells and heterocysts of *N. commune* HUN retain their ultrastructural organization and the integrity of their intra- and extracellular membranes after 2 years of desiccation and subsequent rehydration (Peat *et al.*, 1988). A laboratory study of a non-colony-forming culture of *N. commune* UTEX 584 showed that nitrogenase is more sensitive to water stress than the intracellular ATP pool (Potts *et al.*, 1984). Following the rewetting of dried *N. commune*, the sequence in which physiological activities become apparent is first respiration, then photosynthesis, and finally nitrogenase activity (Scherer *et al.*, 1984). Respiration and photosynthesis preceded growth and were carried out by existing cells, whereas nitrogen fixation depended on newly differentiated heterocysts.

In addition to the occurrence of a number of truly halophilic cyanobacteria in hypersaline environments, numerous forms have been recorded from environments with fluctuating salinities. There is now considerable understanding of the physiological mechanisms by which cyanobacteria are adapted to such environments. Essentially, cells have adopted two different types of mechanism (Reed *et al.*, 1986a,b; Hagemann *et al.*, 1990) and, although few strains have been investigated with respect to both, it seems likely that the occurrence of both is universal. One is the avoidance of toxic internal amounts of inorganic ions such as Na^+, using active export systems (Yopp *et al.*, 1978; Reed *et al.*, 1985; Warr *et al.*, 1984). The other is the synthesis and accumulation of osmoprotective compounds to achieve an equilibrium of osmotic potential.

Three main salt-tolerant groups have been recognized, based on the osmoprotective compounds accumulated (Reed *et al.*, 1986a). In order of their association with increasing tolerance, the compounds are: (1) sucrose (Warr *et al.*, 1984) or trehalose (Reed and Stewart, 1983; Reed *et al.*, 1984); (2) glucosylglycerol; (3) glutamate betaine (Mackay *et al.*, 1983) or glycine betaine (Mohammad *et al.*, 1983). For instance, a survey of some 70 strains (Reed *et al.*, 1984) showed that there was a trend toward sucrose accumulation in freshwater strains and glucosylglycerol accumulation in marine ones.

If salt concentrations are lowered (hypoosmotic shock), the cells have to remove the accumulated osmoprotective compound(s) and Reed *et al.* (1986) found that considerable amounts were released extracellularly. In *Synechocystis* PCC 6714 a hypoosmotic shock from 684 to 2 mM led to 40–50% of the photosynthetically labeled organic material being released to the medium (Fulda *et al.*, 1990), presumably due to transient changes in membrane permeability (Reed *et al.*, 1985). Glucosylglycerol was the main component, but the compounds covered practically the whole range of low-molecular-weight substances in the cells. A hyperosmotic shock led to the release of only one-quarter as much glucosylglycerol as the hypoosmotic shock.

Intertidal *Calothrix scopulorum* (Jones and Stewart, 1969) and *Rivularia atra* (Reed and Stewart, 1983) both showed at high salinity greater inhibition of nitrogenase activity than of photosynthesis. This may reflect the channeling of fixed carbon into an osmoticum, making less available as an energy source for nitrogenase, as apparently also occurs in a laboratory strain of *Nostoc muscorum* (Blumwald and Tel-Or, 1982). Both *N. muscorum* and *Synechococcus* PCC 6301 show an initial period of increased photosynthetic activity following salt upshock, which leads to the accumulation of osmoregulatory sugars (Blumwald and Tel-Or, 1984). An initial inhibition of protein synthesis is followed by adaptation to this process and, in the case of *Nostoc muscorum*, subsequently by nitrogenase activity. Adaptation in *N. muscorum* was most efficient in response to NaCl-induced stress and functioned only partially under stress induced by either KCl or a nonionic osmoticum (Blumwald and Tel-Or, 1984).

The influence of a salt hyperosmotic shock on protein synthesis has been studied in some detail in *Microcystis firma* (Hagemann *et al.*, 1989) and *Synechocystis* sp. PCC 6803 (Hagemann *et al.*, 1990). After a hyperosmotic shock of 684 mM NaCl to the *Synechocystis* strain, total protein synthesis was almost completely blocked. Then, parallel to the accumulation of glycosylglycerol, protein synthesis recovered gradually, but remained diminished. Unlike the situation in water-stressed *Nostoc commune* (see above) or heat-shocked *Synechococcus* sp. PCC 6301 (Borbely *et al.*, 1985), the qualitative protein composition of salt-shocked *Synechocystis* sp. PCC 6803 cells remained unchanged; however, two-dimensional electrophoresis revealed proteins whose rate of synthesis was enhanced.

4.3. Light

Cyanobacterial photosynthesis is treated in Chapter 3, so only a few aspects of the influence of light are discussed in any detail here. Nevertheless some general points are worth bearing in mind when considering the ecological significance of other literature on cyanobacteria. Apart from a few topics like carbohydrate status and nitrogen fixation, most laboratory studies have been made under continuous illumination. Little is known about the influence of light–dark cycles on other factors such as heat, hyper- or hypoosmotic shocks, toxic agents, or interactions with other microbes. Almost nothing is known about the influence on photosynthesis and respiration of the extremely rapid and marked fluctuations in light flux which often occur in nature and are especially pronounced in monsoon regions of the subtropics. Laboratory light sources seldom permit experimental light flux values anywhere near the values for incident radiation in the tropics. Although a moderate number of studies on the influence of UV sources have been reported, only one has grown an organism under the high UV flux that might occur in a natural terrestrial environment (see below). Most materials for experimental studies are grown in the near absence of UV radiation. The influence of dissolved oxygen concentration, another factor which has received relatively little study (Section 4.4), may be expected to interact with UV dosage in determining photooxidation effects. Few colonial cyanobacteria are grown in the laboratory under conditions permitting the formation of the typical morphology found in nature, which in turn is likely to have a marked influence on the light regime for individual cells. All of these differences from the natural environment may not only influence the results of particular experiments, but also suggest the possibility of laboratory selection for modified strains.

Some of the above problems are due to the practical difficulties of simulating natural light regimes, especially those in the aquatic environment, so field experiments have played an important role in present understanding. Ecological aspects of phenotypic responses to differences in the natural light climate have been reviewed by Wyman and Fay (1987) and Whitton (1987b); photomovement, which plays an important role in the ecological response of many filamentous forms, has been reviewed by Häder (1987). In view of the profound influence of light on cyanobacterial functioning, an understanding of the role of light in regulation at the transcriptional level is especially important; several studies have started to provide such information (Schaeffer and Golden, 1989; Bustos *et al.*, 1990; Federspiel and Grossman, 1990).

The UV flux reaching the earth's surface was probably greater when cyanobacteria were first evolving than it is today (Sagan, 1965), so their survival would have necessitated the evolution of very efficient mechanisms for minimizing the impact of UV by absorbing potentially harmful wavelengths or subsequently repairing damage. There is an extensive literature on the effects of UV on cyanobacteria (Whitton, 1987b), but its value for interpreting and predicting the ability for organisms to survive in nature is not always clear; many of the earlier studies used UV-C radiation (<280 nm), yet this does not penetrate the earth's atmosphere. Interest in the impact of UV-B (280–320 nm) and UV-A (>320 nm) radiations has, however, been stimulated in recent years by the apparent decrease in the stratospheric ozone shield. A UV-A/B-absorbing pigment with maxima at 312 and 330 nm has been described (Scherer *et al.*, 1988) from field materials of *Nostoc commune;* in colonies grown under solar UV radiation the pigment concentration reached 10% dry weight. Although some pigment formed in laboratory-grown materials in the absence of UV-B light, its presence led to a much higher concentration.

Photoreactivation, the process by which pyrimidine dimers formed in UV-irradiated DNA are monomerized, has been described in various cyanobacteria: *Agmenellum quadruplicatum* (Van Baalen, 1968), *Anacystis nidulans* (=*Synechococcus* PCC 6301) (Wu *et al.*, 1967; Werbin and Rupert, 1968), *Gloeocapsa alpicola* (=*Synechocystis* PCC 6308) (O'Brien and Houghton, 1982), *Plectonema boryanum* (Saito and Werbin, 1970), and *Anabaena* spp. (Srivastava *et al.*, 1971; Levine and Thiel, 1987). The action spectra for photoreactivation in *Agmenellum quadruplicatum* (Van Baalen and O'Donnell, 1972) and *Anacystis nidulans* (=*Synechococcus* PCC 6301) (Eker *et al.*, 1990) are similar, with maxima at about 440 nm.

In addition to the demonstrated interactions between UV light and various cellular components, it has been suggested (Rambler *et al.*, 1977) that the colored sheaths which are especially widespread in terrestrial forms play a role in photoprotection. As they are such a conspicuous feature in many cyanobacteria, the sparse literature is summarized here.* Yellow and brown tints predominate, although red and violet shades also occur (Fritsch, 1945). Red and blue colors have been ascribed to a substance gloeocapsin, yellow and brown ones to scytonemin; the latter becomes green in the presence of acids and takes on a violet-grayish tint with iodine reagents. Two crystallizable substances (fuxorhodin, fuxochlorin) have

---

*Understanding of this subject has been greatly enhanced by Garcia-Pichel and Castenholz (1991); details are given in a note added in proof to this review.

been isolated from the dark-brown envelopes of *Calothrix scopulorum* (Kylin, 1927, 1943), suggesting that scytonemin may be a mixture of the two.

Coloration of the sheaths of terrestrial forms is often restricted to those parts of the stratum that directly face the incident light (Brand, 1990; Ercegović, 1929); in larger *Nostoc* colonies it is confined to the peripheral regions (Fritsch, 1945). However, other observations indicate that the relationship between presence of a yellow-brown pigment and light is less clearcut. Based on field observations on the typical form of *Scytonema myochrous* and the status *petalonemoides,* Jaag (1945) ascribed the coloration to a combination of high light intensities and low moisture content. A study (Pentecost, 1985) of ten sites with differing water availability also indicated that scytonemin production was negatively correlated with water availability. Some *Calothrix* strains whose pigment has a similar color to that in *Scytonema* (Gemsch, 1943) can produce intense brown sheaths when grown in the dark (B. A. Whitton, unpublished data).

4.4. Dissolved Oxygen

Marked diel changes in ambient oxygen concentration are very widespread in nature, and the extreme situation, with supersaturation by day and anoxic conditions by night, probably occurs quite often in thermal springs (Revsbech and Ward, 1984) and shallow subtropical and tropical waters, such as waste stabilization ponds (Konig, 1985) and deepwater rice fields (Whitton *et al.,* 1988a). A *Microcystis* pond in South India showed a diel range of 26 mg liter^{-1} O_2, though it did not quite become anoxic (Marzolf and Saunders, 1984). Apart from studies on anoxygenic photosynthesis (Padan and Cohen, 1982) and on the influence of dissolved oxygen on nitrogenase activity (Section 4.5), relatively little is known about the physiology of organisms where such changes occur. However, it seems probable that the efficient mechanisms of inorganic carbon acquisition which develop in (many) cyanobacteria grown under conditions of C depletion effectively suppress photorespiration (Colman, 1989). A diel study (Sirenko *et al.,* 1968) of *Aphanizomenon flos-aquae* suggested that it has a special oxidation–reduction system which regulates oxygen concentration; sulfhydryl groups in the mucilaginous coating may be involved. Adaptation of *Anabaena cylindrica* to oxygen supersaturation involves induction of superoxide dismutase and catalase (Mackey and Smith, 1983).

The physiological changes in response to changes in oxygen concentration may also be important in minimizing toxic effects of the H_2O_2 produced in some, but not all, cyanobacteria (Stevens *et al.,* 1973). Although some cyanobacteria form H_2O_2 in the dark (Stevens *et al.,* 1973), it is a strictly light-dependent process in *Anacystis nidulans* SAG 1402-1 (=*Synechococcus* PCC 6301) (Roncel *et al.,* 1989). Because of proposals (Fitzgerald, 1966; Kemp *et al.,* 1966; Kay *et al.,* 1982) to use H_2O_2 as an algicide, it is important to establish how it acts on bloom-forming cyanobacteria and, in particular, whether strains are likely to evolve which are resistant to it. Several cyanobacteria tested by Kay *et al.* were found to be more sensitive than the green alga *Ankistrodesmus*, *Microcystis* being the most sensitive with a toxicity threshold of 1.7 mg liter^{-1} H_2O_2. A somewhat similar result was found by Barroin and Feuillade (1986) in a comparison of *Oscillatoria rubescens* and *Pandorina morum*.

There are a number of studies on organisms undergoing marked seasonal shifts in environmental oxygen concentration. In the case of *Oscillatoria limnetica* in Solar Lake (Israel, Egypt), the principal effect of a shift from an oxygenated to an anoxic environment is an indirect one associated with the increase in S^{2-} content, the replacement of oxygenic by anoxygenic photosynthesis (Padan and Cohen, 1982). *Microcystis* in temperate lakes overwinters on bottom sediments under anoxic or near-anoxic conditions (Fallon and Brock, 1981; Reynolds *et al.*, 1981; Caceres and Reynolds, 1984).

4.5. Nitrogen

Nitrogen fixation is discussed in Chapter 3 and has been the subject of many reviews (e.g., Van Baalen, 1987), including several recent ones on ecological aspects (Howarth *et al.*, 1988a,b; Paerl, 1990). The following are merely a few topics of particular interest to the author.

Cyanobacteria typically contain a higher N content [4–9% (Fogg *et al.*, 1975)] than do eukaryotic phytoplankton [1–3% (Wheeler *et al.*, 1983)]. This is in part due to the presence of cellulose or other types of wall material in many eukaryotes, but it is also influenced by the frequent presence of substantial reserves of macromolecular N and perhaps also in nature (excluding the oceans) by the greater likelihood of cyanobacteria being P-limited rather than N-limited. It is widely accepted that all cyanobacteria can use nitrate- and ammonium-N for growth (Stanier and Cohen-Bazire, 1977; Carr and Wyman, 1986), though the possibility should not be ruled out that strains characteristic of particular environments may eventually be shown to lack the ability to use one of these substrates. A range of other inorganic and organic N sources has been shown to be utilized by various strains (Carr and Wyman, 1986); the poor ability to use urea in some reports may be due to the omission of Ni from trace element stocks, as cyanobacterial urease has been shown to resemble that of other organisms in requiring this element (Daday *et al.*, 1988; Singh and Rai, 1990). Kratz and Myers (1955) found that *Anacystis nidulans* (=*Synechococcus* PCC 6301), *Anabaena variabilis,* and *Nostoc muscorum* were all unable to utilize the mixture of amino acids in casein hydrolysate; a possible explanation may be the absence of specific uptake systems (Carr and Wyman, 1986). Several types of amino acid uptake system have, however, been reported for *Anacystis nidulans* (Lee-Kaden and Simonis, 1982).

The majority of cyanobacteria capable of fixing (di)nitrogen are heterocystous, but an increasing range of nonheterocystous ones are being found (Gallon, 1990). Heterocystous forms show considerable diversity in the morphology, distribution, and frequency of heterocysts, so it seems likely that the different genera and species have also evolved different strategies with respect to nitrogen fixation. Differences have been reported among species with respect to the concentration of combined nitrogen required to suppress heterocyst differentiation, and also of the relative influence of nitrate-N and ammonium-N in doing so. Most strains appear to be like the *Anabaena variabilis* studied by Ogawa and Carr (1969) in that ammonium-N is more effective, but *Anabaena* CA (Bottomley *et al.*, 1979) is much more strongly inhibited by nitrate-N. It is at present difficult to relate such physiological differences to morphological differences. However, organisms living in environments with markedly fluctuating levels of combined nitrogen probably retain their heterocysts at higher concentrations of combined nitrogen. The frequent occurrence of highly pigmented heterocysts in Rivulariaceae (Fritsch, 1945), especially those of many rice fields, may be associated with an environment where nitrogen fixation is important by day and combined nitrogen by night (Whitton, 1987b; Rother *et al.*, 1988).

The morphologically more complex cyanobacteria often show a lower heterocyst frequency (in relation to total cells) in nature than do simpler forms like *Anabaena;* the heterocysts of the former also often persist for longer. Diel measurements of fixation of CO_2 and N_2 (assayed as nitrogenase activity) in a stream population of *Rivularia* showed (Livingstone *et al.*, 1984) a molar ratio of 400:1 by daytime and only a slight further contribution to N_2 fixation by night, suggesting that much of the nitrogen source may have been combined nitrogen. There must presumably be some selective advantage for the dominance of a heterocystous form at the site, such as a much greater dependence on nitrogen fixation for some part of the year; this may represent an alternative strategy to that of an organism which lacks heterocysts, but can maintain a large N reserve. Many examples of morphologically more complex forms (*Gloeotrichia, Tolypothrix, Hapalosiphon*) have been grown in axenic culture with only N_2 as a nitrogen source, so it remains doubtful whether cyanobacterial populations exist in nature which possess heterocysts, yet nevertheless have an obligate requirement for combined nitrogen. The only laboratory examples known to the author are several mutant *Anabaena* strains and a *Calothrix* strain growing heterotrophically in the dark.

Nitrogen fixation has been reported in a number of axenic unicellular cyanobacteria [reviewed by Gallon and Chaplin (1988) and Gallon (1990)]. These strains have been isolated from a variety of habitats, including rocks, rice fields, and the marine intertidal zone, but apparently not the open ocean. Nitrogenase activity in *Gloeocapsa alpicola* (=*Gloeothece*) (Gallon *et al.*, 1975) and some marine *Synechococcus* strains (Leon *et al.*, 1986) occurs only during the dark periods of a light:dark cycle. However, at least some strains supported only by N_2 can continue to grow under continuous light when grown under synchronized conditions, apparently by a temporal separation of photosynthesis and nitrogen fixation during the cell cycle (Mitsui *et al.*, 1986). In *Synechococcus* RF-1 nitrogen fixation is apparently associated with an endogenous rhythm (Grobbelaar *et al.*, 1987; Huang and Chow, 1990; Huang *et al.*, 1990).

There is a relatively large number of records of nitrogen fixation by field populations dominated by nonheterocystous filamentous cyanobacteria; however, unless axenic cultures or a technique demonstrating localization of nitrogenase within the cells are used, it is open to doubt whether this fixation is due to heterotrophic bacteria (Paerl, 1990). Unequivocal results mostly concern sheathed genera, such as *Lyngbya* and *Plectonema*. A number of field studies have reported nitrogen fixation in *Microcoleus* mats (Potts and Whitton, 1977; Stal *et al.*, 1984; Paerl *et al.*, 1989), but so far only one axenic isolate has been shown to fix nitrogen (Pearson *et al.*, 1979); nitrogenase activity was evident whether or not the trichomes were aggregated into bundles. The integrity of the bundle is, however, essential for maintaining high rates of N_2 fixation in *Trichodesmium* (Carpenter and Price, 1976), and it has been suggested that aggregation is important in providing localized N_2-depleted zones (Paerl and Carlton, 1988; Paerl, 1990). Immunogold labelling showed that nitrogenase activity was limited to randomly distributed trichomes (10–40%) within the colonies (Bergman and Carpenter, 1991).

4.6. Phosphorus

In view of the fact that algal growth in unpolluted freshwaters is often P-limited, and that phosphate plays an especially important role in eutrophication, less attention has been paid to phosphate than might have been expected. Healey (1982) summarized key physiological features for cyanobacteria. Uptake of orthophosphate is an apparent hyperbolic function of external phosphate concentration, as in other microbes. Half-saturation values (k) and maximum rates of uptake (V_{max}) are similar to those of eukaryotic algae, but generally lower than those of heterotrophic bacteria. The rate of phosphate uptake is generally greater in the light than the dark, with stimulation ranging from slight (Batterton and Van Baalen, 1968) to as much as eight times (Simonis *et al.*, 1974); however at concentrations of phosphate over 1 mM the rate of uptake by *Anacystis nidulans* (=*Synechococcus* PCC 6301) is greater in the dark, as passive uptake apparently dominates over active uptake (Simonis *et al.*, 1974). The phosphate uptake characteristics of *A. nidulans* (=*Synechococcus* PCC 6301) conform to properties predicted by a linear force–flow relationship (Falkner *et al.*, 1989). Linearity extends to high phosphate concentrations when the organism has been preconditioned to high phosphate levels. Under phosphate-limited growth linearity is confined to a small concentration range, with threshold values decreasing below 10 mM. The authors suggested that the uptake system responds to changes in the external phosphate concentration in the same way as do sensory systems, by amplifying signals and adapting to them.

An anomalous feature of phosphate metabolism which has never been clearly resolved is shown in reports that efflux of inorganic or organic phosphate occurred simultaneously with uptake (Lean and Nalewajko, 1976; Nalewajko and Lean, 1978); gross uptake often greatly exceeded net influx. However, others have found little or no evidence for such efflux (Grillo and Gibson, 1979; see also Healey, 1982).

Addition of phosphate to P-deficient cultures leads to rapid uptake and often proceeds until cellular phosphorus content greatly exceeds that of exponentially growing cells (Healey, 1973; Sicko-Goad and Jensen, 1976; Falkner et al., 1984). This phosphate accumulates rapidly in polyphosphate granules: in Calothrix strains they are usually obvious with the light microscope within a few minutes (Sinclair and Whitton, 1977). The polyphosphate granules can accumulate a range of "heavy metals" (Jensen et al., 1982), which become toxic if increasing P deficiency leads to breakdown of the polyphosphate and hence mobilization of any associated metal (Sicko-Goad and Lazinsky, 1986). The presence of accumulated aluminum in polyphosphate granules of Anabaena cylindrica reduced the subsequent ability to use phosphate from the granules (Pettersson et al., 1988), probably by the binding of the aluminum to ATP (Pettersson and Bergman, 1989).

Most cyanobacteria can also mobilize organic phosphates in their environment by means of cell-bound ("surface") phosphatase, which is usually an inducible activity (Healey, 1982). However, there are considerable differences among strains with respect to their abilities to hydrolyze organic phosphates in their environment. All strains showing cell-bound phosphatase activity hydrolyze monoesters, most hydrolyze diesters, but only some hydrolyze phytic acid (Whitton et al., 1991); most strains also release phosphomonoesterase extracellularly, but none release phosphodiesterase. Marked differences occur in pH optima, with a maximum at pH 10.0 for cell-bound and extracellular phosphomonoesterase in Calothrix parietina D550 (Grainger et al., 1989), but at pH 7.0 in Nostoc commune UTEX 584 (Whitton et al., 1990). In batch culture cell-bound phosphomonoesterase and phosphodiesterase activities become detectable when mean cellular P falls below about 0.8% dry weight, but the strains differ, in that activity continues to increase with age of culture in Calothrix parietina, but reaches a peak and then decreases in Nostoc commune.

An ability to mobilize insoluble forms of inorganic phosphate is apparently also widespread. All but 1 of 18 strains tested by Bose et al. (1971) solubilized tricalcium phosphate; other materials utilized as P sources by cyanobacteria include Mussorie rock phosphate (Roychoudhury and Kaushik, 1989) and hydroxyapatite (H. J. Cameron and Julian, 1988). It has been suggested (Natesan and Shanmugasundaram, 1989) that extracellular phosphatase is involved in mobilization of insoluble inorganic phosphate, but without any direct proof.

Environmental factors may have a marked influence on the P status of cells. In addition to factors such as light, which have a relatively direct effect on energy supply, iron, aluminum, and sulfur are especially important. One important effect of iron occurs in the environment: phosphate is precipitated or complexed (see Section 4.7). In the case of aluminum and *Anabaena cylindrica*, the metal does not influence uptake and polyphosphate granule formation is actually enhanced (Pettersson *et al.*, 1988); aluminum appears to act on intracellular phosphate metabolism (see above), leading to symptoms of P starvation. Sulfur deficiency also leads to increased polyphosphate formation (Lawry and Jensen, 1986), but diminished growth is apparently due to lack of key S-containing amino acids rather than any direct influence on phosphate metabolism.

Past studies were often carried out under phosphate conditions very different from those encountered in nature, so there must be unease that this has influenced present understanding. Media developed up to the early 1960s often used very high levels of phosphate as a buffer. Although concentrations are much lower in most modern media, they are still sufficiently high that many cultures are seldom grown to conditions of P limitation. For instance, the widely used BG-11 medium (Rippka *et al.*, 1979) contains sufficient phosphate that batch cultures need to reach about 0.7 g liter^{-1} dry weight before inducible phosphatase activity develops in cyanobacterial strains. For this reason morphogenetic changes associated with moderate P deficiency, such as the formation of akinetes and hairs, or addition of P to P-deficient material, such as formation of hormogonia, have often been overlooked. It also poses the risk of selection of strains with anomalous phosphorus metabolism, perhaps including the loss of ability to form features such as akinetes.

4.7. Iron

A number of cyanobacteria deposit iron (oxides) in their sheaths (Desikachary, 1946), among which *Lyngbya ochracea* appears to be especially widespread. The ferromanganese nodules on the floor of some Tasmanian lakes may be formed by a member of the Stigonematales (Tyler and Buckney, 1980). Little more is known about the mechanism of deposition than when Naumann in 1925 (quoted in Desikachary, 1959) first speculated whether it is just a physicochemical process or whether bacteria also play a role. In some cases it may occur simply as the result of Fe^{2+} becoming oxidized in the presence of oxygen released by the cyanobacterium. The explanation is presumably more complicated in *Cyanodictyon imperfectum*, a species whose vertical distribution in Lake Kinneret showed two peaks, one near the bottom, and the other in the upper 5 m (Hickel and Pollingher, 1988). Deposition of iron oxide is localized in dense rings between two adjoining cells, perhaps due to binding of colloidal iron oxides, a phenomenon probably universal among microbes with anionic exopolymers (Ghiorse, 1989). In a *Microcystis aeruginosa* population in Blelham Tarn, England, an iron-containing colloid in the water was found to play an important role in seasonal changes of colony buoyancy (Oliver *et al.*, 1985); colonies coprecipitated with the colloid following breakdown of the thermocline in autumn. The presence of ferric "oxides" in sheaths or their vicinity raises the question as to whether organisms can mobilize phosphate associated with the deposit.

Many, but apparently not all, cyanobacteria produce siderophores under iron-limited conditions (Boyer *et al.*, 1987). Examples of both hydroxamate and catechol siderophores have been reported (Trick and Kerry, 1989). The siderophore of *Anabaena* sp. 7120 (schizokinen) has been shown (Clarke *et al.*, 1987) to complex both iron and copper, but only the schizokinen ferric complex is transported by the cell. The growth kinetics of siderophore-producing species differs from traditional nutrient-limited growth kinetics and clearly reflects the presence of a high-affinity, siderophore-mediated iron transport system (Kerry *et al.*, 1988): a reduction in iron initially leads to a reduced growth rate, but further reduction leads to an increased rate, coinciding with initiation of extracellular siderophore production. The level to which the external iron concentration had to drop depended in *A. variabilis* ATCC 29413 on whether the organism utilized nitrate or N_2 as N source.

A number of morphological responses to iron limitation in the laboratory have been reported. In addition to the formation of replacement heterocysts (Section 2.3) in *Calothrix*, hairs form as a response to iron limitation in some strains. In contrast to hairs forming as a response to phosphate limitation, these do not possess cell-bound phosphatase activity (Douglas *et al.*, 1986; Whitton, 1987a); however, hormogonia are formed as a response to addition of the limiting element in both cases. In trichomes of *Anabaena variabilis* UTEX 1444 grown in the presence of nitrate to avoid heterocyst formation there is a differentiational response among the cells (Foster *et al.*, 1988). At a stage when many cells show lysis, parts of the trichome show no disorganization; autoradiography revealed that uptake of ^{55}Fe was restricted to only a few cells and the authors suggested that, under conditions of lysis, iron released by disorganized cells may be taken up by the remaining viable cells.

Many changes in cellular composition and ultrastructure due to iron limitation have been reported (Riethman and Sherman, 1988). The chlorophyll content decreases and glycogen granules increase in *Anacystis nidulans* (Sherman and Sherman, 1983), *Agmenellum quadruplicatum* (Hardie *et al.*, 1983a,b), and *Calothrix parietina* (Douglas *et al.*, 1986). The degradation of some membranes is also a response in all three strains, but they apparently differ in other features. Intrathylakoidal vacuoles are very pronounced in *Agmenellum*, moderate in *Calothrix*, and (judging by the published micrographs) do not occur in *Anacystis*. Carboxysomes increase in quantity in *Anacystis*, but not *Agmenellum* or *Calothrix*.

The levels of ferredoxin decline under moderate iron stress, but have been shown to be replaced functionally by flavodoxin in several cyanobacteria (e.g., Hutber *et al.*, 1977). The gene encoding flavodoxin in *Anacystis nidulans* R2 (=*Synechococcus* sp. PCC 7942) is encoded as part of a 1900-base message induced by iron stress (Laudenbach *et al.*, 1988; Laudenbach and Straus, 1988). Antibodies cross-reactive with specific membrane proteins revealed (Riethman and Sherman, 1988) the presence of three iron-stress-induced polypeptides in *Anacystis* R2 (=*Synechococcus* sp. PCC 7942). K. J. Reddy *et al.* (1988), who characterized a gene *irpA* encoding for one of these, a cytoplasmic protein, suggested that this gene is regulated by iron and that the product is involved in iron acquisition or storage. The presence in *Synechococcus* sp. PCC 7942 and *Synechocystis* sp. PCC 6308 of specific "outer membrane" proteins, which may have a role in Fe transport, has been demonstrated by Scanlan *et al.* (1989). These were associated with the inner wall layer rather than the cytoplasmic membrane, and it seems possible that this was also the location of the other iron-stress-induced "membrane" proteins mentioned above.

Several studies have reported inhibitory effects of iron due to interactions with other essential elements. For instance, Storch and Dunham (1986) reported that the degree of inhibition by iron added to *Anabaena flos-aquae* suspended in Lake Erie water or a low-nutrient artificial medium was reduced by the addition of EDTA or phosphate. The authors concluded that the presence of EDTA reduced the ability of ferric iron to remove soluble phosphate from the medium.

5. TAXONOMY

The development of an effective taxonomic system for cyanobacteria imposes problems for which there is no exact parallel with any other group of organisms and, as mentioned in Section 1, they have been treated under both the International Code of Botanical Nomenclature (as blue-green algae) and the International Code of Bacteriological Nomenclature. In order to understand the present position and consider possible future developments, it is necessary to consider the history of the subject.

The International Code of Botanical Nomenclature takes the starting point for the valid publication of names of heterocystous blue-greens as

Bornet and Flahault (1886a,b, 1987, 1888) and that for nonheterocystous filamentous forms as Gomont (1982). In other cases—the morphologically less complex forms—the starting point is Linnaeus (1753). The publications of Bornet and Flahault in particular still provide an important source of morphological information, but most of these and the other accounts available 40 years later were assembled by Geitler (1932) into a comprehensive flora; although this was nominally a regional account, in practice it was cosmopolitan. At least for freshwaters, many subsequent researchers have taken this flora as their source for accounts of genera and species. Elenkin's floras (1936/1938/1949) are even more comprehensive and have been regarded as a favored source of information by some specialists; overall, however, they have been much less used, mainly because of the language (Russian), but also because Elenkin tended to be a taxonomic "splitter." Among the other major floristic accounts, the following are some of the best known. Frémy's (1929, 1929–1933) studies are still of importance because of his first-hand experience of tropical and intertidal environments. Desikachary's (1959) flora mentions much of the literature since Geitler's account, and is widely quoted because it is the only comprehensive one in English. Bourrelly (1970/1985) uses his account of the genera of freshwater algae to broaden generic limits of some groups, in particular the inclusion of *Phormidium* and *Symploca* within *Lyngbya*. Many other systemic accounts are listed by Anagnostidis and Komárek (1985).

These and other more recent morphology-based systems have shown many improvements with time, but suffer to varying degrees from several problems. Remarkably little consideration has been given to the theoretical concepts underlying the recognition of limits to genera and species, the account by Hoffman (1988a) being one of the few exceptions. Differing numbers of characters are used to separate related taxa, but in many cases only one is used. As a result numerous poorly described species are separated from others based on a single criterion. The characters chosen are frequently not of the "presence or absence" type, but involve subjective decisions about where to split a continuum. However, many experimental studies have now confirmed what the more observant of the older researchers already suspected—that many "blue-greens" show great morphological variability in response to environmental variation (Jeeji-Bai, 1977). Various approaches have been adopted to deal with these problems.

A number of papers by J. Komárek and several allied researchers have made substantial improvements to the "botanical" approach. Relatively little discussion is given to the theoretical principles upon which the studies are based (Anagnostidis and Komárek, 1985), but the studies benefit from a combination of extensive first-hand experience of the organisms in many different regions, often combined with experimental studies. The accounts of *Chamaesiphon* (Kann, 1972a,b, 1973), *Phormidium* (Kann, 1973; Kann and Komárek, 1970) and *Homoeothrix* (Komárek and Kann, 1973) supersede those published previously and include long bibliographies. A recent series of papers provides detailed accounts of the Chroococcales (Komárek and Anagnostidis, 1988), Oscillatoriales (Anagnostidis and Komárek, 1988), and Nostocales (Komárek and Anagnostidis, 1989). The generic limits adopted tend to be quite narrow, leading to smaller and more homogeneous genera than in most other botanical floras. These papers provide the most detailed guides available to the morphological and ecological literature, considerable new insight into morphology, and accounts not only of well-known forms, but ones which are little known, though often widespread. A new flora is in preparation.

An entirely different approach was adopted by F. Drouet to deal with the problem of morphological variability, starting with Drouet and Daily (1956) and followed by a series of monographs by Drouet (see Castenholz and Waterbury, 1989a). Essentially these reversed for "blue-greens" the taxonomic developments made by mainstream bacteriologists since the 1890s, reverting to a few simple morphological characters easily recognized in preserved specimens. The species *sensu* Drouet represents a polymorphic cluster of "ecophenes" in a few broad genera. The system became progressively more simplified with successive monographs, until more than 2000 species were reduced to 62 species. The nomenclatural simplicity of Drouet's system led to many ecological accounts using his names, especially descriptions of marine communities by U. S. researchers. Two of his generic names, *Anacystis* and *Agmenellum*, were given to strains which have subsequently been used widely in research. No extensive critique of his system has been ever published. However, reviews of the various monographs (e.g., Komárek, 1983) almost all acknowledge the valuable contribution Drouet made in bringing together the systematic literature, but are otherwise highly critical of the system. Key flaws are the neglect of characters other than those visible in preserved material and the fact that a series of studies on DNA/DNA hybridization and DNA base composition by W. T. Stam and co-workers (e.g., Stam and Holleman, 1979) have shown the unsatisfactory nature of particular decisions made by Drouet.

A third taxonomic approach was introduced by R. Y. Stanier and his associates in the early 1970s. This involved the long-term strategy of dealing with cyanobacteria in a similar way to other major groups of prokaryotes. The system is based on axenic, clonal cultures and the use of characters from a range of disciplines, rather than morphology alone. An extensive study (Rippka *et al.*, 1979) of a range of strains led to a proposed system for the major taxonomic groups (see Section 2.1) and revised descriptions for some of the most important genera. The accounts (largely by R. W. Castenholz and J. B. Waterbury) in Staley *et al.* (1989) give as detailed an account of genera as possible with the strains in culture, but leave consideration of genera not yet in culture and all species until more information is available.

The proposal by Stanier *et al.* (1978) that cyanobacteria be placed under the International Code of Nomenclature of Bacteria was intended to remove these organisms from the jurisdiction of the Botanical Code. In practice, however, both codes have been employed, with authors choosing one or the other, or in a few cases describing organisms according to both. The various approaches and problems have been aired by Friedmann and Borowitzka (1982), while Castenholz and Waterbury (1989a,b) summarize the different nomenclatural recommendations. One particular problem is that well-known organisms may be given new names under the Bacteriological Code (e.g., Florenzano *et al.*, 1985). This potentially chaotic situation will be removed if the recommendation of the International Committee on Systematic Bacteriology in 1986, " . . that names of cyanobacteria described and validly published as blue-green algae under the International Code of Botanical Nomenclature are recognized as having been validly published under the International Code of Nomenclature of Bacteria," is accepted by the 1990 International Congress of Microbiology. Castenholz and Waterbury (1989a) suggest that a new data for establishing an "Approved List of Names" should be agreed jointly by bacteriologists and botanists and that Geitler (1932) represents a reasonable and practical point of departure.

In view of the fact that a *modus vivendi* has now been reached between supporters of the "bacteriological" and "botanical" systems of classification, the problems for taxonomists are now largely practical ones. There is a need for an extensive body of standardized information on field populations and laboratory strains, including morphological, cytological, ultrastructural, physiological, and biochemical characters, which can be used for assessing similarity. Molecular methods such as DNA–DNA hybridization (Stam and Venema, 1977; Stam, 1980) and restriction fragment length polymorphism (Douglas and Carr, 1988) have also been applied to problems at lower taxonomic levels, but recent molecular sequencing data have mostly been applied to broader questions of evolution, in particular the relationship of cyanobacteria, prochlorophytes, cyanelles, and plastids.

The use of *Prochlorothrix* in a comparison of a range of 16S RNA nucleotide sequences led to the construction of evolutionary trees by Turner *et al.* (1989), which show clear separation of green chloroplasts from *Prochlorothrix.* In contrast, the use of DNA sequences from a photosynthetic membrane protein (*psbA*) from photosystem II led Morden and Golden (1989) to the opposite conclusion. In view of the fact that Lewin (1989) has speculated whether the prochlorophytes may prove to be polyphyletic, it is unfortunate that neither *Prochloron* nor the deep oceanic prochlorophytes were included in these studies. Janssen *et al.* (1989) assessed the evolutionary relationships of *psbA* genes from cyanobacteria (but not a prochlorophyte), the cyanelles of *Cyanophora paradoxa*, and plastids and concluded that the data strongly support the postulated bridge position of cyanelles between chloroplasts and free-living cyanobacteria. The fact that

Prochlorothrix appears to possess the shorter, plastid-type *psbA* gene (Morden and Golden, 1989) suggested to Janssen *et al.* the possible polyphyletic origin of plastids. Use of a wider range of organisms to provide sequence data, together with more than one method to reconstruct the evolutionary trees, should help to resolve these various questions (Penny, 1989).

NOTE ADDED IN PROOF

The rapid progress of research on cyanobacteria made it inevitable that important studies were published while the paper was in press. In particular the characterization of the brown pigment scytonemin (Garcia-Pichel and Castenholz, 1991) has added considerably to the account given in Section 4.3. Scytonemin, which appears to be a single lipid-soluble compound, was identified in more than 30 species of cyanobacteria from cultures and natural populations exposed to intense solar radiation. It has a prominent absorption maximum in the near ultraviolet and was effective in protecting the cells from incoming near-ultraviolet-blue radiation. UV-A radiation was very effective in eliciting scytonemin synthesis. The results strongly suggest that scytonemin formation is an adaptive strategy for photoprotection against short-wavelength irradiance. The authors argue that in contrast to the various repair systems for UV damage, which provide a "remedial" strategy, scytonemin provides a "preventive" strategy. It should be of special value for species that endure long periods of exposure under physiologically inactive or quiescent conditions such as those subject to intermittent desiccation.

REFERENCES

Adams, D. G., and Carr, N. G., 1981, The developmental biology of heterocyst and akinete formation in cyanobacteria, *CRC Crit. Rev. Microbiol.* **9:**45–100.

Anagnostidis, K., and Komárek, J., 1985, Modern approach to the classification systems of cyanophytes 1—Introduction, *Arch. Hydrobiol.* (Suppl.) **71:**291–302.

Anagnostidis, K., and Komárek, J., 1988, Modern approach to the classification system of cyanophytes 3—Oscillatoriales, *Arch. Hydrobiol.* (Suppl.) **80:**327–472.

Barroin, G., and Feuillade, M., 1986, Hydrogen peroxide as potential algicide for *Oscillatoria rubescens* D.C., *Water Res.* **20:**619–623.

Batterton, J. C., Jr., and Van Baalen, C., 1968, Phosphorus deficiency and phosphate uptake in the blue-green alga *Anacystis nidulans, Can. J. Microbiol.* **14:**341–348.

Bauld, J., 1981, Occurrence of benthic microbial mats in saline lakes, *Hydrobiologia* **81:**87–111.

Bergman, B., and Carpenter, E. J., 1991, Nitrogenase confined to randomly distributed trichomes in the marine cyanobacterium *Trichodesmium thiebauti, J. Phycol.* **27:**158–165.

Bloor, S., and England, R. R., 1989, Antibiotic production by the cyanobacterium *Nostoc muscorum, J. Appl. Phycol.* **1:**367–372.

Blumwald, E., and Tel-Or, E., 1982, Osmoregulation and cell composition in salt adaptation of *Nostoc muscorum, Arch. Microbiol.* **132:**168–172.

Blumwald, E., and Tel-Or, E., 1984, Salt adaptation of the cyanobacterium *Synechococcus* 6311 growing in a continuous culture (turbidostat), *Plant Physiol.* **74:**183–185.

Borbely, G., Suranyi, G., Korcz, A., and Palfi, Z., 1985, Effect of heat shock on protein synthesis in the cyanobacterium *Synechococcus* sp. strain PCC 6301, *J. Bacteriol.* **161:**1125–1130.

Bornet, E., and Flahault, C., 1886a, Revision des Nostocacées Heterocystées, A.. *Sci. Nat. Bot. Ser. VII* **3:**323–381.

Bornet, E., and Flahault, C., 1886b, Revision des Nostocacées Heterocystées, A.. *Sci. Nat. Bot. Ser. VII* **4:**343–373.

Bornet, E., and Flahault, C., 1887, Revision des Nostocacées Heterocystées, A.. *Sci. Nat. Bot. Ser. VII* **5:**51–129.

Bornet, E., and Flahault, C., 1888, Revision des Nostocacées Heterocystées, A.. *Sci. Nat. Bot. Ser. VII* **7:**177–262.

Borowitzka, L. A., 1986, Osmoregulation in blue-green algae, *Prog. Phycol. Res.* **4:**243–256.

Bose, P., Nagpal, U. S., Nagpal, Venkataraman, G. S., and Goyal, S. K., 1971, Solubilization of tricalcium phosphate by blue-green algae, *Curr. Sci.* **7:**165–166.

Bottomley, P. J., Grillo, J. F., Van Baalen, C., and Tabita, R., 1979, Synthesis of nitrogenase and heterocysts by *Anabaena* sp. CA in the presence of high levels of ammonia, *J. Bacteriol.* **140:**938–943.

Bourrelly, P., 1970/1985, *Les Algues d'Eau Douce. III Les Algues Bleues et Rouges, les Eugleniens, Peridiniens et Cryptomonadines*, N. Boubée, Paris.

Boyer, G. L., Gillam, A. H., and Trick, C., 1987, Iron chelation and uptake, in: *The Cyanobacteria* (P. Fay and C. Van Baalen, eds.), Elsevier, Amsterdam, pp. 415–436.

Brand, F., 1900, Der Formenkreis von *Gloeocapsa alpina*, *Bot. Centralbl.* **15:**152–159.

Brock, T. D., 1978, *Thermophilic Microorganisms and Life at High Temperatures*, Springer, New York, 633 pp.

Burger-Wiersma, Veenhuis, M., Korthals, H. J., van de Wiel, C. C. M., and Mur, L. R., 1986, A new prokaryote containing chlorophylls *a* and *b*, *Nature* **320:**262–264.

Burger-Wiersma, T., Stal, L. J., and Mur, L. C., 1989, *Prochlorothrix hollandica* gen. nov., sp. nov., a filamentous oxygenic photoautotrophic procaryote containing chlorophylls *a* and *b*: Assignment to *Prochlorotrichaceae* fam. nov. and order *Prochlorales* Florenzano, Balloni, and Materassi 1986, with emendation of the ordinal description, *Int. J. Syst. Bacteriol.* **39:**250–257.

Bustos, S. A., Schaeffer, M. R., and Goulden, S. S., 1990, Different and rapid responses of four cyanobacterial *psbA* transcripts to changes in light intensity, *J. Bacteriol.* **172:**1998–2004.

Caceres, O., and Reynolds, C. S., 1984, Some effects of artificially-enhanced anoxia on the growth of *Microcystis aeruginosa* Kütz. emend. Elenkin, *Phil. Trans. R. Soc. Lond. B* **293:**419–477.

Cameron, H. J., and Julian, G. R., 1988, Utilization of hydroxyapatite by cyanobacteria as their sole source of phosphate and calcium, *Plant Soil* **109:**123–124.

Cameron, R. E., 1969, Abundance of Microflora in Soils of Desert Regions, National Aeronautics and Space Administration Technical Report No. 32-1378, Jet Propulsion Laboratory, Pasadena, California.

Canabaeus, L., 1929, *Über die Heterocysten und Gasvakuolen der Blaualgen und ihre Beziehung zueinander.* Pflanzenforschung 13, Fisher, Jena.

Cano, M. M. S., de Mulé, M. C. Z., de Caire, G. Z., and de Halperin, D. R., 1990, Inhibition of *Candida albicans* and *Staphylococcus aureus* by phenolic compounds from the terrestrial cyanobacterium *Nostoc muscorum*, *J. Appl. Phycol.* **2:**79–82.

Capone, D. G., and Carpenter, E. J., 1982, Nitrogen fixation in the marine environment. *Science* (Wash. D.C.) **21:**1140–1142.

Carpenter, E. J., 1983, Nitrogen fixation by marine *Oscillatoria* (*Trichodesmium*) in the world's oceans, in: *Nitrogen in the Marine Environment* (E. J. Carpenter and D. G. Capone, eds.), Academic Press, New York, pp. 65–104.

Carr, N. G., and Wyman, M., 1986, Cyanobacteria: Their biology in relation to oceanic picoplankton, *Can. Bull. Fish. Aquat. Sci.* **214:**159–204.

Castenholz, R. W., 1982, Motility and taxes, in: *The Biology of Cyanobacteria* (N. G. Carr and B. A. Whitton, eds.), Blackwell, Oxford, and University of California Press, Berkeley, pp. 413–439.

Castenholz, R. W., 1989a, Subsection III. Order Oscillatoriales, in: *Bergey's Manual of Systematic Bacteriology*, Volume 3 (J. T. Staley, M. P. Bryant, N. Pfennig, and J. G. Holt, eds.), Williams and Wilkins, Baltimore, pp. 1771–1780.

Castenholz, R. W., 1989b, Subsection IV. Order Nostocales, in: *Bergey's Manual of Systematic Bacteriology*, Volume 3 (J. T. Staley, M. P. Bryant, N. Pfennig, and J. G. Holt, eds.), Williams and Wilkins, Baltimore, pp. 1780–1793.

Castenholz, R. W., 1989c, Subsection V. Order Stigeonematales, in: *Bergey's Manual of Systematic Bacteriology*, Volume 3 (J. T. Staley, M. P. Bryant, N. Pfennig, and J. G. Holt, eds.), Williams and Wilkins, Baltimore, pp. 1780–1799.

Castenholz, R. W., and Waterbury, J. B., 1989a, Cyanobacteria, in: *Bergey's Manual of Systematic Bacteriology*, Volume 3 (J. T. Staley, M. P. Bryant, N. Pfennig, and J. G. Holt, eds.), Williams and Wilkins, Baltimore, pp. 1710–1727.

Castenholz, R. W., and Waterbury, J. B., 1989b, Taxa of the cyanobacteria, in: *Bergey's Manual of Systematic Bacteriology*, Volume 3 (J. T. Staley, M. P. Bryant, N. Pfennig, and J. G. Holt, eds.), Williams and Wilkins, Baltimore, pp. 1727–1728.

Castenholz, R. W., and Wickstrom, C. E., 1975, Thermal streams, in: *River Ecology* (B. A. Whitton, ed.), pp. 264–284, Blackwell, Oxford.

Chisholm, S. W., Olson, R. J., Zettler, E. R., Goericke, R., Waterbury, J. B., and Welschmeyer, N. A., 1988, A novel free-living prochlorophyte abundant in the oceanic euphotic zone, *Nature* **334:**340–343.

Clarke, S. E., Stuart, J., and Sanders-Loehr, J., 1987, *Appl. Environ. Microbiol.* **53:**917–922.

Cmiech, H. A., Reynolds, C. S., and Leedale, G. F., 1984, Seasonal periodicity, heterocyst differentiation and sporulation in planktonic Cyanophyceae in a shallow lake with special reference to *Anabaena solitaria*, *Br. Phycol. J.* **19:**245–257.

Cohen, Y., 1989, Photosynthesis in cyanobacterial mats and its relation to the sulfur cycle: A model for microbial sulfur interactions, in: *Microbial Mats* (Y. Cohen and E. Rosenberg, eds.), American Society for Microbiology, Washington, D. C., pp. 22–36.

Cohen, Y., and Rosenberg, E. (eds.), 1989, *Microbial Mats*, American Society for Microbiology, Washington, D. C.

Cohen, Y., Padan, E., and Shilo, M., 1975, Facultative anoxygenic photosynthesis in the cyanobacterium *Oscillatoria limnetica*, *J. Bacteriol.* **123:**855–861.

Cohen, Y., Jørgensen, B. B., Revsbech, N. P., and Poplawski, R., 1986, Adaptation to hydrogen sulfide of oxygenic and anoxygenic photosynthesis among cyanobacteria, *Appl. Environ. Microbiol.* **51:**398–407.

Colman, B., 1989, Photosynthetic carbon assimilation and the suppression of photorespiration in the cyanobacteria, *Aquat. Bot.* **34:**211–231.

Daday, A., Mackerras, A. H., and Smith, G. D., 1988, A role for nickel in cyanobacterial nitrogen fixation and growth via cyanophycin metabolism, *J. Gen. Microbiol.* **134:**2659–2663.

De, P. K., 1939, The role of blue-green algae in nitrogen fixation in rice fields, *Proc. R. Soc. Lond.* **127B:**121–139.

Desikachary, T. V., 1946, Germination of the heterocyst in two members of the Rivulariaceae *Gloeotrichia raciborskii* and *Rivularia mangini* Frémy, *J. Indian Bot. Soc.* **25:**3–17.

Desikachary, T. V., 1959, *Cyanophyta*, Indian Council of Agricultural Research, New Delhi.

Dodds, W. K., and Castenholz, R. W., 1987, Effects of grazing and light on the growth of *Nostoc pruniforme* (Cyanobacteria), *Br. Phycol. J.* **23:**219–227.

Douglas, D., Peat, A., and Whitton, B. A., 1986, Influence of iron status on structure of the cyanobacterium (blue-green alga) *Calothrix parietina*, *Cytobios* **47:**155–165.

Douglas, S. E., and Carr, N. G., 1988, Examination of genetic relatedness of marine *Syn-*

echococcus spp. by using restriction fragment length polymorphisms, *Appl. Environ. Microbiol.* **54**:3071–3078.

Drews, G., and Weckesser, J., 1982, Function, structure and composition of cell walls and external layers, in: *The Biology of Cyanobacteria* (N. G. Carr and B. A. Whitton, eds.), Blackwell, Oxford, and University of California Press, Berkeley, pp. 333–357.

Drouet, F., and Daily, W. A., 1956, Revision of the coccoid Myxophyceae, *Butler Univ. Stud.* **10**:1–218.

Dyck, L. A., and Speziale, B. J., 1990, Infestations of *Lyngbya wollei:* Structural and phenological characteristics, *J. Phycol.* **26**(2) Suppl. 10.

Eker, A. P. M., Kooiman, P., Hessels, J. K. C., and Yasui, A., 1990, DNA photoreeactivating enzyme from the cyanobacterium *Anacystis nidulans, J. Biol. Chem.* **265**:8009–8015.

Elenkin, A. A., 1936/1938/1949, *Sinenzelenye vodoroslii SSSR. Monographia Algarum Cyanophycearum Aquidulcium et Terrestrium in Finibus URSS Inventarum, Pars Generalis, Pars Specialis I et II,* Akademii Nauk SSSR, Moscow.

Entsch, B., Sim, R. G., and Hatcher, B. G., 1983, Indications from photosynthetic components that iron is a limiting nutrient in primary producers on coral reefs, *Mar. Biol.* **73**:17–30.

Ercegović, A., 1929, *Dalmatella,* nouvelle genre des cyanophycées lithophytes de la côte Adriatique, *Acta Bot. Inst. Bot. Univ. Zagreb* **4**:35–41.

Ernst, W. H. O., 1974, *Schwermetallvegetation der Erde,* Gustav Fischer Verlag, Stuttgart.

Falkner, G., Falkner, R., Graffius, D., and Strasser, P., 1984, Bioenergetic and ecological aspects of phosphate uptake by blue-green algae, *Arch. Hydrobiol.* **101**:89–99.

Falkner, G., Falkner, R., and Schwab, A. J., 1989, Bioenergetic characterization of transient state phosphate uptake by the cyanobacterium *Anacystis nidulans, Arch. Microbiol.* **152**:353–361.

Fallon, R. D., and Brock, T. D., 1981, Overwintering of *Microcystis* in Lake Mendota, *Freshwater Biol.* **11**:217–226.

Fattom, A., and Shilo, M., 1984, Hydrophobicity as an adhesion mechanism of benthic cyanobacteria, *Appl. Environ. Microbiol.* **47**:135–143.

Fay, P., Stewart, W. D. P., Walsby, A. E., and Fogg, G. E., 1968, Is the heterocyst the site of nitrogen fixation in blue-green algae? *Nature* (London) **220**:810–812.

Federspiel, N. A., and Grossman, A. R., 1990, Characterization of the light-regulated operon encoding the phycoerythrin-associated linker proteins from the cyanobacterium *Fremyella diplosiphon, J. Bacteriol.* **172**:4072–4081.

Fisher, R. W., and Wolk, C. P., 1976, Substance stimulating the differentiation of spores of the blue-green alga, *Cylindrospermum licheniforme, Nature* **259**:394–395.

Fitzgerald, G. P., 1966, Use of potassium permanganate for control of problem algae, *J. Am. Water Works Assoc.* **58**:609–614.

Florenzano, G. C. S., Pelosi, E., and Vincenzini, M., 1985, *Cyanospira rippkae* and *Cyanospira capsulata* (gen. nov. and spp. nov.): New filamentous heterocystous cyanobacteria from Magadi Lake (Kenya), *Arch. Microbiol.* **140**:301–306.

Fogg, G. E., 1975, *Algal Cultures and Phytoplankton Ecology,* 2nd ed., University of Wisconsin Press, Madison, Milwaukee.

Fogg, G. E., 1987, Marine planktonic cyanobacteria, in: *The Cyanobacteria* (P. Fay and C. Van Baalen, eds.), Elsevier, Amsterdam, pp. 393–414.

Fogg, G. E., Stewart, W. D. P., and Fay, P., 1973, *The Blue-Green Algae,* Academic Press, London.

Foster, E. W., Barton, L. L., and Johnson, G. V., 1988, Differential cellular response of *Anabaena variabilis* to iron, *J. Plant Nutr.* **11**:1193–1203.

Fox, G. E., Stackebrandt, E., Hespell, R. B., Gibson, J., Maniloff, J., Dyer, T. A., Wolfe, R. S., Balch, W. E., Tanner, R. S., Magrum, L. J., Zablen, L. B., Blaemore, R. Gupta, R., Bonen, L., Lewis, B. J., Stahl, D. A., Luehrsen, K. R., Chen, K. N., and Woese, C. R., 1980, The phylogeny of procaryotes, *Science* **209**:457–463.

Foy, R. H., 1980, The influence of surface to volume ratio on the growth rates of planktonic blue-green algae, *Br. Phycol. J.* **15**:278–289.

Foy, R. H., and Smith, R. V., 1980, The role of carbohydrate accumulation in the growth of planktonic *Oscillatoria* species, *Br. Phycol. J.* **13:**139–150.

Frémy, P., 1929, *Les Myxophycées de l'Afrique Equatoriale Française*, Archives de Botanique III, Mémoire No. 2.

Frémy, P., 1929–1933, Cyanophycées des Côtes d'Europe, *Mém. Soc. Natl. Sci. Nat. Math. Cherbourg* **41:**1–236.

Friedmann, E. I., 1980, Endolithic microorganisms in the Antarctic desert ecosystem, *Orig. Life* **10:**223–235.

Friedmann, E. I., 1982, Endolithic microorganisms in the Antarctic cold desert, *Am. Assoc. Adv. Sci. Publ.* **215:**1045–1053.

Friedmann, E. I., and Borowitzka, L. J., 1982, The symposium on taxonomic concepts in blue-green algae: Towards a compromise with the Bacteriological Code? *Taxon* **31:**673–683.

Friedmann, E. I., and Galun, M., 1974, Desert algae, lichens and fungi, in: *Desert Biology* (G. W. Brown, ed.), Academic Press, New York, pp. 165–212.

Friedmann, E. I., and Ocampo, R., 1976, Endolithic blue-green algae in the dry valleys: Primary producers in the Antarctic desert ecosystem, *Science* **193:**1247–1249.

Fritsch, F. E., 1945, *The Structure and Reproduction of the Algae*, Volume II, Cambridge University Press, Cambridge.

Fulda, S., Hagemann, M., and Libbert, E., 1990, Release of glycosylglycerol from the cyanobacterium *Synechocystis* spec. SAG 92.79 by hypoosmotic shock, *Arch. Microbiol.* **153:**405–408.

Gallon, J. R., 1990, The physiology and biochemistry of N_2 fixation by nonheterocystous cyanobacteria, *Phykos* **28:**18–46.

Gallon, J. R., and Chaplin, A. E., 1988, Recent studies on N_2 fixation by nonheterocystous cyanobacteria, in: *Nitrogen Fixation: One hundred years After* (H. Bothe, F. J. de Bruyn, and W. E. Newton, eds.) pp. 183–188, Gustav Fisher, Stuttgart.

Gallon, J. R., Kurz, W. G. W., and LaRue, T. A., 1975, The physiology of nitrogen fixation by a *Gloeocapsa* sp., in: *Nitrogen Fixation in Free-Living Microorganisms* (W. D. P. Stewart, ed.), Cambridge University Press, Cambridge, pp. 159–173.

Garcia-Pichel, F., and Castenholz, R. W., 1991, Characterization and biological implications of scytonemin, a cyanobacterial sheath pigment, *J. Phycol.* **27:**395–409.

Geitler, L., 1932, Cyanophyceae, in: *Rabenhorst's Kryptogamen-Flora*, Volume 14, Akademische Verlagsgesellschaft, Leipzig.

Gemsch, W., 1943, Vergleichende Untersuchungen über Membranfarbung und Membran-farbstoffe in der Gattungen *Gloeocapsa* Kütz. und *Scytonema* Ag., *Ber. Schweiz. Ges.* **53:**121–192.

Ghiorse, W. C., 1989, Manganese and iron as physiological electron donors and acceptors in aerobic–anaerobic transition zones, in: *Microbial Mats* (Y. Cohen and E. Rosenberg, eds.), American Society for Microbiology, Washington, D. C., pp. 163–169.

Gibson, C. E., and Smith, R. V., 1982, Freshwater plankton, in: *The Biology of Cyanobacteria* (N. G. Carr and B. A. Whitton, eds.) pp. 463–490, Blackwell, Oxford, and University of California Press, Berkeley.

Giovannoni, S. J., Turner, S., Olsen, G. J., Barns, S., Lane, D. J., and Pace, N. R., 1988, Evolutionary relationships among cyanobacteria and green chloroplasts, *J. Bacteriol.* **170:**3584–3592.

Glover, H. E., 1986, The physiology and ecology of marine cyanobacteria, *Synechococcus* spp., *Adv. Aquat. Microbiol.* **3:**49–107.

Golden, J. W., and Wiest, D. R., 1988, Genome rearrangement and nitrogen fixation in *Anabaena* blocked by inactivation of *xisA* gene, *Science* **242:**1421–1423.

Golden, J. W., Robinson, S. J., and Haselkorn, R., 1985, Rearrangement of nitrogen fixation genes during differentiation in the cyanobacterium *Anabaena*, *Nature* **314:**419–423.

Golubic, S., and Focke, J. A., 1978, *Phormidium hendersonii* Howe: Identity and significance of a modern stromatolite building microorganism, *J. Sediment. Petrol.* **48:**751–764.

Gomont, M., 1892, Monographie des Oscillariées, *Ann. Sci. Nat. Ser. Bot.* **9:**49–53.

Grainger, S. L. J., Peat, A., Tiwari, D. N., and Whitton, B. A., 1989, Phosphomonoesterase activity of the cyanobacterium (blue-green alga) *Calothrix parietina*, *Microbios* **59:**7–17.

Grillo, J. F., and Gibson, J., 1979, Regulation of phosphate accumulation in the unicellular cyanobacterium *Synechococcus*, *J. Bacteriol.* **140:**508–517.

Grobbelaar, N., Huang, T.-C., Lin, H.-Y., and Chow, T.-J., 1987, Dinitrogen-fixing endogenous rhythm in *Synechococcus* RF-1, *FEMS Microbiol. Lett.* **37:**173–177.

Häder, D.-P., 1987, Photomovement, in: *The Cyanobacteria* (P. Fay and C. Van Baalen, eds.), Elsevier, Amsterdam, pp. 325–345.

Hagemann, M., Erdmann, N., and Wittenburg, E., 1989, Studies concerning enzyme activities in salt-loaded cells of the cyanobacterium *Microcystis firma*, *Biochem. Physiol. Pflanz.* **184:**87–94.

Hagemann, M., Wölfel, L., and Krüger, B., 1990, Alterations of protein synthesis in the cyanobacterium *Synechocystis* sp. PCC 4803 after a salt shock, *J. Gen. Microbiol.* **136:**1393–1399.

Hardie, L. P., Balkwill, D. L., and Stevens, S. E., Jr., 1983a, Effects of iron starvation on the physiology of the cyanobacterium *Agmenellum quadruplicatum*, *Appl. Environ. Microbiol.* **45:**999–1006.

Hardie, L. P., Balkwill, D. L., and Stevens, S. E., Jr., 1983b, Effects of iron starvation on the ultrastructure of the cyanobacterium *Agmenellum quadruplicatum*, *Appl. Environ. Microbiol.* **45:**1007–1017.

Hawley, G. R. W., and Whitton, B. A., 1991a, Seasonal changes in chlorophyll-containing picoplankton populations of two lakes in northern England, *Int. Rev. Ges. Hydrobiol.* **76:**545–554.

Hawley, G. R. W., and Whitton, B. A., 1991b, Survey of algal picoplankton from lakes in five continents, *Verh. Int. Ver. Limnol.* **24:**1220–1222.

Healey, F. P., 1973, Characteristics of phosphorus deficiency in *Anabaena*, *J. Phycol.* **9:**383–394.

Healey, F. P., 1982, in: *The Biology of Blue-Green Algae* (N. G. Carr and B. A. Whitton, eds.), Blackwell, Oxford, and University of California Press, Berkeley, pp. 105–124.

Herdman, M., 1987, Akinetes: Structure and function, in: *The Cyanobacteria* (P. Fay and C. Van Baalen, eds.), Elsevier, Amsterdam, pp. 227–250.

Herdman, M., and Rippka, R., 1988, Hormogonia and baeocytes, *Meth. Enzymol.* **167:**232–242.

Hickel, B., and Pollingher, U., 1988, Mass development of an iron precipitating cyanophyte (*Cyanodictyon imperfectum*) in a subtropical lake (Lake Kinneret, Israel), *Phycologia* **27:**291–297.

Hirosawa, T., and Wolk, C. P., 1979, Isolation and characterization of a substance which stimulates the formation of akinetes in the cyanobacterium *Cylindrospermum licheniforme* Kütz., *J. Gen. Microbiol.* **114:**433–441.

Hoffmann, L., 1988a, Criteria for the classification of blue-green algae (cyanobacteria) at the genus and at the species level, *Arch. Hydrobiol.* **80:**131–139.

Hoffmann, L., 1988b, The development of hormogonia, a possible taxonomic criterium in false branching blue-green algae (Cyanophyceae, Cyanobacteria), *Arch. Protistenk.* **135:**41–43.

Hoffman, L., 1988c, Algae of terrestrial habitats, *Bot. Rev.* **55:**77–105.

Howarth, R. W., Marino, R., Lane, J., and Cole, J. J., 1988a, Nitrogen fixation in freshwater, estuarine, and marine ecosystems. 1. Rates and importance, *Limnol. Oceanogr.* **33:**669–687.

Howarth, R. W., Marino, R., and Cole, J. J., 1988b, Nitrogen fixation in freshwater, estuarine, and marine ecosystems. 2. Biogeochemical controls, *Limnol. Oceanogr.* **33:**688–701.

Huang, T.-C., and Chow, T.-J., 1990, Characterization of the rhythmic nitrogen-fixing activity of *Synechococcus* sp. RF-1 at the transcription level, *Curr. Microbiol.* **20:**23–26.

Huang, T.-C., Tu, J., Chow, T.-J., and Chen, T.-H., 1990, Circadian rhythm of the prokaryote *Synechococcus* sp. RF-1, *Plant Physiol.* **92:**531–533.

Huntsman, S. A., and Sunda, W. G., 1980, The role of trace metals in regulating phytoplanton growth, in: *The Physiological Ecology of Phytoplankton* (I. Morris, ed.), Blackwell, Oxford, pp. 285–328.

Hutber, G. N., Hutson, K. G., and Rogers, L. J., 1977, Effect of iron deficiency on levels of two ferredoxins and flavodoxins in a cyanobacterium, *J. Bacteriol.* **170:**258–265.

Islam, R. I., and Whitton, B. A., in press, Cell composition and nitrogen-fixing activity of the deepwater rice-field cyanobacterium (blue-green alga) *Calothrix* D764, *Microbios.*

Jaag, O., 1945, Untersuchungen über die Vegetation und Biologie der Algen des nackten Gesteins in den Alpen, im Jura und im schweizerischen Mittelland, *Beitr. Kryptogamen flora Schweiz* **9:**1–560.

Janssen, I., Jakowitsch, J., Michalowski, C. B., Bohnert, H. J., and Löffelhardt, W., 1989, Evolutionary relationship of *psbA* genes from cyanobacteria, cyanelles and plastids, *Curr. Genet.* **15:**335–340.

Jeeji-Bai, N., 1977, Morphological variation of some species of *Calothrix* and *Fortiea*, *Arch. Protistenk.* **119:** 367–387.

Jensen, T. E., 1985, Cell inclusions in the cyanobacteria, *Arch. Hydrobiol.* (Suppl.) **71:**33–73.

Jensen, T. E., and Clark, R. L., 1969, Cell wall and coat of the developing akinete of a *Cylindrospermum* species, *J. Bacteriol.* **97:**1494–1495.

Jensen, T. E., Baxter, M., Rachlin, J. W., and Jani, V., 1982, Uptake of heavy metals by *Plectonema boryanum* (Cyanophyceae) into cellular compartments, especially polyphosphate bodies: An X-ray energy dispersive study, *Environ. Pollut. A* **27:**119–127.

Johnson, P. W., and Sieburth, J. Mc. N., 1979, Chroococcoid cyanobacteria in the sea: a ubiquitous and diverse phototrophic biomass, *Limnol. Oceanogr.* **24:**928–935.

Jones, K., and Stewart, W. D. P., 1969, Nitrogen turnover in marine and brackish habitats. III. The production of extracellular nitrogen by *Calothrix scopulorum*, *J. Mar. Biol. Assoc. U. K.* **49:**475–483.

Jørgensen, B. B., and Nelson, D. C., 1988, Bacterial zonation, photosynthesis, and spectral light distribution in hot spring microbial mats of Iceland, *Microb. Ecol.* **16:**133–147.

Jüttner, F., 1987, Volatile organic substances, in: *The Cyanobacteria* (P. Fay and C. Van Baalen, eds.), Elsevier, Amsterdam, pp. 453–469.

Jüttner, F., Höflacher, B., and Wurster, K., 1986, Seasonal analysis of volatile organic bigenic substances (VOBS) in freshwater phytoplankton populations dominated by *Dinobryon*, *Microcystis* and *Aphanizomenon*, *J. Phycol.* **22:**169–175.

Kann, E., 1972a, Zur Systematik und Ökologie der Gattung *Chamaesiphon* (Cyanophyceae) 1. Systematik, *Arch. Hydrobiol.* (Suppl.) **41:**117–171.

Kann, E., 1972b, Zur Systematik und Ökologie der Gattung *Chamaesiphon* (Cyanophyceae) 2. Ökologie, *Arch. Hydrobiol.* (Suppl.) **41:**243–282.

Kann, E., 1973, Bemerkungen zur Systematik und Ökologie einiger mit Kalk inkrustierter *Phormidium*arten, *Schweiz. Z. Hydrol.* **35:**141–151.

Kann, E., and Komarek, J., 1970, Systematish-Ökologishe Bererkungen zu den Arten des Formenkreis *Phormidium autumnale, Schweiz. Z. Hydrol.* **32:**495–518.

Kay, S. H., Quimbu, P. C., Jr., and Ouzts, J. D., 1982, H_2O_2: A potential algicide for aquaculture, in: *Proceedings 35th Annual Meeting of the Southern Weed Science Society*, Atlanta, Georgia, pp. 275–289.

Kemp, H. T., Fuller, R. G., and Davidson, R. S., 1966, Potassium permanganate as an algicide, *J. Am. Water Works Assoc.* **58:**255–263.

Kerby, N. W., Rowell, P., and Stewart, W. D. P., 1989, The transport, assimilation and production of nitrogenous compounds by cyanobacteria and microalgae, in: *Algal and Cyanobacterial Biotechnology* (R. C. Cresswell, T. A. V., Rees, and N. Shah, eds.), Longman, Avon, England, pp. 50–90.

Kerry, A., Laudenbach, D. E., and Trick, C. G., 1988, Influence of iron limitation and nitrogen source on growth and siderophore production by cyanobacteria, *J. Phycol.* **24:**566–571.

Knoll, A. H., 1985, The distribution and evolution of microbial life in the late Proterozoic era, *Annu. Rev. Microbiol.* **39**:391–417.

Komarek, J., 1983, Review (of F. Drouet, 1981, Revision of the Stigonemataceae with a Summary of the Classification of Blue-green Algae), *Arch. Hydrobiol.* (Suppl.) **67**:113–114.

Komarek, J., and Anagnostidis, K., 1988, Modern approach to the classification system of cyanophytes, *Arch. Hydrobiol.* **73**:157–226.

Komarek, J., and Anagnostidis, K., 1989, Modern approach to the classification system of cyanophytes 4—Nostocales, *Arch. Hydrobiol.* **82**:247–345.

Komarek, J., and Kann, E., 1973, Zur Taxonomie und Ökologie der Gattung *Homoeothrix*, *Arch. Protistenk.* **115**:173–233.

Konig, A., 1985, Ecophysiological studies on some algae and bacteria of waste stabilization ponds, Ph. D. thesis, University of Liverpool, England.

Kratz, W. A., and Myers, J., 1955, Nutrition and growth of several blue-green algae, *Am. J. Bot.* **42**:275–280.

Kylin, H., 1927, Uber die karotinoiden Farbstoffe der Algen, *Hoppe-Seyler's Z. Physiol. Chem.* **166**:39–77.

Kylin, H., 1943, Zur Biochemie der Cyanophyceen, *Kunglina Fysiografiska Sällskapets i Lund Förhandlinger* **13**:64–77.

Lamont, H. C., 1969, Sacrificial cell death and trichome breakage in an Oscillatoriacean blue-green alga—The role of murein, *Arch. Mikrobiol.* **69**:237–259.

Laudenbach, D. E., and Straus, N. A., 1988, Characterization of a cyanobacterial iron stress-induced gene similar to *psbC*, *J. Bacteriol.* **170**:5018–5026.

Laudenbach, D. E., Reith, M. E., and Straus, N. A., 1988, Isolation, sequence analysis, and transcriptional studies of the flavodoxin gene from *Anacystis nidulans* R2, *J. Bacteriol.* **170**:258–265.

Lawry, N. H., and Jensen, T. E., 1986, Condensed phosphate deposition, sulfur amino acid use, and unidirectional transsulfuration in *Synechococcus leopoliensis*, *Arch. Microbiol.* **144**:317–323.

Lazaroff, N., 1973, Photomorphogenesis and Nostocacean development, in: *The Biology of Blue-Green Algae* (N. G. Carr and B. A. Whitton, eds.), Blackwell, Oxford, and University of California Press, Berkeley, pp. 279–319.

Lean, D. R. S., and Nalewajko, C., 1976, Phosphate exchange and organic phosphorus excretion by freshwater algae, *J. Fish. Res. Board Can.* **33**:1312–1323.

Lee-Kaden, J., and Simonis, W., 1982, Amino acid uptake and energy coupling dependent on photosynthesis in *Anacystis nidulans*, *J. Bacteriol.* **151**:229–236.

Leon, C., Kumazawa, S., and Mitsui, A., 1986, Cyclic appearance of aerobic nitrogenase activity during synchronous growth of unicellular cyanobacteria, *Curr. Microbiol.* **13**:149–153.

Levine, E., and Thiel, T., 1987, UV-inducible DNA repair in the cyanobacteria *Anabaena* spp., *J. Bacteriol.* **169**:3988–3993.

Lewin, R. A., 1976, Naming the blue-greens, *Nature* **259**:360.

Lewin, R. A., 1989, Order Prochlorales, in: *Bergey's Manual of Systematic Bacteriology*, Volume 3 (J. T. Staley, M. P. Bryant, N. Pfennig, and J. G. Holt, eds.), Williams and Wilkins, Baltimore, pp. 1799–1806.

Linnaeus, C., 1753, *Species Plantarum, Exhibentes Plantas Rite Cognitas, et Genera Relatas, cum Differentus Specificis, Nominibus Trivialibus, Synonymis Selectis, Locis Natalibus, Secundum Systema Sexuale Digestas II*, Stockholm, pp. 561–1200.

Livingstone, D., Pentecost, A., and Whitton, B. A., 1984, Diel variations in nitrogen and carbon dioxide fixation by the blue-green alga *Rivularia* in an upland stream, *Phycologia* **23**:125–133.

Mackay, M. A., Norton, R. S., and Borowitzka, L. J., 1983, Marine blue-green algae have a unique osmoregulatory system, *Mar. Biol.* **73**:301–307.

Mackey, E. J., and Smith, G. D., 1983, Adaptation of the cyanobacterium *Anabaena cylindrica* to high oxygen tension, *FEBS Lett.* **156**:108–112.

Mahasneh, I. A., Grainger, S. L. J., and B. A. Whitton, 1990, Influence of salinity on hair formation and phosphatase activities of the blue-green alga (cyanobacterium) *Calothrix viguieri* D253, *Br. Phycol. J.* **25**:25–32.

Mann, N., and Carr, N. G., 1974, Control of macromolecular composition and cell division in the blue-green alga *Anacystis nidulans, J. Gen. Microbiol.* **83**:399–405.

Martin, T. C., and Wyatt, J. T., 1974, Extracellular investments in blue-green algae with particular emphasis on the genus *Nostoc, J. Phycol.* **10**:204–210.

Marzolf, G. R., and Saunders, G. W., 1984, Patterns of diel oxygen changes in ponds of tropical India, *Verh. Int. Ver. Limnol.* **22**:1722–1726.

Mathijs, H. C. P., Burger-Wiersma, T., and Mur, L. R., 1989, A status report on *Prochlorothrix hollandica* a free-living prochlorophyte, in: *Prochloron A Microbial Enigma* (R. A. Lewin and L. Cheng, eds.), Chapman & Hall, New York, pp. 83–87.

Matsumoto, A., and Tsuchiya, Y., 1988, Earth-musty odor-producing cyanophytes isolated from five water areas in Tokyo, *Water Sci. Technol.* **20**:179–183.

May, V., 1989, Long-term observations on *Anabaena circinalis* Rabenhorst (Cyanophyta), *Hydrobiologia* **179**:237–244.

Mitsui, A., Kumazawa, S., Takahashi, A., Ikemoto, H., Cao, S., and Arai, T., 1986, Strategy by which nitrogen-fixing unicellular cyanobacteria grow photoautotrophically, *Nature* **323**:720–722.

Mohammad, M. A., Reed, R. H., and Stewart, W. D. P., 1983, The halophilic cyanobacterium, *Synechocystis* DUN 52 and its osmotic responses, *FEMS Microbiol. Lett.* **16**:287–290.

Monty, C. L. V., 1965, Recent algal stromatolites in the Windward Lagoon, Andros Island, Bahamas, *Ann. Soc. Géol. Belg. Bull.* **96**:585–624.

Monty, C. L. V., 1979, Scientific reports of the Belgian expedition in the Australian Great Barrier Reefs, 1967. Sedimentology: 2. Monospecific stromatolites from the Great Barrier Reef Tract and their palaeontological significance, *Ann. Soc. Géol. Belg. Bull.* **101** (1978):163–171.

Moore, R. E., 1981, Toxins from marine blue-green algae, in: *The Water Environment. Algae and Health* (W. W. Carmichael, ed.), Plenum Press, New York, pp. 15–23.

Morden, C. W., and Golden, S. S., 1989, *psnA* genes indicate common ancestry of prochlorophytes and chloroplasts, *Nature* **337**:382–385.

Naes, H., Utkilen, H. C., and Post, A. F., 1988, Factors influencing geocosmin production by the cyanobacterium *Oscillatoria brevis, Water Sci.* **20**:125–131.

Nair, V. R., Devassy, V. P., and Qasim, S. Z., 1980, Zooplankton & *Trichodesmium* phenomenon, *Indian J. Mar. Sci.* **9**:1–6.

Nalewajko, C., and Lean, D. R. S., 1978, Phosphorus kinetics—algal growth relationships in batch cultures, *Mitt. Int. Ver. Theor. Angew. Limnol.* **21**:184–192.

Natesan, R., and Shanmugasundaram, S., 1989, Extracellular phosphate solubilization by the cyanobacterium *Anabaena* ARM310, *J. Biosci.* **14**:203–208.

Nichols, J. M., and Adams, D. G., 1982, Akinetes, in: *The Biology of Cyanobacteria* (N. G. Carr and B. A. Whitton, eds.), Blackwell, Oxford, and University of California Press, Berkeley, pp. 387–412.

O'Brien, P. A., and Houghton, J. A., 1982, Photoreactivation and excision repair on UV induced pyrimidine dimers in the unicellular cyanobacterium *Gloeocapsa alpicola* (*Synechocystis* PCC 6803), *Photochem. Photobiol.* **36**:417–422.

Ogawa, R. E., and Carr, J. E., 1969, The influence of nitrogen on heterocyst production in blue-green algae, *Limnol. Oceanogr.* **14**:342–351.

Oliver, R. L., Thomas, R. H., Reynolds, C. S., and Walsby, A. E., 1985, The sedimentation of buoyant *Microcystis* colonies caused by precipitation with an iron-containing colloid, *Proc. R. Soc. Lond. B* **223**:511–528.

Paasche, E., 1960, On the relationship between primary production and standing stock of phytoplankton, *J. Cons. Perm. Int. Explor. Mer.* **26**:33–48.

Padan, E., and Cohen, Y., 1982, Anoxygenic photosynthesis, in: *The Biology of Cyanobacteria* (N.

G. Carr and B. A. Whitton B. A., eds.), Blackwell, Oxford, and University of California Press, Berkeley, pp. 215–235.

Paerl, H. W., 1988, Nuisance phytoplankton blooms in coastal, estuarine, and inland waters, *Limnol. Oceanogr.* **333**:823–847.

Paerl, H. W., 1990, Physiological ecology and regulation of N_2 fixation in natural waters, *Adv. Microb. Ecol.* **11**:305–344.

Paerl, H. W., and Carlton, R. G., 1988, Control of N_2 fixation by oxygen depletion in surface-associated microzones, *Nature* (London) **332**:260–262.

Paerl, H. W., Bebout, B. M., and Prufert, L. E., 1989, Naturally occurring patterns of oxygenic photosynthesis and N_2 fixation in a marine microbial mat: physiological and ecological ramifications, in: *Microbial Mats* (Y. Cohen and E. Rosenberg, eds.) pp. 326–241, American Society for Microbiology, Washington, D. C.

Palmer, R. J., Jr., and Friedmann, E. I., 1990, Water relations and photosynthesis in the cryptoendolithic microbial habitat of hot and cold deserts, *Microb. Ecol.* **19**:111–118.

Pandey, K. D., and Kashyap, A. K., 1987, Factors affecting formation of spores (akinetes) in cyanobacterium *Anabaena doliolum* (AdS strain), *J. Plant Physiol.* **127**:123–134.

Pandey, R. K., and Talpasayi, E. R. S., 1980, Control of sporulation in a blue-green alga *Nodularia spumigena* Mertens, *Indian J. Bot.* **3**:128–133.

Pankow, H., 1986, Über endophytische und epiphytische Algen in bzw. auf der Gallerthulle von *Microcystis*-Kolonien, *Arch. Protistenk.* **132**:377–380.

Pardy, R. L., 1989, *Prochloron* in symbiosis, in: *Prochloron A Microbial Enigma* (R. A. Lewin and L. Cheng, eds.), Chapman and Hall, New York, pp. 19–29.

Pearson, H. W., Howsley, R., Kjeldson, C. K., and Walsby, A. E., 1979, Aerobic nitrogenase activity by axenic cultures of the blue-green alga *Microcoleus chthonoplastes*, *FEMS Microbiol. Lett.* **5**:163–167.

Peat, A., Powell, N., and Potts, M., 1988, Ultrastructural analysis of the rehydration of desiccated *Nostoc commune* HUN (Cyanobacteria) with particular reference to the immunolabelling of *NifH*, *Protoplasma* **146**:72–80.

Penny, D., 1989, What, if anything, is *Prochloron? Nature* **337**:304–305.

Pentecost, A., 1985, Investigation of variation in heterocyst numbers, sheath development and false-branching in natural populations of Scytonemataceae (Cyanobacteria), *Arch. Hydrobiol.* **102**:343–353.

Peterson, R. B., and Wolk, C. P., 1978, High recovery of nitrogenase activity and of [55]Fe-labeled nitrogenase in heterocysts isolated from *Anabaena variabilis*, *Proc. Natl. Acad. Sci. USA* **75**:6271–6275.

Pettersson, A., and Bergman, B., 1989, Effects of aluminum on ATP pools and utilization in the cyanobacterium *Anabaena cylindrica:* A model for the *in vivo* toxicity, *Physiol. Plant.* **76**:527–534.

Pettersson, A., Hällbom, L., and Bergman, B., 1988, Aluminium effects on uptake and metabolism of phosphorus by the cyanobacterium *Anabaena cylindrica, Plant Physiol.* **86**:112–116.

Pierce, J., and Omata, T., 1988, Uptake and utilization of inorganic carbon by cyanobacteria, *Photosynth. Res.* **16**:141–154.

Potts, M., and Friedmann, E. I., 1981, Effects of water stress on cryptoendolithic cyanobacteria from hot desert rocks, *Arch. Microbiol.* **130**:267–271.

Potts, M., and Whitton, B. A., 1977, Nitrogen fixation by blue-green algal communities in the intertidal zone of the lagoon of Aldabra Atoll, *Oecologia* (Berl.) **27**:275–283.

Potts, M., Bowman, M. A., and Morrison, N. S., 1984, Control of matrix potential (πm) in immobilized cultures of cyanobacteria, *FEMS Microbiol. Lett.* **24**:193–196.

Procter, L. M., and Fuhrman, J. A., 1990, Viral mortality of marine bacteria and cyanobacteria, *Nature* **343**:60–62.

Rambler, M., Margulis, L., and Barghoorn, E. S., 1977, Natural mechanisms of protection of a blue-green alga against ultra-violet light, in: *Chemical Evolution of the Early Precambrian* (C. Ponnamperuma, ed.), Academic Press, New York, pp. 133–141.

Rana, B. C., Gopal, T., and Kumar, H. D., 1971, Studies on the biological effects of industrial wastes on the growth of algae, *Environ. Health* **13**:138–143.

Reddy, K. J., Bullerjahn, G. S., Sherman, D. M., and Sherman, L. A., 1988, Cloning, nucleotide sequence, and mutagenesis of a gene (*irpA*) involved in iron-deficient growth of the cyanobacterium *Synechococcus* sp. strain PCC7942, *J. Bacteriol.* **170**:4466–4476.

Reddy, P. M., 1983, Changes in polyphosphate bodies during sporulation and spore germination in cyanobacteria, *Biochem. Physiol. Pflanz.* **178**:77–79.

Reed, R. H., and Stewart, W. D. P., 1983, Physiological responses of *Rivularia atra* to salinity: Osmotic adjustment in hyposaline medium, *New Phytol. Reed,* **95**:595–603.

Reed, R. H., Richardson, D. L., Warr, S. R. C., and Stewart, W. D. P., 1984, Carbohydrate accumulation and osmotic stress in cyanobacteria, *J. Gen. Microbiol.* **130**:1–4.

Reed, R. H., Richardson, D. L., and Stewart, W. D. P., 1985, Na$^+$ uptake and extrusion in the cyanobacterium *Synechocystis* PCC 6714 in response to hypersaline treatment. Evidence for transient changes in plasmalemma, *Biochim. Biophys. Acta* **814**:347–355.

Reed, R. H., Borowitzka, L. J., Kackay, M. A., Chudek, J. A., Foster, R., Warr, S. R. C., Moore, D. J., and Stewart, W., D. P., 1986a, Organic solute accumulation in osmotically stessed cyanobacteria, *FEMS Microbiol. Rev.* **39**:51–56.

Reed, R. H., Warr, S. R. C., Kerby, N. W., and Stewart, W. D. P., 1986b, Osmotic shock-induced release of low molecular weight metabolites from free-living and immobilized cyanobacteria, *Enzyme Microb. Technol.* **8**:101–104.

Revsbech, N. P., and Ward, D. M., 1984, Microelectrode studies of interstitial water chemistry and photosynthetic activity in a hot spring microbial mat, *Appl. Environ. Microbiol.* **48**:270–275.

Reynolds, C. S., 1987, Cyanobacterial water-blooms, *Adv. Bot. Res.* **13**:68–143.

Reynolds, C. S., Jaworski, G. H. M., Cmiech, H. A., and Leedale, F. F., 1981, On the annual cycle of the blue-green *Microcystis aeruginosa* Kütz. emend Elenkin, *Phil. Trans. R. Soc. Lond. B* **293**:419–477.

Riethman, H. C., and Sherman, L. A., 1988, Immunological characterization of iron-regulated membrane proteins in the cyanobacterium *Anacystis nidulans* R2, *Plant Physiol.* **88**:497–505.

Rippka, R., 1988, Recognition and identification of cyanobacteria, *Meth. Enzymol.* **167**:28–67.

Rippka, R., Deruelles, J. B., Waterbury, J. B., Herdman, M., and Stanier, R. Y., 1979, Generic assignments, strain histories and properties of pure cultures of cyanobacteria, *J. Gen. Microbiol.* **111**:1–61.

Robinson, B. L., and Miller, J. H., 1970, Photomorphogenesis in the blue-green alga *Nostoc commune* 584, *Physiol. Plant.* **23**:461–472.

Roger, P. A., and Kulasooriya, S. A., 1980, *Blue-Green Algae and Rice,* International Rice Research Institute, Los Baños, Philippines.

Roman, M. R., 1978, Ingestion of the blue-green alga *Trichodesmium* by the harpactacoid copepod, *Macrosetella gracilis, Limnol. Oceanogr.* **23**:1245–1248.

Roncel, M., Navarro, J. A., and de la Rosa, M. A., 1989, Coupling of solar energy to hydrogen peroxide production in the cyanobacterium *Anacystis nidulans, Appl Environ. Microbiol.* **55**:483–487.

Rother, J. A., and Fay, P., 1979, Some physiological characteristics of planktonic blue-green algae during bloom formation in three Salopian meres, *Freshwater Biol.* **9**:369–370.

Rother, J. A., Aziz, A., Hye Karim, N., and Whitton, B. A., 1988, Ecology of deepwater rice-fields in Bangladesh. 4. Nitrogen fixation by blue-green algal communities, *Hydrobiologia* **169**:43–56.

Roychoudhury, P., and Kaushik, B. D., 1989, Solubilization of Mussorie rock phosphate by cyanobacteria, *Curr. Sci.* **58**:569–570.

Rueter, J. G., 1988, Iron stimulation of photosynthesis and nitrogen fixation in *Anabaena* 7120 and *Trichodesmium* (Cyanophyceae), *J. Phycol.* **24**:249–254.

Rueter, J. G., Ohki, K., and Fujita, Y., 1990, The effect of iron nutrition on photosynthesis and nitrogen-fixation of *Trichodesmium* (Cyanophyceae), *J. Phycol.* **26**:30–35.

Sagan, C., 1965, in: *The Origins of Prebiological Systems and of Their Molecular Matrices* (S. W. Fox, ed.), Academic Press, New York.

Saito, N., and Werbin, H., 1970, Purification of a blue-green algal deoxyribonucleic acid photoreactivation enzyme. An enzyme requiring light as a physical cofactor to perform its catalytic function, *Biochemistry* (N.Y.) **9:**2610–2620.

Sarma, T. A., and Khattar, J. I. S., 1986, Accumulation of cyanophycin and glycogen during sporulation in the blue-green alga *Anabaena torulosa, Biochem. Physiol. Pflanz.* **181:**155–164.

Scanlan, D. J., Mann, N. H., and Carr, N. G., 1989, Effect of iron and other nutrient limitations on the pattern of outer membrane proteins in the cyanobacterium *Synechococcus* PCC7942, *Arch. Microbiol.* **152:**224–228.

Schaefer, M. R., and Golden, S. S., 1989, Differential expression of members of a cyanobacterial *psbA* gene family in response to light, *J. Bacteriol.* **171:**3973–3981.

Scherer, S., and Potts, M., 1989, Novel water stress protein from a desiccation-tolerant cyanobacterium, *J. Bacteriol.* **264:**12546–12553.

Scherer, S., Ernst, A., Chen, T.-W., and Böger, P., 1984, Rewetting of drought-resistant blue-green algae: Time course of water uptake and reappearance of respiration, photosynthesis, and nitrogen fixation, *Oecologia* (Berl.) **62:**418–423.

Scherer, S., Chen, T. W., and Böger, P., 1988, A new UV A/B protecting pigment in the terrestrial cyanobacterium *Nostoc commune, Plant Physiol.* **88:**1055–1057.

Schopf, J. W., and Walter, M. R., 1982, Origin and evolution of cyanobacteria: The geological evidence, in: *The Biology of Cyanobacteria* (N. G. Carr and B. A. Whitton, eds.), Blackwell, Oxford, and University of California Press, Berkeley, pp. 543–564.

Schwabe, G. H., and Mollenhauer, R., 1967, Über den Einfluss der Begleitbakterien auf das Lagerbild von *Nostoc sphaericum, Nova Hedwigia* **13:**77–80.

Severina, I. I., Skulachev, V. P., and Fedorova, N. D., 1989, Specific features of the energetics and motility of cyanobacterium *Synechococcus* WH 8113, *Biol. Membrany* **6**(4):434–436 [in Russian, with English abstract].

Sherman, D. A., and Sherman, L. A., 1983, The effects of iron deficiency and iron restoration on the ultrastructure of the cyanobacterium, *Anacystis nidulans, J. Bacteriol.* **156:**393–401.

Sherman, L., Bricker, T., Guikema, J., and Pakrasi, H., 1987, The protein composition of the photosynthetic complexes from the cyanobacterial thylakoid membrane, in: *The Cyanobacteria* (P. Fay and C. Van Baalen, eds.), Elsevier, Amsterdam, pp. 1–33.

Sherr, E. B., and Sherr, B. F., 1987, High rates of consumption of bacteria by pelagic ciliates, *Nature* **325:**710–711.

Shilo, M., 1989, The unique characteristics of benthic cyanobacteria, in: *Microbial Mats* (Y. Cohen and E. Rosenberg, eds.), American Society for Microbiology, Washington, D. C., pp. 207–213.

Sicko-Goad, L., and Jensen, T. E., 1976, Phosphate metabolism in blue-green algae. II. Changes in phosphate distribution during starvation and the 'polyphosphate' overplus phenomenon in *Plectonema boryanum, Can. J. Microbiol.* **24:**105–108.

Sicko-Goad, L., and Lazinsky, D., 1986, Quantitative ultrastructural changes associated with lead-coupled luxury phosphate uptake and polyphosphate utilization, *Arch. Environ. Contam. Taxicol.* **15:**617–627.

Simon, R. D., 1987, Inclusion bodies in the cyanobacteria: Cyanophycin, polyphosphate, polyhedral bodies, in: *The Cyanobacteria* (P. Fay and C. Van Baalen, eds.), Elsevier, Amsterdam, pp. 199–225.

Simon, R. D., and Weathers, P., 1976, Determination of the structure of the novel polypeptide containing aspartic acid and arginine which is found in cyanobacteria, *Biochim. Biophys. Acta* **420:**165–176.

Simonis, W., Bornefeld, T., Lee-Kaden, J., and Majumdar, K., 1974, Phosphate uptake and photophosphorylation in the blue-green alga *Anacystis nidulans,* in: *Membrane Transport in Plants* (U. Zimmerman and J. Dainty, eds.) pp. 220–225, Springer, Berlin.

Sinclair, C., and Whitton, B. A., 1977, Influence of nutrient deficiency on hair formation in the Rivulariaceae, *Br. Phycol. J.* **12:**297–313.

Singh, R. N., and Tiwari, D. N., 1970, Frequent heterocyst germination in the blue-green alga *Gloeotrichia ghosei* Singh, *J. Phycol.* **6:**172–176.

Singh, S., and Rai, A. K., 1990, Nickel-dependent growth and urea uptake in the cyanobacteria *Anabaena doliolum* and *Anacystis nidulans, Indian J. Exp. Biol.* **28:**80–82.

Sirenko, L. A., Stetsenko, N. M., Arendarchuk, V. V., and Kuz'menko, M. I., 1968, Role of oxygen conditions in the vital activity of certain blue-green algae, *Mikrobiol. Transl.* **37**(2):199–202.

Srivastava, B. S., Kumar, H. D., and Singh, H. N., 1971, The effect of caffeine and light on killing of the blue-green alga *Anabaena doliolum* by ultraviolet radiation, *Arch. Microbiol.* **78:**139–144.

Stacey, G. C., Van Baalen, C., and Tabita, F. R., 1977, Isolation and characterization of a marine *Anabaena* sp. capable of rapid growth on molecular nitrogen, *Arch. Microbiol.* **114:**197–201.

Stal, L. J., Grossberger, S., and Krumbein, W. E., 1984, Nitrogen fixation associated with the cyanobacterial mat of a marine laminated microbial ecosystem, *Mar. Biol.* **82:**217–224.

Stal, L. J., Heyer, H., Bekker, S., Villbrandt, M., and Krumbein, W. E., 1989, Aerobic–anaerobic metabolism in the cyanobacterium *Oscillatoria limosa*, in: *Microbial Mats* (Y. Cohen and E. Rosenberg, eds.), American Society for Microbiology, Washington, D. C., pp. 255–276.

Staley, J. T., Bryant, M. P., Pfennig, N., and J. G. Holt (eds.), 1989, *Bergey's Manual of Systematic Bacteriology*, Volume 3, Williams and Wilkins, Baltimore, pp. 1710–1727.

Stam, W. T., 1980, Relationships between a number of filamentous blue-green algal strains (Cyanophyceae) revealed by DNA–DNA hybridization, *Arch. Hydrobiol.* (Suppl.) **56:**351–374.

Stam, W. T., and Holleman, H. C., 1979, Cultures of *Phormidium, Plectonema, Lyngbya* and *Synechococcus* (Cyanophyceae) under different conditions: Their growth and morphological variability, *Acta Bot. Neerl.* **26:**327–342.

Stam, W. T., and Venema, G., 1977, The use of DNA–DNA hybridization for determination of the relationship between some blue-green algae (Cyanophyceae), *Acta Bot. Neerl.* **26:**327–342.

Stanier, R. Y., and Cohen-Bazire, G., 1977, Phototrophic prokaryotes: The cyanobacteria, *Annu. Rev. Microbiol.* **31:**225–274.

Stanier, R. Y., Sistrom, W. R., Hansen, T. A., Whitton, B. A., Castenholz, R. W., Pfennig, N., Gorlenko, V. N., Kondratieva, E., M. N., Eimhjellen, K. E., Whittenbury, R., Gherna, R. L., and Trüper, H. G., 1978, Proposal to place the nomenclature of the cyanobacteria (blue-green algae) under the rules of the International Code of Nomenclature of Bacteria, *Int. J. Syst. Bacteriol.* **28:**335–336.

Steinberg, C. E. W., and Hartmann, H. M., 1988, Planktonic bloom-forming cyanobacteria and the eutrophication of lakes and rivers, *Freshwater Biol.* **20:**279–287.

Stevens, S. R., Jr., Patterson, C. O. P., and Myers, J., 1973, The production of hydrogen peroxide by blue-green algae: a survey, *J. Phycol.* **9:**427–430.

Stockner, J. G., 1988, Phototrophic picoplankton: An overview from marine and freshwater ecosystems, *Limnol. Oceanogr.* **33:**765–775.

Stockner, J. G., and Antia, N. J., 1986, Algal picoplankton from marine and freshwater ecosystems: A multidisciplinary approach, *Can. J. Fish. Aquat. Sci.* **43:**2472–2503.

Stockner, J. G., and Shortreed, K. S., 1988, Response of *Anabaena* and *Synechococcus* to manipulation of nitrogen–phosphorus ratios in a lake fertilization experiment, *Limnol. Oceanogr.* **33:**1348–1361.

Storch, T. A., and Dunham, V. L., 1986, Iron-mediated changes in the growth of Lake Erie phytoplankton and axenic algal cultures, *J. Phycol.* **22:**109–117.

Sutherland, J. M., Herdman, M., and Stewart, W. D. P., 1979, Akinetes of the cyanobacterium *Nostoc* PCC 7524: Macromolecular composition, structure and control of differentiation, *J. Gen. Microbiol.* **115**:273–287.

Thomas, J., 1972, Relationship between age of culture and occurrence of the pigments of photosystem II of photosynthesis in heterocysts of a blue-green alga, *J. Bacteriol.* **110**:92–95.

Trick, C. G., and Kerry, A., 1989, Isolation and purification of siderophores produced by cyanobacteria, *Anacystis nidulans* R2, *Anabaena variabilis*, and *Spirulina maxima*, *J. Phycol.* (Suppl.) **25**:(2):11.

Turner, S., Burger-Wiersma, T., Giovannoni, S. J., Mur, L. R., and Pace, N. R., 1989, The relationship of a prochlorophyte *Prochlorothrix hollandica* to green chloroplasts, *Nature* **337**:380–382.

Tyler, P. A., and Buckney, R. T., 1980, Ferromanganese concretions in Tasmanian lakes, *Aust. J. Mar. Freshwater Res.* **31**:525–531.

Van Baalen, C., 1968, The effects of ultraviolet irradiation on a coccoid blue-green alga: Survival, photosynthesis, and photoreactivation, *Plant Physiol.* **43**:1689–1695.

Van Baalen, C., 1987, Nitrogen fixation, in: *The Cyanobacteria* (P. Fay and C. Van Baalen, eds.), Elsevier, Amsterdam, pp. 187–198.

Van Baalen, C., and O'Donnell, R., 1972, Action spectra for ultraviolet killing and photoreactivation in the blue-green alga *Agmenellum quadruplicatum*, *Photochem. Photobiol.* **35**:359–364.

Van de Water, S. D., and Simon, R. D., 1982, Induction and differentiation of heterocysts in the filamentous cyanobacterium *Cylindrospermum licheniforme*, *J. Gen. Microbiol.* **128**:917–925.

Van Liere, L., and Walsby, A. E., 1982, Interactions of cyanobacteria with light, in: *The Biology of Cyanobacteria* (N. G. Carr and B. A. Whitton, eds.), Blackwell, Oxford, and University of California Press, Berkeley, pp. 9–45.

Vincent, W. (ed.), 1987, Dominance of bloom-forming cyanobacteria (blue-green algae), *N. Z. J. Mar. Freshwater Res.* **21**:361–542.

Walsby, A. E., 1972, Gas vacuoles, in: *The Biology of Blue-Green Algae* (N. G. Carr and B. A. Whitton, eds.), Blackwell, Oxford, and University of California Press, Berkeley, pp. 340–352.

Walsby, A. E., 1987, Mechanisms of buoyancy regulation by planktonic cyanobacteria with gas vesicles, in: *The Cyanobacteria* (P. Fay and C. Van Baalen, eds.) pp. 377–392, Elsevier, Amsterdam, New York, Oxford.

Ward, A. K., Dahm, C. N., and Cummins, K. W., 1985, *Nostoc* (Cyanophyta) productivity in Oregon stream ecosystems: Invertebrate influences and differences between different morphological types, *J. Phycol.* **21**:223–227.

Ward, D. M., Weller, R. Shiea, J., Castenholz, R. W., and Cohen, Y., 1989, Hot spring microbial mats: Anoxygenic and oxygenic mats of possible evolutionary significance, in: *Microbial Mats* (Y. Cohen and E. Rosenberg, eds.), American Society for Microbiology, Washington, D. C., pp. 3–15.

Warr, S. R. C., Reed, R. H., and Stewart, W. D. P., 1984, Physiological responses of *Nodularia harveyana* to osmotic stress, *Mar. Biol.* **79**:21–26.

Watanabe, M., Koyosawa, H., and Hayashi, H., 1985, Studies on planktonic blue-green algae 1. *Anabaena macrospora* Klebahn from Kizaki, *Bull. Natl. Sci. Mus. B Tokyo* **11**:69–76.

Waterbury, J. B., 1989, Subsection II. Order Pleurocapsales Geitler 1925, emend. Waterbury and Stanier 1978, in: *Bergey's Manual of Systematic Bacteriology*, Volume 3 (J. T. Staley, M. P. Bryant, N. Pfennig, and J. G. Holt, eds.), Williams and Wilkins, Baltimore, pp. 1747–1770.

Waterbury, J. B., Watson, S. W. Guillard, R. R. L., and Brand, L. E., 1979, Widespread occurrence of a unicellular marine planktonic cyanobacterium, *Nature* **277**:293–294.

Waterbury, J. B., Willey, J. M., Franks, D. G., Valois, F. W., and Watson, S. W., 1985, A cyanobacterium capable of swimming motility, *Science* **230**:74–76.

Werbin, H., and Rupert, C. S., 1968, Presence of photoreactiving enzyme in blue-green algal cells, *Photochem. Photobiol.* **7**:225–230.

Wheeler, P. A., 1983, Phytoplankton nitrogen metabolism, in: *Nitrogen in the Marine Environment* (E. Carpenter, ed.), Academic Press, New York.

Whitton, B. A., 1987a, The biology of Rivulariaceae, in: *The Cyanobacteria* (P. Fay and C. Van Baalen, eds.), Elsevier, Amsterdam, pp. 513–534.

Whitton, B. A., 1987b, Survival and dormancy of blue-green algae, in: *Survival and Dormancy of Microorganisms* (Y. Henis, ed.), Wiley, New York, pp. 109–167.

Whitton, B. A., 1980, Zinc and plants in rivers and streams, in: *Zinc in the Environment*, Part II: *Health Effects* (J. O. Nriagu, ed.), Wiley-Interscience, New York, pp. 415–437.

Whitton, B. A., 1988, Hairs in eukaryotic algae, in: *Algae and the Aquatic Environment* (F. E. Round, ed.), Biopress, Bristol, England, pp. 446–460.

Whitton, B. A., and Potts, M., 1982, Marine littoral, in: *The Biology of Cyanobacteria* (N. G. Carr and B. A. Whitton, eds.) pp. 515–542, Blackwell, Oxford, and University of California Press, Berkeley.

Whitton, B. A., Rother, J. A., and Paul, A. R., 1988a, Ecology of deepwater rice-fields in Bangladesh 2. Chemistry of sites at Manikganj and Sonargaon, *Hydrobiologia* **169**:21–30.

Whitton, B. A., Aziz, A., Kawecka, B., and Rother, J. A., 1988b, Ecology of deepwater rice-fields in Bangladesh 3. Associated algae and macrophytes, *Hydrobiologia* **169**:31–42.

Whitton, B. A., Potts, M., Simon, J. W., and Grainger, S. L. J., 1990, Phosphatase activity of the blue-green alga (cyanobacterium) *Nostoc commune* UTEX 584, *Phycologia* **29**:139–145.

Whitton, B. A., Grainger, S. L. J., Hawley, G. R. W., and Simon, J. W., 1991, Cell-bound and extracellular phosphatase activities of cyanobacterial isolates, *Microb. Ecol.* **21**:85–98.

Willey, J. M., and Waterbury, J. B., 1989, Chemotaxis toward nitrogenous compounds by swimming strains of marine *Synechococcus* spp., *Appl. Environ. Microbiol.* **55**:1888–1894.

Woese, C. R., 1987, Bacterial evolution, *Microbiol. Rev.* **51**:221–271.

Wolk, C. P., 1982, Heterocysts, in: *The Biology of Cyanobacteria* (N. G. Carr and B. A. Whitton, eds.), Blackwell, Oxford, and University of California Press, Berkeley, pp. 359–386.

Wu, J. H., Lewin, R. A., and Werbin, H., 1967, Photoreactivation of UV-irradiated blue-green algal virus LPP-1, *Virology* **31**:657–664.

Wyatt, J. T., and Silvey, J. K. G., 1969, Nitrogen fixation by *Gloeocapsa*, *Science* **165**:908–909.

Wyman, M., and Fay, P., 1987, Acclimation to the natural light climate, in: *The Cyanobacteria* (P. Fay and C. Van Baalen, eds.), Elsevier, Amsterdam, pp. 347–376.

Yopp, J. H., Miller, D. M., and Tindall, D. R., 1978, Regulation of introcellular water potential in the halophilic blue-green alga *Aphanothece halophytica* (Chroococcales), in: *Energetics and Structure of Halophilic Microorganisms* (S. R. Caplan and M. Ginzburg, eds.), Elsevier/North-Holland Biomedical Press, New York, pp. 619–624.

Zohary, T., 1985, Hyperscums of the cyanobacterium *Microcystis aeruginosa* in a hypertrophic lake (Harteespoort Dam, South Africa), *J. Plankton Res.* **7**:399–409.

Zohary, T., 1989, Cyanobacterial hyperscums of hypertrophic water bodies, in: *Microbial Mats* (Y. Cohen and E. Rosenberg, eds.), American Society for Microbiology, Washington, D. C., pp. 52–63.

Taxonomy, Phylogeny, and General Ecology of Anoxygenic Phototrophic Bacteria

2

JOHANNES F. IMHOFF

1. INTRODUCTION

Phototrophic bacteria, including oxygenic and anoxygenic phototrophic bacteria, can transform light energy into metabolically useful chemical energy by chlorophyll- or bacteriochlorophyll-mediated processes. Major differences between oxygenic and anoxygenic phototrophic bacteria relate to their photosynthetic pigments and the structure and complexity of the photosynthetic apparatus (Stanier *et al.*, 1981). Photosynthesis in anoxygenic phototrophic bacteria depends on oxygen-deficient conditions, because synthesis of the photosynthetic pigments is repressed by oxygen (bacteria like *Erythrobacter longus* are exceptions to this rule); in contrast to photosynthesis in plants and cyanobacteria (including *Prochloron* and related forms), oxygen is not produced. Unlike the cyanobacteria and eukaryotic algae, anoxygenic phototrophic bacteria are unable to use water as an electron donor. Most characteristically, sulfide and other reduced sulfur compounds, but also hydrogen and a number of small organic molecules, are used as photosynthetic electron donors. [Anoxygenic photosynthesis with sulfide, an inhibitor of photosystem II, as electron donor is also car-

Since the manuscript for this article was completed a number of new species of phototrophic bacteria have been described. These are the purple non-sulfur bacteria *Rhodopseudomonas julia* (Kompantseva, 1989), *Rhodospirillum centenum* (Favinger *et al.*, 1989), *Rhodopseudomonas cryptolactis* (Stadtwald-Demnick *et al.*, 1990), *Rhodopseudomonas rosea* (Janssen and Harfoot, 1991) and *Rhodoferax fermentans* (Hiraishi *et al.*, 1991), the purple sulfur bacterium *Thiocapsa halophila* (Caumette *et al.*, 1991), the green sulfur bacterium *Pelodictyon phaeoclathratiforme* (Overmann and Pfennig, 1989), and the aerobic bacteriochlorophyll containing bacteria *Erythrobacter sibiricus* (Yurkov and Gorlenko, 1990) and *Roseobacter litoralis* (Shiba, 1991).

JOHANNES F. IMHOFF ● Institut für Mikrobiologie und Biotechnologie, Rheinische Friedrich-Wilhelms-Universität, D-5300 Bonn, Germany.
Photosynthetic Prokaryotes, edited by Nicholas H. Mann and Noel G. Carr. Plenum Press, New York, 1992.

ried out by some cyanobacteria using photosystem I only (Cohen *et al.*, 1975; Garlick *et al.*, 1977).] As a consequence, the ecological niches of anoxygenic phototrophic bacteria are anoxic parts of waters and sediments, which receive light of sufficient quantity and quality to allow phototrophic development. Representatives of this group are widely distributed in nature and found in freshwater, marine, and hypersaline environments, hot springs, and arctic lakes, as well as elsewhere. They live in all kinds of stagnant water bodies, in lakes, waste water ponds, coastal lagoons, stratified lakes, and other aquatic habitats, but also in marine coastal sediments, in moist soils, and in paddy fields.

The anoxygenic phototrophic bacteria are an extremely heterogeneous eubacterial group, on the basis of both structural and physiological properties. They are treated taxonomically in a number of well-distinct families and groups, and also appear to be phylogenetically quite diverse. The various species of these bacteria contain several types of bacteriochlorophylls and a variety of carotenoids as pigments, which function in the transformation of light into chemical energy and give the cell cultures a distinct coloration varying with the pigment content from various shades of green, yellowish-green, brownish-green, brown, brownish-red, red, pink, purple, and purple-violet to even blue (carotenoidless mutants of some species containing bacteriochlorophyll *a*).

For methods of enrichment, isolation, and characterization of anoxygenic phototrophic bacteria the reader may consult Imhoff (1988c). These aspects are not treated here.

2. TAXONOMY

Traditionally two major groups of anoxygenic phototrophic bacteria are distinguished, the phototrophic green and the purple bacteria. A clear separation of these two groups is possible on the basis of pigment composition and structure of the photosynthetic apparatus. The various types of photosynthetic pigments are listed in Tables I and II.

The *phototrophic purple bacteria* [purple sulfur bacteria (Chromatiaceae and Ectothiorhodospiraceae) and purple non-sulfur bacteria (Rhodospirillaceae)] have intracytoplasmic membranes that are formed by invagination of the cytoplasmic membrane and are continuous with this membrane. These intracytoplasmic membranes consist of small fingerlike intrusions, vesicles, tubules, or lamellae parallel or at an angle to the cytoplasmic membrane. They carry the photosynthetic apparatus, the reaction centers and light-harvesting pigment−protein complexes surrounding the reaction center in the plane of the membrane. The pigments are bacteriochlorophyll *a* and *b* and carotenoids of the spirilloxanthin, okenone, or rhodopinal series (Schmidt, 1978).

Table I. Characteristic Absorption Maxima of Different Bacteriochlorophylls in Living Cells[a]

Bacteriochlorophyll	Esterifying alcohol[b]	Characteristic absorption maxima (nm)
a	P, Gg	375, 590, 800–810, 830–890
b[c]	P, Gg	400, 605, 835–850, 1015–1035
c[d]	F	335, 460, 745–760
d	F	325, 450, 725–745
e	F	345, 450–460, 715–725
g	Gg	370, 419, 575, 670, 770–790

[a]Data are collected from Brockmann and Lipinski (1983), Gloe et al. (1975), and Pfennig (1978).
[b]Esterifying alcohol: P, phytol; Gg, geranylgeraniol; F, farnesol.
[c]In *Ectothiorhodospira halochloris* and *Ectothiorhodospira abdelmalekii* a phytadienol was found (Steiner et al., 1981; also R. Steiner, personal communication).
[d]In *Chloroflexus aurantiacus* a straight-chain aliphatic stearyl alcohol (C-18) is esterified to the propionic acid side chain (Gloe and Risch, 1978).

The *phototrophic green bacteria* [green sulfur bacteria (Chlorobiaceae) and multicellular filamentous green bacteria (Chloroflexaceae)] do not have intracytoplasmic membrane systems. They possess specialized structures, the chlorobium vesicles or chlorosomes, that are located in the cytoplasm and attached to the surface of the cytoplasmic membrane (Cohen-Bazire et al., 1964; Staehelin et al., 1978, 1980). These chlorosomes contain large amounts of bacteriochlorophyll c, d, or e and are potent light-harvesting organelles. The reaction centers of phototrophic green bacteria are located in the cytoplasmic membrane at the attachment sites of the

Table II. Major Carotenoid Groups of Anoxygenic Phototrophic Bacteria[a]

Biosynthetic group	Major components
Normal spirilloxanthin series	Lycopene, rhodopin, spirilloxanthin
Rhodopinal series	Lycopene, lycopenal, lycopenol, rhodopin, rhodopinal, rhodopinol
Alternative spirilloxanthin series	Hydroxyneurosporene, spheroidene, spheroidenone (spirilloxanthin)
Okenone series	Okenone
Isorenieratene series	β-Carotene, isorenieratene
Chlorobactene series	γ-Carotene, chlorobactene

[a]Data are taken from Schmidt (1978).

chlorosomes. The localization of the small amounts of bacteriochlorophyll *a* that are present in these bacteria is mainly restricted to the reaction centers and protein complexes of the so-called baseplate, which connects the chlorosome with the cytoplasmic membrane (Staehelin *et al.*, 1980; Amesz and Knaff, 1988). A small amount (about 1% of the total) of bacteriochlorophyll *a* has been found inside the chlorosomes (Gerola and Olson, 1986). Carotenoids of the green bacteria are of the chlorobactene and isorenieratene series (Schmidt, 1978).

For careful taxonomic characterization of a species and for identification of new isolates many more properties are required and all available information on physiological and structural properties, including chemical structures of cell components, should be used for this purpose.

Significant features of dissimilatory sulfur metabolism, such as the ability to oxidize sulfide, the preference to grow photolithotrophically with sulfide as electron donor (compared to photoorganotrophic growth), the formation of sulfur globules inside or outside the cells, and the oxidation of extracellular elemental sulfur are properties that have been used as criteria to distinguish between major groups of the phototrophic purple bacteria (Molisch, 1907; Bavendamm, 1924; Pfennig and Trüper, 1971; Imhoff, 1984a) and among the phototrophic green bacteria (Pfennig, 1989a,b; Gibson *et al.*, 1984). In addition, differentiation of the genera and species is based on a number of morphological properties, such as cell form and size and flagellation and intracytoplasmic membrane structures, on pigment composition, on DNA base ratio, and also on physiological properties such as carbon and nitrogen substrate utilization and ability to respire aerobically and anaerobically in the dark, among others (Pfennig and Trüper, 1974; Trüper and Pfennig, 1981; Imhoff and Trüper, 1989).

More recently, chemotaxonomic methods have also been applied to the taxonomy of phototrophic bacteria and have supported the heterogeneity of this group of bacteria. The cellular composition of polar lipids, fatty acids, and quinones was found to be useful for the taxonomic characterization of phototrophic bacteria. Principal differences in the composition of these compounds were found among the major groups, but the recognition of species is also possible by comparing the composition of these membrane constituents (Imhoff *et al.*, 1982; Imhoff, 1982, 1984b, 1988b; Hansen and Imhoff, 1985). Also, other molecular information, such as size and sequence of cytochromes *c*, lipopolysaccharide structure, and 16S rRNA oligonucleotide catalogues and sequence data are available and can be used to determine similarity and taxonomic relations among species. The lipid A structure of the lipopolysaccharides shows significant differences among members of the purple non-sulfur bacteria (Weckesser *et al.*, 1974, 1979; Mayer, 1984); it is quite similar among all the Chromatiaceae (Meissner *et al.*, 1988b), but absent from *Chloroflexus* (Meissner *et al.*, 1988a).

2.1. The Phototrophic Purple Bacteria

Molisch (1907) first considered the pigmentation of purple sulfur and purple non-sulfur bacteria as a criterion to combine these groups taxonomically in a new order [Rhodobacteria (Molisch, 1907)] with the families Thiorhodaceae and Athiorhodaceae. This order and these families have been renamed Rhodospirillales, Chromatiaceae, and Rhodospirillaceae, respectively, by Pfennig and Trüper (1971). Presently we distinguish three groups of phototrophic purple bacteria, Chromatiaceae, Ectothiorhodospiraceae, and purple non-sulfur bacteria (Imhoff, 1984a; Imhoff *et al.*, 1984). It has been proposed to abandon the use of the family name Rhodospirillaceae because of the extreme heterogeneity of this group, which is not in accord with the properties of a taxonomic unit of a family (see also Section 3, on phylogenetic considerations).

The *Chromatiaceae* Bavendamm 1924 [emended description Imhoff (1984a)] comprise those phototrophic bacteria that, under the proper growth conditions, deposit globules of elemental sulfur inside their cells (Imhoff, 1984a). This definition agrees with that of the Thiorhodaceae by Molisch (1907). This family is a quite coherent group, as shown by the similarity of 16S rRNA molecules (Fowler *et al.*, 1984) and lipopolysaccharides (Meissner *et al.*, 1988b). All but one species (*Thiocapsa pfennigii*) have the vesicular type of intracytoplasmic membranes. All species are able to grow photoautotrophically under anaerobic conditions in light using sulfide or elemental sulfur as an electron donor. Several species are able to grow under photoheterotrophic conditions, some species grow as chemoautotrophs, and a few species also grow chemoheterotrophically (Gorlenko, 1974; Kondratieva *et al.*, 1976; Kämpf and Pfennig, 1980). Many species are motile by means of flagella, some have gas vesicles, and only *Thiocapsa* species completely lack motility. Some properties of the Chromatiaceae species are shown in Table III.

We distinguish two major physiological groups of Chromatiaceae, metabolically specialized and versatile species. Among the specialized species are *Chromatium okenii*, *Chromatium weissei*, *Chromatium warmingii*, *Chromatium buderi*, *Thiospirillum jenense*, *Thiocapsa pfennigii*, and the gas-vacuolated species of the genera *Lamprocystis* and *Thiodictyon*. These species depend on strictly anaerobic conditions and are obligately phototrophic. Sulfide is required; thiosulfate and hydrogen are not used as electron donors. Only acetate and pyruvate (or propionate) are photoassimilated in the presence of sulfide and CO_2. They do not grow with organic electron donors and chemotrophic growth is not possible. None of the investigated species assimilates sulfate as a sulfur source.

The versatile species photoassimilate a greater number of organic substrates and also grow in the absence of reduced sulfur sources with organic

Table III. Some Characteristic Properties of Chromatiacene[a]

Species	Cell shape	Cell diameter (μm)	Major carotenoids	Growth factor	Flagella	Gas vesicles
Thiospirillum						
jenense	Spiral	2.5–4.5	rh, ly	B_{12}	+	−
Chromatium						
okenii	Rod	4.5–6.0	ok	B_{12}	+	−
weissei	Rod	3.5–4.0	ok	B_{12}	+	−
warmingii	Rod	3.5–4.0	ra, ro	B_{12}	+	−
buderi	Rod	3.5–4.5	ra	B_{12}	+	−
tepidum	Rod	1.2	sp, rv	−	+	−
minus	Rod	2.0	ok	−	+	−
salexigens	Rod	2.0–2.5	sp	B_{12}	+	−
vinosum	Rod	2.0	sp, ly, rh	−	+	−
violascens	Rod	2.0	rh, ro, ra	−	+	−
gracile	Rod	1.0–1.3	sp, ly, rh	−	+	−
minutissimum	Rod	1.0–1.2	sp, ly, rh	−	+	−
purpuratum	Rod	1.2–1.7	ok	o	+	−
Thiocystis						
violacea	Sphere	2.5–3.5	ra, ro, rh	−	+	−
gelatinosa	Sphere	3.0	ok	−	+	−
Lamprocystis						
roseopersicina	Sphere	3.0–3.5	la, lo	−	+	+
Lamprobacter						
modestohalophilus	Rod	2.0–2.5	o	B_{12}	+	+
Thiodictyon						
elegans	Rod	1.5–2.0	ra, rh	−	−	+
bacillosum	Rod	1.5–2.0	ra	−	−	+
Amoebobacter						
roseus	Sphere	2.0–3.0	sp	B_{12}	−	+
pendens	Sphere	1.5–2.5	sp	B_{12}	−	+
pedioformis	Sphere	2.0	sp	(B_{12})	−	+
purpureus	Sphere	1.9–2.3	ok	o	−	+
Thiopedia						
rosea	Ovoid	1.0–2.0	ok	B_{12}	−	+
Thiocapsa						
roseopersicina	Sphere	1.2–3.0	sp	−	−	−
pfennigii	Sphere	1.2–1.5	ts	−	−	−

[a]Abbreviations: o, not determined; bchl, bacteriochlorophyll; (B_{12}), vitamin B_{12} strongly enhancing growth, but not absolutely required; −1, hydrogenase present, but growth with H_2 not demonstrated. Utilization of organic carbon: −, only acetate and pyruvate (or propionate) are photoassimilated; +, other carbon sources are used as well. Carotenoids: la, lycopenal; lo, lycopenol; ly, lycopene; ok, okenone; ra,

| Chemoautotrophy | Thiosulfate | Utilization of | | | G + C content (mole %) |
		Hydrogen	Sulfate	Organic carbon	
−	−	−	−	−	45.5
−	−	−	−	−	48.0–50.0
−	−	−	−	−	48.0–50.0
−	−	−	−	−	55.1–60.2
−	−	−	−	−	62.2–62.8
−	−	−	−	−	61.5
+	+	−1	−	−	62.2
+	+	−1	−	−	64.6
+	+	+	+	+	61.3–66.3
+	+	+	+	+	61.8–64.3
+	+	+	+	+	68.9–70.4
+	+	+	+	+	63.7
o	+	o	o	+	68.9
+	+	−1	+/−	+	62.8–67.9
+	−	−1	o	−	61.3
−	o	o	o	−	63.8
+	+	+	−	+	64.0
−	−	o	o	−	65.3
−	−	o	o	−	66.3
+	+	−	−	+	64.3
−	+	−	−	+	65.3
+	+	−	−	+	65.5
+	+	−	−	+	63.4–64.1
−	−	−	−	+	o
+	+	+	+	+	63.3–66.3
−	−	−1	o	−	69.4–69.9

rhodopinal; rh, rhodopin; ro, rhodopinol, rv, rhodovibrin; sp, spirilloxanthin; ts, 3,4,3′,4′-tetrahydrospirilloxanthin. Data are collected from Caumette *et al.* (1988), Eichler and Pfennig (1986, 1988), Gorlenko *et al.* (1979), Kämpf and Pfennig (1980), Madigan (1986), Schmidt (1978), and Trüper and Pfennig (1981).

substrates as electron donors for photosynthesis. Most of these species are able to grow with sulfate as the sole sulfur source. They also can grow under chemoautotrophic conditions. Among these species are the small-celled *Chromatium* species, *Thiocystis violacea*, *Thiocapsa roseopersicina*, *Lamprobacter modestohalophilus*, and *Amoebobacter* species.

The *Ectothiorhodospiraceae* (Imhoff, 1984a) are phototrophic sulfur bacteria that, during oxidation of sulfide, deposit elemental sulfur outside the cells. They are distinguished from the Chromatiaceae by lamellar intracytoplasmic membrane structures, by significant differences of the polar lipid composition (Imhoff *et al.*, 1982), and by the dependence on saline and alkaline growth conditions (Imhoff, 1989a). *Ectothiorhodospira halophila* is the most halophilic eubacterium known and even grows in saturated salt solutions. All species are motile by polar flagella and *Ectothiorhodospira vacuolata* in addition forms gas vesicles. In a phylogenetic tree based on 16S rRNA data Ectothiorhodospiraceae form a separate branch, close to the branch of the Chromatiaceae (Stackebrandt *et al.*, 1984; Woese *et al.*, 1985b). Some properties of *Ectothiorhodospira* species are shown in Table IV.

The group of the *purple non-sulfur bacteria* [Rhodospirillaceae, Pfennig and Trüper (1971)] is by far the most diverse group of the phototrophic purple bacteria (Imhoff and Trüper, 1989). This diversity is reflected in greatly varying morphology, internal membrane structure, carotenoid composition, utilization of carbon sources, and electron donors, among other features. The intracytoplasmic membranes are small, fingerlike intrusions, vesicles, or different types of lamellae. Most species are motile by flagella; gas vesicles are not formed by any of the known species. Some properties of purple non-sulfur bacteria are shown in Table V.

The preferred growth mode of all species is photoheterotrophic, under anaerobic conditions in the light with various organic substrates. Many species also are able to grow photoautotrophically with either molecular hydrogen or sulfide as electron donor and CO_2 as the sole carbon source. Most of these species oxidize sulfide to elemental sulfur only (Hansen and van Gemerden, 1972). In *Rhodobacter sulfidophilus* and *Rhodopseudomonas palustris* sulfate is the final oxidation product and is formed without accumulation of elemental sulfur as an intermediate (Hansen, 1974; Hansen and Veldkamp, 1973). Three species, *Rhodobacter veldkampii*, *Rhodobacter adriaticus*, and *Rhodobacter euryhalinus*, are known to deposit elemental sulfur outside the cells, while oxidizing sulfide to sulfate (Hansen *et al.*, 1975; Neutzling *et al.*, 1984; Hansen and Imhoff, 1985; Kompantseva, 1985).

Most representatives can grow under microaerobic to aerobic conditions in the dark as chemoheterotrophs, a few also as chemoautotrophs. Some species are very sensitive to oxygen, but most are quite tolerant of oxygen and grow well under aerobic conditions in the dark. Under these conditions synthesis of photosynthetic pigments is repressed and the cultures are faintly colored or colorless. Even under phototrophic growth

Table IV. Some Characteristic Properties of Ectothiorhodospiraceae[a]

Species	Cell shape	Cell diameter (μm)	Color of culture	Major carotenoid	Major bchl	Gas vesicles	Optimum salinity range (%)	G + C content (mole %)
Ectothiorhodospira								
mobilis	Rod-spiral	0.7–1.0	Red	sp, rh	a	–	3–15	62.0–69.9
shaposhnikovii	Rod-spiral	0.8–0.9	Red	sp, rh	a	–	1–3	61.2–62.8
vacuolata	Rod	1.5	Red	sp	a	+	1–6	61.4–63.6
halophila	Spiral	0.8–0.9	Red	sp	a	–	15–30	64.3–69.7
halochloris	Spiral	0.5–0.6	Green	rhg*, rh	b	–	14–27	50.5–52.9
abdelmalekii	Spiral	0.9–1.2	Green	b	b	–	12–20	63.3–63.8
marismortui	Pleomorphic Rods	0.9–1.3	Red	(sp)	a	–	3–8	65

[a] Abbreviations: bchl, bacteriochlorophyll. Carotenoids: rh, rhodopin; rhg*, rhodopin glucoside and derivatives; sp, spirilloxanthin; (sp), most probably spirilloxanthin. Data are collected from Imhoff (1984b, 1989a); Imhoff and Trüper (1981); Oren et al., 1989; Schmidt (1978).
[b] Most probably similar to *E. halochloris*.

Table V. Some Characteristic Properties of Purple Non-Sulfur Bacteria (Rhodospirillaceae)[a]

Species	Cell shape	Cell diameter (μm)	ICM	Major carotenoid	Sulfide oxidation to	Growth factors	Salt response	G + C content (mole %)	Other significant properties
Rhodospirillum									
rubrum	Spiral	0.8–1.0	V	sp, rv	S°	b	F	63.8–65.8	—
photometricum	Spiral	1.2–1.5	S	rv, rh	—	YE	F	65.8	—
molischianum	Spiral	0.7–1.0	S	ly, rh	—	AA	F	61.7–64.8	—
fulvum	Spiral	0.5–0.7	S	ly, rh	—	paba	F	64.3–65.3	—
salexigens	Spiral	0.6–0.7	L	sp	—	Glutamate	H	64.0	NaCl required
salinarum	Spiral	0.8–0.9	V	sp	—	YE	H	67.4–68.1	NaCl required
mediosalinum	Spiral	0.8–1.0	V	sp	S°	t, paba, n	H	66.6	NaCl required
Rhodopila									
globiformis	Sphere	1.6–1.8	V	kts	—	b, paba	F	66.3	Acidic pH
Rhodomicrobium									
vannielii	Ovoid-rod	1.0–1.2	L	rh, ly, sp	+	None	F	61.8–63.8	Peritr. flagella, exposore formation, acidic pH
Rhodobacter									
capsulatus	Rod	0.5–1.2	V	sn, se	S°	t, (b, n)	F	65.5–66.8	—
veldkampii	Rod	0.6–0.8	V	sn, se	S°/Sul	b, t, paba	F	64.4–67.5	Nonmotile
sphaeroides	Ovoid-rod	0.7	V	sn, se	S°	b, t, n	F	68.4–69.9	—

Species	Shape	Size	ICM	Carotenoids	Sulfide oxidation	Growth factors	Salt response	%GC	Remarks
sulfidophilus	Rod	0.6–0.9	V	sn, se	Sul	b, t, n, paba	M	67.0–71.0	NaCl required
euryhalinus	Rod	0.7–1.0	V	se	S°/Sul	b, t, n, paba	M	62.1–68.6	NaCl required
adriaticus	Rod	0.5–0.8	V	sn, se	S°/Sul	b, t	M	64.9–66.7	NaCl required, nonmotile
Rhodocyclus									
purpureus	Half circle	0.6–0.7	T	ra, rh	–	B₁₂	F	65.3	Nonmotile
gelatinosus	Rod	0.4–0.5	T	sn, se	–	b, t	F	70.5–72.4	Gelatin liquefied
tenuis	Spiral	0.3–0.5	T	ly, rh, ra	–	None	F	64.8	—
Rhodopseudomonas									
palustris	Rod	0.6–0.9	L	sp, rv, rh	Sul	paba, (b)	F	64.8–66.3	—
viridis	Rod	0.6–0.9	L	neu*, ly*	–	paba, b	F	66.3–71.4	bchl *b*
sulfoviridis	Rod	0.5–0.9	L	neu, sp	+	b, p, paba	F	67.8–68.4	bchl *b*
blastica	Rod	0.6–0.8	L	sn, se	–	B₁₂, b, n, t	F	65.3	—
acidophila	Rod	1.0–1.3	L	rh, rg, rag	–	None	F	62.2–66.8	Acidic pH
rutila	Rod	0.4–1.0	L	sp, rv	–	None	F	67.6–69.4	—
marina	Rod	0.7–0.9	L	sp	+	o	M	61.5–63.8	NaCl required

[a]Abbreviations: o, not determined. bchl, bacteriochlorophyll. ICM, structure of intracytoplasmic membrane system: V, vesicles; L, lamellae; S, stacks; T tubes. Salt response: F, freshwater species; M, marine species; H, halophilic species. Sulfide oxidation to: S°, elemental sulfur only; S°/Sul, sulfate with intermediate formation of extracellular elemental sulfur globules; –, sulfide not oxidized; +, sulfide oxidized to various products. Growth factors: b, biotin; n, niacin; t, thiamine; paba, p-aminobenoic acid; YE, yeast extract; AA, amino acids; vitamins in parentheses are required only by some strains. Carotenoids: kts, ketocarotenoids; ly, lycopene; ly*, 1,2-dihydrolycopene; neu, neurosporene; neu*, 1,2-dihydroneurosporene; ra, rhodopinal; rag, rhodopinal glucoside; rg, rhodopin glucoside; rh, rhodopin; rv,rhodovibrin; se, spheroidene; sn, spheroidenone; sp, spirilloxanthin. Data are collected from Akiba *et al.* (1983), Drews (1981), Eckersley and Dow (1980), Hansen (1974), Hansen and Imhoff (1985, Imhoff (1983), Kompantseva (1985), Kompantseva and Gorlenko (1984), Neutzling *et al.* (1984), Nissen and Dundas (1984), Schmidt (1978), Schmidt and Bowien (1983), and Trüper and Pfennig (1981).

conditions (anaerobic/light) many species exhibit considerable respiratory capacity. Respiration under these conditions is inhibited, however, by light. The fact that considerable respiratory activity is expressed also under phototrophic growth conditions enables these bacteria (e.g., *Rhodobacter capsulatus, Rhodobacter sphaeroides, Rhodocyclus gelatinosus, Rhodospirillum rubrum*) to switch immediately from phototrophic to respiratory metabolism when environmental conditions change.

Some species also may perform a respiratory metabolism anaerobically in the dark with sugars and either nitrate, dimethylsulfoxide, or trimethylamine-*N*-oxide as electron sink, or—though poorly—with metabolic intermediates as electron acceptors (fermentation).

One or more vitamins are generally required as growth factors, most commonly biotin, thiamine, niacin, and *p*-aminobenzoic acid; these compounds are rarely needed by species of the Chromatiaceae and Ectothiorhodospiraceae, which may require vitamin B_{12} as sole growth factor. Growth of most purple non-sulfur bacteria is enhanced by small amounts of yeast extract, and some species have a complex nutrient requirement.

A clear separation of the purple non-sulfur bacteria from the purple sulfur bacteria as well as their own diversity are well evidenced in a number of chemotaxonomic observations, such as cytochrome *c* structures, lipid composition, quinone composition, lipopolysaccharide structure, and DNA/rRNA hybridization (Imhoff *et al.*, 1984). Also, oligonucleotide patterns (and, as far as available, nucleotide sequences) of 16S rRNA molecules support this picture (Woese *et al.*, 1984a,b). The similarity coefficients derived from these patterns are higher than 0.6 within the Chromatiaceae (Fowler *et al.*, 1984) and higher than 0.5 within the Ectothiorhodospiraceae (Stackebrandt *et al.*, 1984), but considerably lower between most of the species of the purple non-sulfur bacteria. Binary similarity coefficients higher than 0.6 have been obtained only for the couples *Rhodobacter capsulatus* and *Rhodobacter sphaeroides, Rhodocyclus tenuis* and *Rhodocyclus gelatinosus,* as well as *Rhodospirillum rubrum* and *Rhodospirillum photometricum* (Gibson *et al.*, 1979; Woese *et al.*, 1984a). Similarity coefficients between species of the purple non-sulfur bacteria and the purple sulfur bacteria were between 0.3 and 0.4. Furthermore, the purple non-sulfur bacteria are not only heterogeneous in themselves, but on the basis of 16S rRNA analyses, representatives of this group are much more similar to certain non-phototrophic, purely chemotrophic bacteria than to other phototrophic species (see also Section 3 on phylogenetic considerations).

2.2. The Phototrophic Green Bacteria

Although both the green sulfur bacteria and the multicellular and filamentous green bacteria (sometimes also called green non-sulfur bacteria) contain chlorosomes, their physiological properties are significantly different and they apparently belong to completely different lines of descent (see below).

All *green sulfur bacteria* (Chlorobiaceae) have highly similar physiological capacities. They are metabolic specialists, strictly anaerobic and obligately phototrophic. All species grow photolithotrophically with CO_2 as sole carbon source. In the presence of sulfide and CO_2 only acetate is assimilated as an organic carbon source. CO_2 is assimilated via reactions of the reductive tricarboxylic acid cycle (Evans *et al.*, 1966; Fuchs *et al.*, 1980a,b; Ivanovsky *et al.*, 1980). Sulfide is required; it is used as electron donor and sulfur source and is oxidized to sulfate with the intermediate accumulation of elemental sulfur globules outside the cells. Some strains also use thiosulfate and molecular hydrogen as photosynthetic electron donors.

We distinguish green and brown species. The green species contain bacteriochlorophyll *c* or *d* and the carotenoids chlorobactene and OH-chlorobactene as light-harvesting pigments. The brown species have bacteriochlorophyll *e* and the carotenoids isorenieratene and β-isorenieratene as light-harvesting pigments (Liaaen-Jensen, 1965). The different pigment content is responsible for differences in the light absorption properties. The brown species have a broader absorption range, between 480 and 550 nm, which apparently is of ecological significance (see below). All species lack flagella; some have gas vesicles. Only *Chloroherpeton thalassium* is motile by gliding and in addition forms gas vesicles; a further difference from all other species of this group is the carotenoid composition, with γ-carotene as major component. Physiological properties and results of 16S rRNA analyses led to the inclusion of *Chloroherpeton* into this group (Gibson *et al.*, 1985). Some properties of the green sulfur bacteria are shown in Table VI.

The only well-studied bacterium of the *multicellular filamentous green bacteria* that is presently in pure culture is *Chloroflexus aurantiacus* (Pierson and Castenholz, 1974a). Other representatives are *Heliothrix oregonensis*, *Oscillochloris chrysea*, *Oscillochloris trichoides*, and *Chloronema giganteum*. Cells are arranged in filaments and move by gliding. Some species form gas vesicles. These bacteria are facultatively aerobic and grow preferably with organic substrates under phototrophic and chemotrophic conditions. Sulfide and other reduced sulfur compounds are not important as electron donors. Photosynthetic pigments are bacteriochlorophyll *c* and *a* (in a similar ratio as in green sulfur bacteria) and the carotenoids γ- and β-carotene. Some properties of this group are shown in Table VII.

The recently described *Heliothrix oregonensis* contains bacteriochlorophyll *a*, but lacks other bacteriochlorophylls and does not contain chlorosomes or intracytoplasmic membranes (Pierson *et al.*, 1985). The 5S rRNA nucleotide sequence of *Heliothrix* is similar to that of *Chloroflexus*

Table VI. Some Characteristic Properties of Green Sulfur Bacteria[a]

Species	Cell shape	Cell diameter (μm)	Color	Major carotenoid	Major bchl	Flagella	Gas vesicles	Salinity	Sulfide	G + C content (mole %)	Other significant properties
Chlorobium											
limicola	Rod	0.7–1.0	Green	cl	c or d	–	–	F	H	51.0–58.1	—
vibrioforme	Vibrio	0.5–0.7	Green	cl	c or d	–	–	2%	H	52.0–57.1	—
phaeobacteroides	Rod	0.6–0.8	Brown	irt, β-irt	e	–	–	F	H	49.0–50.0	—
phaeovibrioides	Vibrio	0.3–0.4	Brown	irt, β-irt	e	–	–	2%	H	52.0–53.0	—
chlorovibrioides	Vibrio	0.3–0.4	Green	cl	c or d	–	–	2–3%	H	o	—
Prosthecochloris											
aestuarii	Sphere	0.5–0.7	Green	cl	c	–	–	2–5%	H	50.0–56.0	Prosthecae
phaeoasteroides	Sphere	0.3–0.6	Brown	irt	e	–	–	0.5–2%	H	52.2	Prosthecae
Ancalochloris											
perfilievii	Sphere	0.5–1.0	Green	o	o	–	+	F	L	o	Prosthecae, not pure
Pelodictyon											
luteolum	Ovoid	0.6–0.9	Green	cl	c or d	–	+	F	L	53.5–58.1	—
clathratiforme	Rod	0.7–1.2	Green	cl	c or d	–	+	F	L	48.5	Nets formed, not pure
phaeum	Vibrio	0.6–0.9	Brown	irt	e	–	+	3%	L	o	
Chloroherpeton											
thalassium	Rod	1.0	Green	γ-c	c	gl	+	1–2%	L	45.0–48.2	Flexible, gliding

[a] Abbreviations: o, no data available; bchl, bacteriochlorophyll; gl, gliding motility. Carotenoids: cl, chlorobactene; irt, isorenieratene; β-irt, β-isorenieratene; γ-c, γ-carotene. Salt response: F, freshwater isolates; numbers give optimum salinity of isolates from brackish or marine habitats. Sulfide: H, high sulfide concentrations favorable; L, low sulfide concentrations favorable. Data are collected from Gibson *et al.* (1984), Gorlenko and Lebedeva (1971), and Trüper and Pfennig (1981).

Table VII. Some Characteristic Properties of Multicellular Filamentous Green Bacteria[a]

Species	Cell shape	Cell diameter (μm)	Color	Major carotenoid	Major bchl	Motility	Gas vesicles	Preferred salinity	Sulfide	G + C content (mole %)	Other significant properties
Chloroflexus											
aurantiacus	Filaments	0.5–1.0	Green-orange	β-c, γ-c	*c*	gl	–	F	LO	53.1–54.9	Flexible, thermophilic
Heliothrix											
oregonensis	Filaments	1.5	Orange	γ-cg	*b*	gl	–	F	o	o	Flexible, thermophilic, not pure
Chloronema											
giganteum	Trichomes	2.0–2.5	Green-yellow	o[b]	*d*	gl	+	F	LO	o	Not pure
Oscillochloris											
chrysea	Filaments	4.5–5.5	Green-yellow	o	*c*	gl	+	F	LO	o	Not pure
trichoides	Filaments	1.0–1.4	Green	o	*c*	gl	+	F	LO	o	Not pure

[a] Abbreviations: o, no data available; bchl, bacteriochlorophyll; gl, gliding motility; F, freshwater species. Relations to sulfide: LO, preferable photoorganotrophic growth, sulfide may be used, but only low concentrations are tolerated. Carotenoids: β-c, β-carotene; γ-c, γ-carotene; γ-cg, γ-carotene glucoside, γ-carotene, and oxygenated derivatives. Data are collected from Dubinina and Gorlenko (1975), Gorlenko and Pivovarova (1977), Pierson and Castenholz (1974a), Pierson *et al.* (1985), and Trüper and Pfennig (1981).

[b] Neither bchl *c*, *d*, nor *e* present, and chlorosomes as well as intracytoplasmic membranes lacking, bchl *a* present.

aurantiacus (Pierson *et al.*, 1985). It may be considered as a flexing green bacterium lacking chlorosomes and has been included in this group (Pfennig, 1989b).

2.3. Genera of Uncertain Affiliation

Heliobacterium chlorum (Gest and Favinger, 1983) is unique among all known phototrophic prokaryotes in several aspects. It has a hitherto unknown bacteriochlorophyll *g*, which resembles chlorophylls *a* and *b* in the substitution of the tetrapyrrole ring in position 1 with a vinyl group (Brockmann and Lipinski, 1983). *Heliobacterium chlorum* is an obligately anaerobic bacterium and is extremely oxygen sensitive. It is unable to grow under autotrophic conditions; its metabolism is strictly photoheterotrophic. Sulfide inhibits growth and biotin is required as growth factor. Other bacteria with bacteriochlorophyll *g*, *Heliobacillus mobilis* (Beer-Romero and Gest, 1987; Beer-Romero *et al.*, 1988), *Heliobacterium gestii*, and *Heliobacterium fasciculum* (Ormerod *et al.*, 1990), have been isolated recently.

A quite remarkable group of bacteria, containing bacteriochlorophyll, but unable to grow phototrophically under anaerobic conditions, is represented by a number of Gram-negative aerobic marine bacteria (Sato, 1978; Shiba *et al.*, 1979; Nishimura *et al.*, 1981; Trüper, 1989). The best studied of these bacteria are *Erythrobacter longus* (Shiba and Simidu, 1982) and *Roseobacter denitrificans* (Shiba, 1991), which can synthesize bacteriochlorophyll *a*, form intracytoplasmic membranes, and have reaction center complexes similar to those of other purple bacteria (Harashima *et al.*, 1980; Shimada *et al.*, 1985; Iba *et al.*, 1988). In contrast to all previously known phototrophic purple bacteria, synthesis of bacteriochlorophyll *a* and carotenoids is stimulated by oxygen (Harashima *et al.*, 1980). Shiba (1984) demonstrated that *Roseobacter denitrificans* (formerly designated as *Erythrobacter* strain OCH 114) effectively uses light to increase the cellular ATP level and the incorporation rate of CO_2. According to 16S rRNA analyses, *Erythrobacter longus* belongs to the alpha subgroup of the Proteobacteria (see below), but appears only distantly related to other bacteria of this group, such as *Rhodobacter* and *Rhodopseudomonas* species (Woese *et al.*, 1984a). Recently, it was reported that a *Rhizobium* strain contains bacteriochlorophyll and photosynthetic reaction centers (Evans *et al.*, 1990).

3. PHYLOGENETIC CONSIDERATIONS

The best record of phylogeny is conserved in the primary sequences of nucleic acids and proteins, and attempts to study phylogenetic relatedness have been made by quantitative comparison of sequence data of both kinds of macromolecules. The ideal molecular structure for phylogenetic analyses shows universal distribution, is phylogenetically conservative, has a constant rate of evolution in various lines of descent, and is experimentally accessible to sequence analysis. Principal difficulties for a general concept of prokaryote phylogeny arose from lack of available methods to sequence proteins and nucleic acids and the limited distribution of many proteins. Because of the available methods, proteins were the first macromolecules considered for phylogenetic relationships of various bacteria. Among the phototrophic prokaryotes, *c*-type cytochromes and ferredoxins are two examples of such proteins. On the basis of primary sequence and tertiary structure of the *c*-type cytochromes a first "phylogenetic tree" of the phototrophic purple bacteria was constructed (Dickerson, 1980).

More recently comparative information on 16S rRNA molecules from a great number of diverse bacteria have been worked out, including representatives of all groups of phototrophic prokaryotes (Gibson *et al.*, 1979, 1985; Fowler *et al.*, 1984; Stackebrandt *et al.*, 1984; Woese *et al.*, 1984a,b, 1985b). The comparison of the nucleotide sequence of 16S rRNA is presently the best approach for tracing bacterial phylogeny because analytical procedures are available, the molecule shows universal distribution among prokaryotes, and it, apparently more than others, is phylogenetically conservative. Although the general picture derived from 16S rRNA analyses, as presented below, is now widely accepted, discrepancies between the results obtained and those of other approaches require attention. For the true bacterial phylogeny, other, 16S rRNA-independent properties have to be considered as well.

The method has developed from a comparison of oligonucleotide catalogues derived from 16S rRNA by digestion with T1 RNase (Zablen and Woese, 1975). A large number of representative bacterial strains have been analyzed by this method and as an additional criterion for the comparison, so-called "signature sequences" have been used. These signature sequences are small parts of the total sequence that are regarded as characteristic for the group of bacteria where they are found, but are not (or rarely) present in bacteria outside the considered group. With the development of techniques to completely sequence the 16S rRNA molecule the total information of its nucleic acid sequence (about 1500–1600 bases) is now available for comparison.

On the basis of the data ten major eubacterial groups have been defined (Woese *et al.*, 1985a; Woese, 1987). One of these is represented by the cyanobacteria, one by the green sulfur bacteria, one by the multicellular filamentous green bacteria (*Chloroflexus*) and chemotrophic "relatives," and one by the phototrophic purple bacteria and their chemotrophic "relatives." On the basis of their 16S rRNA, the recently discovered *Heliobacterium chlorum* (Gest and Favinger, 1983) and *Heliobacillus mobilis* (Beer-Romero *et al.*, 1988) do not fit into the aforementioned groups, but are quite similar to Gram-positive bacteria of the genus *Bacillus* (Woese *et al.*, 1985c; Beer-Romero and Gest, 1987). Thus, four major groups of the eubacteria contain representatives of anoxygenic phototrophic bacteria.

These four groups appear phylogenetically only distantly related to each other. In three of these groups, purely chemotrophic bacteria which lack bacteriochlorophyll are found together with the phototrophic bacteria.

The *green sulfur bacteria* form a tight phylogenetic group, separated from other phototrophic and also from known chemotrophic bacteria (Gibson *et al.*, 1985).

Chloroflexus is found related to some nonphototrophic gliding bacteria, such as *Herpetosiphon aurantiacum* and *Thermomicrobium roseum* (Oyaizu *et al.*, 1987).

Deep branching among different groups of the *phototrophic purple bacteria* as well as the close relationship of some of them with purely chemotrophic bacteria were demonstrated on the basis of 16S rRNA similarity (Gibson *et al.*, 1979; Stackebrandt and Woese, 1981). All these related bacteria, whether phototrophic or purely chemotrophic, were called purple bacteria and separated into four subgroups, alpha, beta, gamma, and delta (at present without a phototrophic representative) (Woese *et al.*, 1984a, 1985b). All phototrophic purple bacteria were found together with nonphototrophic representatives in the subgroups. Chromatiaceae and Ectothiorhodospiraceae are in the gamma subgroup, while the purple nonsulfur bacteria are in the alpha and beta subgroups. A new class, the *Proteobacteria*, has been proposed for the purple bacteria and their relatives (Stackebrandt *et al.*, 1988).

The alpha subgroup contains in one branch most of the *Rhodospirillum* species and *Rhodopila globiformis*, in another one *Rhodopseudomonas acidophila*, *Rhodopseudomonas palustris*, *Rhodopseudomonas viridis*, and *Rhodomicrobium vannielii*, and in a third one the *Rhodobacter* species. Specific great similarities of chemotrophic to phototrophic representatives exist between *Paracoccus denitrificans* and the *Rhodobacter* group [*Rhodobacter capsulatus* and *Rhodobacter sphaeroides* (Gibson *et al.*, 1979)], between *Nitrobacter winogradskyi* and *Rhodopseudomonas palustris* (Seewaldt *et al.*, 1982), and among *Aquaspirillum itersonii*, *Azospirillum brasiliense*, and the *Rhodospirillum* group (Woese *et al.*, 1984a).

The beta subgroup contains the species that have been recently combined in the genus *Rhodocyclus* (Imhoff *et al.*, 1984). *Rhodocyclus gelatinosus* appears specifically related to *Sphaerotilus natans;* and *Rhodocyclus tenuis* and *Rhodocyclus purpureus* to *Alcaligenes eutrophus* (Woese *et al.*, 1984b; Woese, 1987).

In *the gamma subgroup*, different clusters of related species were found containing Chromatiaceae and Ectothiorhodospiraceae, respectively (Fowler *et al.*, 1984; Stackebrandt *et al.*, 1984; Woese *et al.*, 1985b). In contrast to the phototrophic bacteria of the alpha and beta subgroups, representatives of Chromatiaceae and Ectothiorhodospiraceae are quite isolated and not intermixed with nonphototrophic representatives of this subgroup.

These data support the idea that ancestors of present-day phototrophic prokaryotes are among the most ancient eubacteria. They further point to early divergences within the phototrophic prokaryotes and to several lines of development of nonphototrophic bacteria from phototrophic ancestors at different times of evolution.

4. GENERAL ECOLOGY

The physiological potency of a bacterium and environmental conditions both determine the development and successful competition of a particular species in nature. In order to explain the natural abundance and to predict possible developments of the phototrophic bacteria it is therefore important to know their ecologically relevant properties. Under natural conditions, temperature, salinity, pH value, concentrations of sulfide and oxygen, stability of anoxic conditions, and availability of essential nutrients (including trace elements and vitamins) are important decisive factors for the development of a particular phototrophic bacterium (Pfennig, 1967, 1977, 1989d). In the following we will discuss some of the bacterial properties and environmental conditions that allow the development of anoxygenic phototrophic bacteria. We first consider some extremes of physical and chemical environmental parameters, such as temperature, salinity, and pH, and then pay attention in particular to the relations of phototrophic bacteria to reduced sulfur compounds and oxygen, and to the selective properties of their photosynthetic pigments. We will also consider the motility of cells as an important property that enables them to adapt their position under changing conditions in a stratified environment. Finally, their position in carbon and sulfur cycles and their contribution to the productivity of lakes will be discussed. Instead of discussing a number of well-investigated lake and sedimental habitats in detail, we will consider the data obtained in many laboratory and *in situ* experiments in more general terms.

Some ecological aspects of phototrophic bacteria have been discussed in more detail by Pfennig (1978, 1989d), van Gemerden and Beeftink (1983), Gorlenko *et al.* (1983), and Madigan (1988).

4.1. Extremes of Temperature, Salinity, and pH

Compared to the moderate conditions of temperature (up to 35°C), salinity (freshwater or marine), and pH (near neutrality, pH 6.5–7.5) which favor many species, some species have adapted to more extreme conditions, which then become highly selective for these species.

4.1.1. High Temperatures

Successful adaptation to high temperatures is found in *Chloroflexus aurantiacus* (optimum 50–60°C) (Pierson and Castenholz, 1974a), *Heliothrix oregonensis* (optimum 40–55°C) (Pierson *et al.*, 1985), and *Chromatium tepidum* (optimum 48–50°C) (Madigan, 1986), In addition, some of the extremely halophilic *Ectothiorhodospira* species prefer elevated temperatures of 35–45°C (Imhoff, 1989a).

Chloroflexus aurantiacus is the most thermophilic of these bacteria and has been isolated from alkaline hot springs all over the world, from sites in Japan, the United States, Iceland, New Zealand (Pierson and Castenholz, 1974a,b; Brock, 1978), and southern France (J. F. Imhoff, unpublished results). *Chloroflexus aurantiacus* preferably grows under photoheterotrophic conditions, although photoautotrophic growth with sulfide as an electron donor is possible (Madigan and Brock, 1975). Under aerobic conditions chemoorganotrophic, respiratory growth occurs, the synthesis of bacteriochlorophyll is repressed, and the color of the cells changes from green to orange. No growth occurs under anaerobic conditions in the dark. This species is well adapted to the high temperatures and high light intensities that are characteristic of these hot springs. Dense mats of *Chloroflexus* are found in the effluent channels and pools of Yellowstone hot springs. The development is at its maximum between 50 and 55°C, but extends down to 40°C and up to 73°C (Bauld and Brock, 1973). In the mat system, *Chloroflexus* may coexist with the unicellular cyanobacterium *Synechococcus lividus*. Underneath a thin, photosynthetically active surface layer, a thick, orange layer of *Chloroflexus* is usually present (Bauld and Brock, 1973). In effluents of springs that are rich in sulfide, mats consisting of dark-green, obligately phototrophic strains of *Chloroflexus* have been found (Giovannoni *et al.*, 1987).

In addition, mesophilic *Chloroflexus* strains have been found as benthic forms in stratified freshwater lakes (Gorlenko, 1975; Pivovarova and Gorlenko, 1977).

4.1.2. High Salinity

Although a number of typical marine phototrophic bacteria are known, some of which also can tolerate higher salt concentrations, full adaptation to elevated salinities is found only in a few specialized halophilic bacteria, such as *Chromatium salexigens* (Caumette *et al.*, 1988), *Rhodospirillum salexigens*, *Rhodospirillum salinarum*, *Ectothiorhodospira halophila*, *Ectothiorhodospira halochloris*, and *Ectothiorhodospira abdelmalekii* (Imhoff, 1988a). While the two *Rhodospirillum* species have been found in cell masses in marine salterns and other habitats with a mineral salts composition similar to seawater (Rodriguez-Valera *et al.*, 1985; see Imhoff, 1988a), the extremely halophilic *Ectothiorhodospira* species bloom in concentrated brines of alkaline soda lakes in various sites around the world (Imhoff *et al.*, 1979; see Imhoff, 1988a).

Ectothiorhodospira species are characterized by their distinct and obligate requirement for salt and alkaline pH (Imhoff, 1989a). On the basis of their salt requirement we distinguish two groups of strains. The first group has salt optima of 1–7% (some strains up to 10%) total salinity and has been isolated from marine environments, but also from hypersaline lakes. The second group of species requires higher salt concentrations. These species form conspicuous mass developments in alkaline soda lakes (Jannasch, 1957; Imhoff *et al.*, 1979) and are well adapted to the high temperature, high pH, high light intensity, and high salinity of these environments. *Ectothiorhodospira halophila* is the most common of the extremely halophilic species and has been isolated from many hypersaline environments (Imhoff, 1988a). *E. halophila* strains isolated from alkaline soda lakes of the Wadi Natrun (Egypt) represent the most halophilic eubacteria known and show optimum growth at higher salinities than most of the known archaeo-bacterial "extreme halophiles." A detailed discussion of the biology of marine and halophilic phototrophic bacteria is given by Imhoff (1988a).

4.1.3. Low pH and High pH

A preference for low pH values is found in *Rhodopila globiformis* (pH optimum 4.8–5.0), *Rhodopseudomonas acidophila* (pH optimum 5.8), and *Rhodomicrobium vannielii* (good growth at neutral and acidic pH down to pH 5.2) (Pfennig, 1969, 1974).

High pH values are preferred by *Chloroflexus aurantiacus* (optimum at pH 7.6–8.4) and by *Ectothiorhodospira* species (pH optima between 7.5 and 9.5) (Pierson and Castenholz, 1974a; Imhoff, 1989a).

4.2. The Stratified Environment

Stratification is a general property of the habitats of phototrophic sulfur bacteria, whether these are stagnant water bodies or top layers of anoxic sediments. Blooms of phototrophic sulfur bacteria usually develop in complex and dynamic multigradient systems, in which the conditions vary strongly with the depth. The development of a particular species usually depends on several of the gradient-forming parameters. The most important properties are the concentrations of sulfide and oxygen and the light intensity, which form countercurrent gradients. The depth at which phototrophic sulfur bacteria may develop is largely restricted to the concomitant presence of light and sulfide. A schematic presentation of a lake environment is shown in Fig. 1. Due to light limitation, layers of phototrophic bacteria are compressed and restricted to the top millimeters in sediments, but may extend down to 30 m depth and deeper in certain lakes.

Excessive concentrations of both sulfide (near the bottom) and oxygen

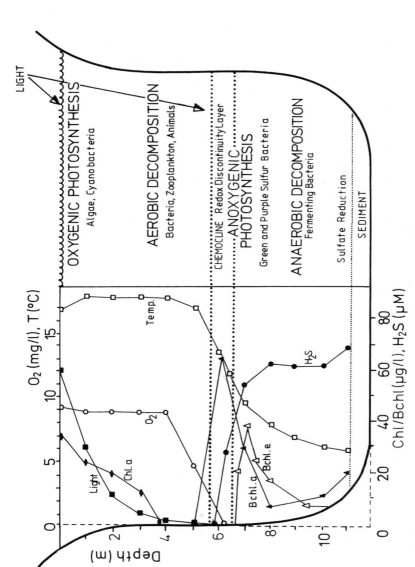

Figure 1. Schematic representation of a typical stratified lake environment. **Right:** major microbial processes together with representative physiological groups of the microbes involved. **Left:** corresponding major environmental factors, such as temperature and light profiles, and oxygen and sulfide concentration gradients. Concentration profiles of chlorophylls indicate separate layers of oxygenic phototrophic organisms and also of phototrophic purple bacteria and brown-colored green sulfur bacteria. Information on the left-hand part is deduced from data on the Schleinsee (Pfennig, 1989d) and the Rotsee (Kohler *et al.*, 1984) and slightly modified for this presentation. Light penetrates down to about 7.5–8 m; its intensity is indicated as percent of

(near the surface) have to be avoided by sensitive bacteria. Low sulfide concentrations do increase, however, the risk of being confronted with oxygen. Therefore, tolerance to oxygen is required by bacteria living in the uppermost layer of the sulfide-containing zone. This top layer of the sulfide horizon should be selective for bacteria that can make use of both sulfide and oxygen, growing as a phototroph during daytime and performing a respiratory metabolism during the night. The lowermost layer favors the development of obligately phototrophic bacteria that are most sensitive to oxygen, most tolerant to sulfide, and also most sensitive to light.

Some of the microbial activities that determine slope and position of these gradients, such as oxygen production by algae and cyanobacteria and sulfide oxidation by phototrophic bacteria, show diurnal, light-dependent fluctuations. Others, such as sulfide production by sulfate-reducing bacteria and oxygen consumption by respiratory bacteria, do not. Due to these processes the sulfide horizon rises during the night and goes down again during the day. The motile phototrophic purple bacteria are able to follow the moving sulfide horizon. Under favorable conditions separate layers of green sulfur bacteria are formed underneath layers of phototrophic purple bacteria and/or of algae and cyanobacteria. Frequently, however, mixed populations of purple and green sulfur bacteria are observed, and purple non-sulfur bacteria regularly accompany the mass development of phototrophic sulfur bacteria. In many investigations of blooming phototrophic sulfur bacteria, the presence of purple non-sulfur bacteria certainly has been overlooked [for discussion see Imhoff (1988a)]. Several observations in natural habitats point to the possible coexistence of these bacteria in the environment.

4.2.1. Sulfide

The ability of phototrophic sulfur bacteria to oxidize reduced sulfur compounds under anoxic conditions is one of their most characteristic and ecologically important properties. Sulfide is not only used as the electron donor and sulfur source, but also serves to maintain anoxic conditions. In a few locations that offer favorable conditions for the development of phototrophic sulfur bacteria the source of sulfide may be geochemical; in most cases, however, it is of biological origin and depends on microbial activity during anaerobic decomposition of organic matter within the sediments of aquatic habitats. One of the possible final stages of anaerobic decomposition of organic matter is performed by sulfate-reducing bacteria, strictly anaerobic bacteria which are involved in the decomposition of the major part of organic carbon in marine sediments and also for the major part of biological sulfide production (Jörgensen and Fenchel, 1974; Widdel, 1988). The close interdependence between sulfate-reducers and phototrophic sulfur bacteria is of particular importance in freshwater habitats. While sulfate is normally abundant in marine habitats (approximately 28 mM) and high rates of sulfide production are found when suitable organic substrates are present, in freshwater habitats usually the concentration of sulfate (approximately 0.2 mM) limits sulfate reduction. Under these conditions blooms of phototrophic sulfur bacteria may develop whenever a balanced coexistence of sulfate-reducing bacteria and sulfide-oxidizing phototrophic bacteria effectively recycles the available sulfur sources.

Whenever the activity of sulfate-reducing bacteria in a sediment is high enough to raise the sulfide horizon into the photic zone, development of purple and green sulfur bacteria is possible. Conspicuous mass developments of these bacteria have been observed in all kinds of aquatic environments where sulfide accumulates and large amounts of sulfide are oxidized by these bacteria. The oxidation of sulfide by purple non-sulfur bacteria, some of which also may use sulfide as photosynthetic electron donor, is of less ecological importance.

4.2.1a. Relations of Phototrophic Bacteria to Sulfide.   Two important factors govern the relation of phototrophic bacteria to sulfide: (i) their tolerance to it, which is a quantitative relation between the concentration of sulfide and unfavorable effects on growth, and (ii) their requirement for it, which reflects the need for a reduced environment and for sulfide as a photosynthetic electron donor. Toxic effects of sulfide are found in all phototrophic bacteria, whether they can use sulfide or not and the concentrations that are tolerated vary considerably among species. Most of the purple non-sulfur bacteria are more sensitive to sulfide than are the phototrophic sulfur bacteria. But also within the purple and green sulfur bacteria, groups of species are distinguished on the basis of low and high sulfide tolerance. The highest tolerance is found in the green sulfur bacteria, which agrees well with their development in the lowermost layer of phototrophic microorganisms in stratified natural environments.

Purple non-sulfur bacteria in general have a low sulfide tolerance (0.4–2.0 mM) which is increased in the presence of small amounts of yeast extract [0.8–3.2 mM (Hansen, 1974)]. Many of these bacteria are dominant in anoxic, but sulfide-poor freshwater habitats. Some purple non-sulfur bacteria also accompany mass developments of purple and green sulfur bacteria in sedimental and lake habitats enriched in sulfide, but seldom form conspicuous mass developments in the absence of phototrophic sulfur bacteria. Several purple nonsulfur bacteria, however, are able to oxidize sulfide either to elemental sulfur or to sulfate (Hansen and Veldkamp, 1973; Hansen and van Gemerden, 1972; Imhoff *et al.*, 1984). Some of these species, in particular typical marine ones, exhibit a sulfide tolerance comparable to that of *Chromatium vinosum*.

Two physiological–ecological groups of the Chromatiaceae can be distinguished on the basis of sulfide and light tolerance (Pfennig, 1978). Moderate to high sulfide concentrations (2–8 mM) and high light intensities (1000–2000 lux) are selective for species without gas vesicles, such as small-celled *Chromatium* species, *Thiocystis*, and *Thiocapsa*. The large-celled *Chromatium* species, *Thiospirillum jenense*, and the gas-vacuolated species of the genera *Amoebobacter*, *Lamprocystis*, *Thiodictyon*, and *Thiopedia* have a selective advantage at low sulfide concentrations (0.4–1.0 mM), low light intensities, and temperatures below 20°C. They are regularly found in blooms in the uppermost layer of the sulfide-containing hypolimnion of stratified lakes.

We can also distinguish two ecological–physiological groups among the green sulfur bacteria. The first group grows well at high sulfide concentrations (4–8 mM) and light intensities of 700–1500 lux. It includes the green species of the genera *Chlorobium* and *Prosthecochloris*. They inhabit sediments and shallow waters enriched in sulfide of freshwater and marine environments. In particular, *Prosthecochloris aestuarii* is common to sediments and shallow waters of marine, but also hypersaline environments (Puchkova, 1984; and J. F. Imhoff, unpublished results). It has been isolated regularly from unstable intertidal mud flats from the German Waddensea close to the island of Sylt (J. F. Imhoff, unpublished results). The second group has a selective advantage at low light intensities (50–100 lux), low sulfide concentrations (0.4–2.0 mM), and low temperatures (below 20°C). This group contains the species of the genera *Pelodictyon* and *Ancalochloris* containing gas vesicles and represents the planktonic species that are found in the upper layers of the sulfide-containing hypolimnion in stratified lakes (Gorlenko and Lebedeva, 1971; Gorlenko, 1972; see also Pfennig, 1978).

4.2.1b. Role of Sulfide and Elemental Sulfur in the Competition between Phototrophic Bacteria.   Concentrations of sulfide and elemental sulfur and the relations of the different phototrophic bacteria to these compounds are significant factors in determining patterns of natural dominance and successful competition. Of particular importance are the affinities for these two sulfur compounds, their oxidation rates, the ability to utilize external elemental sulfur, and the ability to store elemental sulfur inside or outside the cells. The oxidation of sulfide to elemental sulfur yields only two electrons. The oxidation of sulfur to sulfate yields six electrons and consequently renders possible a threefold higher energy yield.

Therefore, several purple non-sulfur bacteria, such as *Rhodobacter capsulatus* and *Rhodospirillum rubrum*, which oxidize sulfide to extracellular elemental sulfur exclusively certainly have a disadvantage compared to other phototrophic bacteria which can also use elemental sulfur as an electron donor. The intracellular storage of elemental sulfur by Chromatiaceae gives these bacteria an additional advantage over the previously mentioned purple non-sulfur bacteria and also over green sulfur bacteria, which store elemental sulfur outside the cells, although both types of sulfur bacteria can use extracellular elemental sulfur and oxidize it to sulfate. All elemental sulfur formed by Chromatiaceae is inaccessible to other bacteria (e.g., green sulfur bacteria), while the elemental sulfur formed by green sulfur bacteria is also available for Chromatiaceae. Intracellular stored sulfur globules are of inestimable value for these bacteria. They serve as a reservoir of photosynthetic electron donors under conditions of external sulfide depletion in the light. Under dark conditions and in the presence of oxygen, intracellular sulfur may support endogenous respiration, even in species that are unable to grow chemotrophically (Breuker, 1964; Kämpf and Pfennig, 1986). Under anaerobic conditions in the dark this sulfur may serve as an electron acceptor during endogenous fermentation of stored carbohydrates (Hendley, 1955; van Gemerden, 1968).

Nevertheless, a number of conditions have been explored by the continuous culture studies of van Gemerden and his co-workers in which the stable coexistence of phototrophic bacteria is possible during autotrophic growth with sulfide and elemental sulfur as electron donors. Three examples of such experiments and possible coexistence between *Rhodobacter capsulatus* and *Chromatium vinosum*, between *Chromatium vinosum* and *Chlorobium limicola*, and between a fast-growing and a slow-growing *Chromatium* species will be shortly discussed. The additional availability of organic substrates further complicates the situation and will not be considered here (van Gemerden and Beeftink, 1983; Veldhuis and van Gemerden, 1986; Hofman *et al.*, 1985; Wijbenga and van Gemerden, 1981).

The purple non-sulfur bacterium *Rb. capsulatus* has a very high affinity for sulfide and is able to grow rapidly in sulfide media with sulfide concentrations below 2 mM and CO_2 as the sole carbon source (Hansen, 1974; van Gemerden and Beeftink, 1983). If one only compares the kinetic parameters for growth on sulfide to those of *Chromatium vinosum*, the latter would seem to be outcompeted at all noninhibitory concentrations of sulfide (van Gemerden and Beeftink, 1983). However, *Rb. capsulatus* oxidizes sulfide to elemental sulfur only, which is deposited outside the cells (Hansen, 1974), whereas *Chromatium vinosum* is able to use the sulfide (in competition with *Rb. capsulatus*), its intracellularly stored elemental sulfur formed during sulfide oxidation, and also the extracellular elemental sulfur, which is formed but not used by *Rb. capsulatus*. Apparently *Rb. capsulatus* is able to coexist under sulfide-limited chemostat conditions due to its high affinity for sulfide. It forms about 5% of the total produced biomass, which indicates that *Chromatium vinosum* had oxidized not only all of the elemental sulfur, but also most of the sulfide (van Gemerden and Beeftink, 1983).

In mixed continuous cultures with sulfide as the sole electron donor, stable coexistence was also observed between *Chromatium vinosum* and *Chlorobium limicola;* at lower dilution rates *Chlorobium limicola* dominated by 90%

(at higher dilution rates the *Chromatium vinosum* was dominant). It was concluded that *Chlorobium limicola,* due to its much higher affinity for sulfide, must have oxidized most of the sulfide and also a large amount of the extracellular sulfur (van Gemerden and Beeftink, 1981). In fluctuating natural ecosystems, the obligately anaerobic green sulfur bacteria may suffer more severely from the presence of oxygen than do many purple sulfur bacteria. The oxygen and light regime may be much more important than sulfide concentrations in governing competition between these two groups of bacteria in nature.

A special case of competition between large-celled and small-celled *Chromatium* species, which in principle have an identical sulfur metabolism, was found to be based on differences in growth rates as well as affinities for and oxidation rates of sulfide (van Gemerden, 1974). Only the oxidation of sulfide is important in the competition of *Chromatium* species, because elemental sulfur is accumulated inside their cells. Small-celled species, such as *Chromatium vinosum,* usually outgrow large-celled species, such as *Chromatium weissei,* rapidly in enrichment cultures. In sulfide-limited continuous mixed cultures, *Chromatium vinosum,* which has the higher affinity for sulfide, grew faster at all sulfide concentrations and outgrew the *Chromatium weissei* under continuous illumination. If the culture was exposed to alternating light and dark periods, however, the bacteria exhibited a balanced coexistence (van Gemerden, 1974). This coexistence was explained by the higher oxidation rate of *Chromatium weissei* compared to the higher growth rate and higher affinity for sulfide of *Chromatium vinosum.* After accumulation of sulfide during a dark period, upon illumination the greater part of it was oxidized by *Chromatium weissei* and stored as elemental sulfur in the cells for further utilization. After prolonged illumination and decrease of the sulfide concentration, however, *Chromatium vinosum* had an advantage. Consequently, fluctuating concentrations during intermittent light/dark cycles, as they occur in the natural environments, enable the large-celled species to exist together with small-celled species and eventually even become dominant.

4.2.2. Oxygen

Sulfide and oxygen are direct competitors in the life of phototrophic sulfur bacteria. Oxygen cannot exist in the presence of sulfide, nor can sulfide exist under conditions of oxygen surplus. They react chemically, though slowly, with each other. Because of this slow chemical reaction, in nature sulfide-containing and oxygen-containing parts of a sedimental or lake environment are separated, but overlap and create a highly dynamic boundary zone, where both sulfide and oxygen are present. This transition zone is quite common to many lakes of the moderate climatic zone and is called the chemocline. The chemocline is characterized by high bacterial populations and high metabolic activity. Dense populations of phototrophic sulfur bacteria often develop in or just below this zone.

For many obligately phototrophic bacteria oxygen is toxic. Most species of the purple sulfur bacteria, however, tolerate low oxygen concentrations and many can grow chemoautotrophically with sulfide, thiosulfate, and elemental sulfur as electron donors in respiratory reactions, though organic substrates are not, or are only poorly, oxidized during respiration (Kämpf and Pfennig, 1980, 1986; De Wit and van Gemerden, 1987). The capacity to grow as chemotrophs under aerobic dark conditions is certainly of selective advantage for purple sulfur bacteria in their highly dynamic environment.

4.2.3. Light

Not only the quantity, but also the quality of light is of major importance for the development of phototrophic bacteria, and due to the different pigment contents of the various phototrophic bacteria, light is also a selective environmental factor. The required light quality is revealed in the absorption spectra of the photosynthetic pigment–protein complexes and is a characteristic property for a particular species.

In all groups of phototrophic bacteria the quantitative content of pigments (bacteriochlorophylls and carotenoids) and the extension of the photosynthetic apparatus (number of chlorosomes in green bacteria and dimensions of the intracytoplasmic membranes in purple bacteria) are regulated by the light intensity (Holt *et al.,* 1966; Cohen-Bazire, 1963; Oelze and Drews, 1972). The highest pigment content is found in cells grown in dim light under anaerobic conditions. Major differences in the sensitivity of the light-harvesting apparatus exist between the phototrophic green and the purple bacteria. The chlorosomes of the green bacteria are unique light-harvesting organelles that are extraordinarily effective in collecting light and allow growth even at light intensities of 5–10 lux, which do not allow growth of any purple bacterium (Biebl and Pfennig, 1978); comparable growth rates are achieved with much lower light intensities than that required by purple bacteria. Both the quality and the quantity of light required by green sulfur bacteria are selective for their development as the lowermost layer of phototrophic organisms in stratified lakes and sediments.

In addition to these quantitative differences in light sensitivity between green and purple bacteria, great variation exists with respect to the chemical structures and the light-absorbing properties of bacteriochlorophylls and carotenoids in different species (see Tables I and II). Most of the purple bacteria have bacteriochlorophyll *a,* with long-wavelength absorption maxima between 800 and 900 nm. Quite a few have bacteriochlorophyll *b*, with an absorption maximum at 1015–1035 nm. The light-harvesting organelles of the phototrophic green bacteria, the chlorosomes, contain either bacteriochlorophyll *c, d,* or *e,* which all have their long-wavelength absorption maximum between 700 and 760 nm. (The chlorophyll *a* of cyanobacteria and eukaryotic algae absorbs below 700 nm.) These spectral properties demonstrate that on the basis of (bacterio) chlorophyll absorption there is no competition among green bacteria, purple bacteria, and oxygenic phototrophic organisms.

Bacteriochlorophyll absorption is of major importance in sediments, because infrared radiation penetrates particularly deep into sandy sediments (Hoffman, 1949). In deeper layers of water, however, the use of bacteriochlorophylls for light harvesting is prevented by the strong absorption of infrared radiation by water, in particular above 800 nm. Therefore, bacteria with bacteriochlorophyll *b* are the least suited to develop in deeper layers of lakes. They appear to be well adapted to shallow waters and in particular to sedimental habitats.

Light absorption by carotenoids is of major ecological significance in deeper layers of lakes, because radiation between 450 and 550 nm penetrates deepest into water (Pfennig, 1967; Culver and Brunskill, 1969). Light of this quality, however, is absorbed by the carotenoids of all kinds of phototrophic microorganism. In an environment with several layers of phototrophic activity, the top layers may select—due to their light absorbance—the bacteria developing in the lower layers. The carotenoids of phototrophic purple bacteria strongly absorb light between 480 and 550 nm, but leave a window for the short-wavelength maximum of the carotenoids of green sulfur bacteria around 450 nm. Underneath a layer of phototrophic purple bacteria the green species with their major absorption peak at this wavelength apparently have a selective advantage over the brown species, as deduced from observations from Lake Ciso (Spain) and laboratory experiments using light filters (Montesinos *et al.,* 1983). In lakes with a dense population of green algae and/or cyanobacteria in the upper layers of the water column, maximum light transmission is near 540 nm and presents highly selective conditions for the brown species. The carotenoids of these bacteria, isorenieratene and β-isorenieratene (Liaaen-Jensen, 1978), have a broad absorption maximum between 450 and 550 nm. The maximum at 550 nm (which is absent from the green species) coincides with maximum light transmission under these conditions.

At extremely low light intensities, the carotenoid/bacteriochlorophyll ratio is strongly increased in the brown *Chlorobium phaeobacteroides,* but not in the green *Chlorobium limicola* (Montesinos *et al.,* 1983). This is an additional property that makes the brown species the fittest at the low light intensities of deep waters, when light absorption by bacteriochlorophyll is not possible. Many observations show that brown species are dominant in lakes where the chemocline is between 10 and 25 m deep and penetrating light intensities are too low for the development of other phototrophic organisms. In a number of freshwater lakes from around the world with blooms of green sulfur bacteria, green species are dominant when these blooms occur near the surface (2–4 m), brown species dominate in blooms below 10 m, and both types are found together in blooms between 5 and 10 m (Montesinos *et al.*, 1983).

It has been pointed out also that among phototrophic purple bacteria the carotenoids are of importance for their development at low light intensities. Many purple sulfur bacteria living at low light intensities in stratified lakes have okenone as major carotenoid (Guerrero *et al.*, 1987). Pure cultures of these species have a broad and high carotenoid absorption maximum and are more sensitive to light than species with, e.g., spirilloxanthin as major carotenoid (Eichler and Pfennig, 1986, 1988). A similarity of the chemical structure between okenone and chlorobactene and isorenieratene is the presence of an aromatic ring at one or both ends of the carbon chain (Schmidt, 1978). This aromatic ring structure combined with the conjugated double bond system of the carotenoid chain may be much more effective in collecting light compared to other carotenoid structures lacking this terminal aromatic ring.

4.2.4. Motility

The ability of a bacterium to actively move to a place where optimal development is possible is certainly of advantage in nature. Such a reaction requires that the bacterium possess mechanisms which sense the environmental conditions, discriminate between pleasant and unpleasant signals, and react to the sensed signals either positively or negatively by movement. Several mechanisms of movement have evolved in bacteria. (i) In planktonic bacteria this may be achieved by changes of cell buoyancy with the formation or destruction of gas vesicles (only movement up and down in stagnant water is possible). (ii) Gliding motility requires a solid surface. (iii) The most advanced and active movement in the aquatic environment is possible by the use of flagella.

Most of the planktonic purple and green sulfur bacteria have gas vesicles that apparently enable them to adapt their buoyancy according to the sulfide concentration and light intensity of this stratified environment. Their adaptation to the planktonic life is further shown by their optimum development at temperatures below 20°C, in low sulfide concentrations, and under low light intensities. With a few exceptions, these bacteria do not have flagella.

Only a few anoxygenic phototrophic bacteria with gliding motility are known. *Chloroflexus aurantiacus* is a typical gliding bacterium living on sediment surfaces of hot spring effluents. Several gliding phototrophic green bacteria contain gas vesicles in addition. These bacteria, therefore, are suited for planktonic development, but also can move on sedimental surfaces.

Most species of the phototrophic purple bacteria are motile by the use of flagella and are able to respond to environmental conditions by phototactic and chemotactic mechanisms. Quite a few species of the purple bacteria completely lack motility: *Thiocapsa roseopersicina, Thiocapsa pfennigii, Rhodocyclus purpureus, Rhodobacter veldkampii,* and *Rhodobacter adriaticus.*

Flagellar movement is not possible by any of the green sulfur bacteria. In nature, however, green sulfur bacteria are often observed as consortia together with motile heterotrophic bacteria (Pfennig, 1989c). A close association between phototrophic and heterotrophic bacteria is observed in the "Chlorochromatium" and "Pelochromatium" consortia. These consortia apparently are highly cooperative relationships between the partners. Growth depends on the presence of sulfide and light. Motility is provided by the heterotrophic partner. The motile consortium is phototactically active (Pfennig, 1989c).

4.3. Position in the Cycle of Matter

Phototrophic bacteria play a significant role in both the carbon and the sulfur cycles of their habitats and are particularly suited to utilize end products of the anaerobic degradation of organic matter, such as hydrogen, sulfide, carbonate, and a number of small organic compounds that are formed during fermentation by various fermenting bacteria and during anaerobic respiration by sulfate-reducing bacteria. Phototrophic bacteria usually cannot degrade macromolecules directly and even sugars are not used or are poor substrates for most of them.

Due to their ability to utilize light as an energy source, anoxygenic phototrophic bacteria are able to produce biomass from these substrates under anaerobic conditions. They constitute an important and indispensable part of an anaerobic cycle of matter, in which sulfate and sulfide perform the function taken by the couple oxygen/water in the aerobic cycle. The phototrophic sulfur bacteria act—similarly to plants, algae, and cyanobacteria in the aerobic cycle—as primary producers and provide external electron acceptors (sulfate and elemental sulfur) for an important component of the anaerobic consumers, the sulfur- and sulfate-reducers.

The function of sulfate-reducing bacteria in the anaerobic cycle of matter is of considerable importance for the establishment of anoxic conditions and for the provision of sulfide. Products of fermenting bacteria are used as electron donors for anaerobic respiration of the sulfate-reducers, with sulfate as electron acceptor being reduced to sulfide. The major carbon compounds released by these bacteria, acetate and carbonate, are both excellent carbon sources for almost all phototrophic bacteria. Although most of the phototrophic sulfur bacteria can grow photoautotrophically, the additional availability of simple organic substrates considerably enhances their development.

As a simple model of such an anaerobic system, a mixed culture of *Escherichia coli, Desulfovibrio desulfuricans,* and a *Chlorobium* species was grown with glucose as substrate. Only *E. coli* is able to use the glucose, which is degraded to several fermentation products. Some of the fermentation products can only be used by *D. desulfuricans,* which in turn produced sulfide and acetate. The products of the former two bacteria, acetate, sulfide, and CO_2 served as substrates for *Chlorobium* (Matheron and Baulaigue, 1976).

A close syntrophic relationship has also been demonstrated between green sulfur bacteria and sulfur-reducers such as *Desulfuromonas acetoxidans.* In this relationship, not sulfate, but elemental sulfur is the oxidized sulfur compound formed extracellularly by the green sulfur bacterium and reduced by the sulfur-reducer, which has a higher affinity for elemental sulfur (Biebl and Pfennig, 1978). In mixed cultures of *D. acetoxidans* and *Chlorobium limicola* only catalytic amounts of sulfur were required for a stable coexistence of the syntrophic partners and resulted in excellent development of the two bacteria.

Evidence for a complete cycle of matter under strictly anoxic conditions has been obtained in a number of Winogradsky columns that were prepared in 1951 and 1952 by the Yugoslavian microbiologist Vlaho Cviic and found untouched and perfectly sealed 22 years later (11 years after he died) on a shelf in his laboratory. These columns, upon opening, revealed ongoing activity of sulfate reduction (*Desulfovibrio* species) and photosynthesis (*Chlorobium* and *Chromatium* species). These bacteria started to grow immediately after inoculation of appropriate media and several strains were isolated (J. F. Imhoff, unpublished results).

The activity of phototrophic sulfur bacteria contributes to the productivity of their habitats. Their productivity depends on the availability of sulfide as the photosynthetic electron donor and consequently is higher in lakes rich in sulfide compared to those poorer in sulfide. Usually the production in a bloom of phototrophic bacteria is $100-200$ mg C m^{-3} day^{-1}, but values of 1600 (Fayetteville Green Lake) and 5700 mg C m^{-3} day^{-1} (Smith Hole) have been reported (Culver and Brunskill, 1969; Wetzel, 1973). On the basis of the total annual productivity, the organic matter produced by phototrophic sulfur bacteria is usually about $3-5\%$ in lakes poor in sulfide and about $9-25\%$ in lakes rich in sulfide, but may reach values of more than 80% of the total production of organic matter [in Solar Lake and Fayetteville Green Lake (Cohen *et al.,* 1977; Culver and Brunskill, 1969; Takahashi and Ichimura, 1968; Pfennig, 1978)]. It is particularly high in permanently stratified lakes. In lakes that are not permanently stratified the contribution to the daily productivity under stratified conditions is higher compared to the annual contribution. On the basis of the annual productivity and the lake area, Pfennig (1978) calculated a total production by phototrophic bacteria of 60 tons year^{-1} of organic carbon in Green Lake (0.25 km^2), a lake with high productivity by phototrophic bacteria, and 3.75 tons of organic carbon in Medicine Lake (0.125 km^2), a lake with a normal productivity by phototrophic bacteria. Accordingly, 84 and 5.25 tons, respectively, of toxic sulfide were oxidized per year by the phototrophic bacteria in these lakes (Pfennig, 1978).

Due to their development at the borderline between anoxic and oxic parts of their environment, the productivity of phototrophic bacteria contributes to both the anaerobic and the aerobic food chains (Fenchel, 1969). Planktonic predators, which generally are obligate aerobes, are often found just above blooms of phototrophic bacteria and many authors have reported on predation of phototrophic bacteria by zooplanktonic species of various kinds of Copepoda and Cladocera. It is sometimes assumed that they periodically dive into the anaerobic layer inhabited by their prey organisms. Their intestines often are colored as a result of ingestion of phototrophic bacteria (van Gemerden and Beeftink, 1983). The activity of predators can be very high and, besides other factors, must be considered as a decisive factor in establishing the population density of phototrophic sulfur bacteria.

REFERENCES

Akiba, T., Usami, R., and Horikoshi, K., 1983, *Rhodopseudomoas rutila*, a new species of non-sulfur purple photosynthetic bacteria, *Int. J. Syst. Bacteriol.* **33**:551–556.

Amesz, J., and Knaff, D. B., 1988, Molecular mechanism of bacterial photosynthesis, in: *Biology of Anaerobic Microorganisms* (A. J. B. Zehnder, ed.), Wiley, Chichester, pp. 113–178.

Bauld, J., and Brock, T. D., 1973, Ecological studies of *Chloroflexus*, a gliding photosynthetic bacterium, *Arch. Mikrobiol.* **92**:267–284.

Bavendamm, W., 1924, *Die farblosen und roten Schwefelbakterien des Süss- und Salzwassers*, Fischer Verlag, Jena.

Beer-Romero, P., and Gest, H., 1987, *Heliobacillus mobilis*, a peritrichously flagellated anoxyphototroph containing bacteriochlorophyll g, *FEMS Microbiol. Lett.* **41**:109–114.

Beer-Romero, P., Favinger, J. L., and Gest, H., 1988, Distinctive properties of bacilliform photosynthetic heliobacteria, *FEMS Microbiol. Lett.* **49**:451–454.

Biebl, H., and Pfennig, N., 1978, Growth yields of green sulfur bacteria in mixed cultures with sulfur and sulfate reducing bacteria, *Arch. Microbiol.* **117**:9–16.

Breuker, E., 1964, Die Verwertung von intrazellulärem Schwefel durch *Chromatium vinosum* im aeroben und anaeroben Licht- und Dunkelstoffwechsel, *Zentralbl. Bakteriol. Parasitenkd. Hyg. Abt. 2* **118**:561–568.

Brock, T. D., 1978, *Thermophilic Microorganisms and Life at High Temperature*, Springer-Verlag, New York.

Brockmann, H., Jr., and Lipinski, A., 1983, Bacteriochlorophyll *g*. A new bacteriochlorophyll from *Heliobacterium chlorum*, *Arch. Microbiol.* **136**:17–19.

Caumette, P., Baulaigue, R., and Matheron, R., 1988, Characterization of *Chromatium salex-*

igens sp. nov., a halophilic Chromatiaceae isolated from Mediterranean salinas, *Syst. Appl. Microbiol.* **10:**284–292.

Caumette, P., Baulaigue, R., and Matheron, R., 1991, *Thiocapsa halophila* sp. nov., a new halophilic phototrophic purple sulfur bacterium, *Arch. Microbiol.* **155:**170–176.

Cohen, Y., Jorgensen, B. B., Padan, E., and Shilo, M., 1975, Sulfide-dependent anoxygenic photosynthesis in the cyanobacterium *Oscillatoria limnetica*, *Nature* **257:**486–492.

Cohen, Y., Krumbein, W. E., and Shilo, M., 1977, Solar Lake (Sinai). II. Distribution of photosynthetic microorganisms and primary production, *Limnol. Oceanogr.* **22:**609–620.

Cohen-Bazire, G., 1963, Some observations on the organization of the photosynthetic apparatus in purple and green bacteria, in: *Bacterial Photosynthesis* (H. Gest, A. San Pietro, and L. P. Vernon, eds.), Antioch Press, Yellow Springs, Ohio, pp. 89–110.

Cohen-Bazire, G., Pfennig, N., and Kunizawa, R., 1964, The fine structure of green bacteria, *J. Cell Biol.* **22:**207-225.

Culver, D. A., and Brunskill, G. J., 1969, Fayetteville Green Lake, New York. V. Studies of primary production and zooplankton in a meromictic lake, *Limnol. Oceanogr.* **14:**862–873.

De Wit, R., and van Gemerden, H., 1987, Chemolithotrophic growth of the phototrophic sulfur bacterium *Thiocapsa roseopersicina*, *FEMS Microbiol. Ecol.* **45:**117–126.

Dickerson, R. E., 1980, Evolution and gene transfer in purple photosynthetic bacteria, *Nature* **283:**210–212.

Drews, G., 1981, *Rhodospirillum salexigens*, spec. nov., an obligatory halophilic phototrophic bacterium, *Arch. Microbiol.* **130:**325–327.

Dubinina, G. A., and Gorlenko, V. M., 1975, New filamentous photosynthetic green bacteria containing gas vacuoles, *Mikrobiology* **44:**452–458.

Eckersley, K., and Dow, C. S., 1980, *Rhodopseudomonas blastica* sp. nov.: A member of the Rhodospirillaceae, *J. Gen. Microbiol.* **119:**465–473.

Eichler, B., and Pfennig, N., 1986, Characterization of a new platelet-forming purple sulfur bacterium, *Amoebobacter pedioforis* sp. nov., *Arch. Microbiol.* **146:**295–300.

Eichler, B., and Pfennig, N., 1988, A new purple sulfur bacterium from stratified fresh-water lakes, *Amoebobacter purpureus* sp. nov., *Arch. Microbiol.* **149:**395–400.

Evans, M. C. W., Buchanan, B. B., and Arnon, D. I., 1966, A new ferredoxin-dependent carbon reduction cycle in a photosynthetic bacterium, *Proc. Natl. Acad. Sci. USA* **55:**928–934.

Evans, W. R., Fleischmann, D. E., Calvert, H. E., Pyati, P. V., Alter, G. M., and Rao, N. S. S., 1990, Bacteriochlorophyll and photosynthetic reaction centers in *Rhizobium* strain BTAi 1, *Appl. Environm. Microbiol.* **56:**3445–3449.

Favinger, J., Stadtwald, R., and Gest, H., 1989, *Rhodospirillum centenum*, sp. nov., a thermotolerant cyst-forming anoxygenic photosynthetic bacterium, *Ant. Leeuwenh.* **55:**291–296.

Fenchel, T., 1969, The ecology of marine microbenthos. IV. Structure and function of the benthic ecosystem, its chemical and physical factors and the microfauna communities with special reference to the ciliated protozoa, *Ophelia* **6:**1–182.

Fowler, V. J., Pfennig, N., Schubert, W., and Stackebrandt, E., 1984, Towards a phylogeny of phototrophic purple sulfur bacteria—16S rRNA oligonucleotide cataloguing of 11 species of Chromatiaceae, *Arch. Microbiol.* **139:**382–387.

Fuchs, G., Stupperich, E., and Jaenchen, R., 1980a, Autotrophic CO_2 fixation in *Chlorobium limicola*. Evidence against the operation of the Calvin cycle in growing cells, *Arch. Microbiol.* **128:**56–63.

Fuchs, G., Stupperich, E., and Eden, G., 1980b, Autotrophic CO_2 fixation in *Chlorobium limicola*. Evidence for the operation of a reductive tricarboxylic acid cycle in growing cells, *Arch. Microbiol.* **128:**64–71.

Garlick, S., Oren, A., and Padan, E., 1977, Occurrence of facultative anoxygenic photosynthesis among filamentous and unicellular cyanobacteria, *J. Bacteriol.* **129:**623–629.

Gerola, P. D., and Olson, J. M., 1986, A new bacteriochlorophyll *a*–protein complex associated with chlorosomes of green sulfur bacteria, *Biochim. Biophys. Acta* **848:**69–76.

Gest, H., and Favinger, J. F., 1983, *Heliobacterium chlorum*, an anoxygenic brownish-green bacterium containing a "new" form of bacteriochlorophyll, *Arch. Microbiol.* **136:**11–16.

Gibson, J., Stackebrandt, E., Zablen, L. B., Gupta, R., and Woese, R. W., 1979, A phylogenetic analysis of the purple photosynthetic bacteria, *Curr. Microbiol.* **3:**59–64.

Gibson, J., Pfennig, N., and Waterbury, J. B., 1984, *Chloroherpeton thalassium* gen. nov. et spec. nov., a nonfilamentous, flexing, and gliding green sulfur bacterium, *Arch. Microbiol.* **138:**96–101.

Gibson, J., Ludwig, W., Stackebrandt, E., and Woese, C. R., 1985, The phylogeny of the green photosynthetic bacteria: Absence of a close relationship between *Chlorobium* and *Chloroflexus*, *Syst. Appl. Microbiol.* **6:**152–156.

Giovannoni, S. J., Revsbech, N. P., Ward, D. M., and Castenholz, R. W., 1987, Obligately phototrophic *Chloroflexus:* Primary production in anaerobic hot spring microbial mats, *Arch. Microbiol.* **147:**80–87.

Gloe, A., and Risch, N., 1978, Bacteriochlorophyll C_S, a new bacteriochlorophyll from *Chloroflexus aurantiacus*, *Arch. Microbiol.* **118:**153–156.

Gloe, A., Pfennig, N., Brockmann, H., Jr., and Trowitsch, W., 1975, A new bacteriochlorophyll from brown-colored Chlorobiaceae, *Arch. Microbiol.* **102:**103–109.

Gorlenko, V. M., 1972, Phototrophic brown sulfur bacteria *Pelodictyon phaeum* non. sp., *Microbiologia* **41:**370–371 [in Russian].

Gorlenko, V. M., 1974, Oxidation of thiosulphate by *Amoebobacter roseus* in darkness under microaerobic conditions, *Microbiologia* **43:**729–731 [in Russian].

Gorlenko, V. M., 1975, Characteristics of filamentous phototrophic bacteria from freshwater lakes, *Microbiology* **44:**682–684.

Gorlenko, V. M., and Lebedeva, E. V., 1971, New green sulphur bacteria with apophyses, *Microbiologia* **40:**1035–1039 [in Russian].

Gorlenko, V. M., and Pivovarova, T. A., 1977, On the belonging of blue-green alga *Oscillatoria coerulescens* Gickelhorn, 1921 to a new genus of Chlorobacteria *Oscillochloris* nov. gen., *Izv. Akad. Nauk SSSR Ser. Biol.* **3:**396–409 [in Russian].

Gorlenko, V. M., and Krasilnikova, E. N., Kikina, O. G., and Tatarinova, N. Ju., 1979, The new motile purple sulphur bacteria *Lamprobacter modestohalophilus* nov. gen., nov. spec. with gas vacuoles, *Biol. Bull. Acad. Sci. USSR* **6:**631–642 [in Russian].

Gorlenko, V. M., Dubinina, G. A., and Kusnetzov, S. I., 1983, *The Ecology of Aquatic Microorganisms*, Schweitzbart'sche Verlagsbuchhandlung, Stuttgart.

Guerrero, R., Pedros-Alio, C., Esteve, I., and Mas, J., 1987, Communities of phototrophic sulfur bacteria in lakes of the Spanish Mediterranean region, *Acta Acad. Aboensis* **47:**125–151.

Hansen, T. A., 1974, Sulfide als electronendonor voor Rhodospirillaceae, Doctoral thesis, University of Groningen, The Netherlands.

Hansen, T. A., and Imhoff, J. F., 1985, *Rhodobacter veldkampii*, a new species of phototrophic purple nonsulfur bacteria, *Int. J. Syst. Bacteriol.* **35:**115–116.

Hansen, T. A., and van Gemerden, H., 1972, Sulfide utilization by purple nonsulfur bacteria, *Arch. Mikrobiol.* **86:**49–56.

Hansen, T. A., and Veldkamp, H., 1973, *Rhodopseudomonas sulfidophila* nov. spec., a new species of the purple nonsulfur bacteria, *Arch. Mikrobiol.* **92:**45–58.

Hansen, T. A., Sepers, A. B. J., and van Gemerden, H., 1975, A new purple bacterium that oxidizes sulfide to extracellular sulfur and sulfate, *Plant Soil* **43:**17–27.

Harashima, K., Hayashi, J.-I., Ikari, T., and Shiba, T., 1980, O_2-stimulated synthesis of bacteriochlorophyll and carotenoids in marine bacteria, *Plant Cell Physiol.* **21:**1283–1294.

Hendley, D. D., 1955, Endogenous fermentation in Thiorhodaceae, *J. Bacteriol.* **70:**625–634.

Hiraishi, A., Hoshino, Y., and Satoh, T., 1991, *Rhodoferax fermentans* gen. nov., sp. nov., a phototrophic purple nonsulfur bacterium previously referred to as the "Rhodocyclus gelatinosus-like" group. *Arch. Microbiol.* **155**:330–336.

Hoffmann, C., 1949, Über die Durchlässigkeit dünner Sandschichten für Licht, *Planta* **37**:48–56.

Hofman, P. A. G., Veldhuis, M. J. W., and van Gemerden, H., 1985, Ecological significance of acetate assimilation by *Chlorobium phaeobacteroides, FEMS Microbiol. Lett.* **31**:271–278.

Holt, S. C., Conti, S. F., and Fuller, R. C., 1966, Effect of light intensity on the formation of the photochemical apparatus in the green bacterium *Chloropseudomonas ethylicum, J. Bacteriol.* **91**:349–355.

Iba, K., Takamiya, K.-I., Toh, Y., and Nishimura, M., 1988, Roles of bacteriochlorophyll and carotenoid synthesis in formation of intracytoplasmic membrane systems and pigment–protein complexes in an aerobic photosynthetic bacterium, *Erythrobacter* sp. strain OCh114, *J. Bacteriol.* **170**:1843–1847.

Imhoff, J. F., 1982, Taxonomic and phylogenetic implications of lipid and quinone compositions in phototrophic microorganisms, in: *Biochemistry and Metabolism of Plant Lipids* (J. F. G. M. Wintermans and P. J. C. Kuiper, eds.), Elsevier Biomedical Press, Amsterdam, pp. 541–544.

Imhoff, J. F., 1983, *Rhodopseudomonas marina* sp. nov, a new marine phototrophic purple bacterium, *Syst. Appl. Microbiol.* **4**:512–521.

Imhoff, J. F., 1984a, Reassignement of the genus *Ectothiorhodospira* Pelsh 1936 to a new family Ectothiorhodospiraceae fam. nov., and emended description of the Chromatiaceae Bavendamm 1924, *Int. J. Syst. Bacteriol.* **34**:338–339.

Imhoff, J. F., 1984b, Quinones of phototrophic purple bacteria, *FEMS Microbiol. Lett.* **25**:85–89.

Imhoff, J. F., 1988a, Halophilic phototrophic bacteria, in: *Halophilic Bacteria* (F. Rodriguez-Valera, ed.), CRC Press, Boca Raton, Florida, pp. 85–108.

Imhoff, J. F., 1988b, Lipids, fatty acids and quinones in taxonomy and phylogeny of anoxygenic phototrophic bacteria, in: *Green Photosynthetic Bacteria* (J. M. Olson, J. G. Ormerod, J. Amesz, E. Stackebrandt, and H. G. Trüper, eds.), Plenum Press, New York, pp. 223–232.

Imhoff, J. F., 1988c, Anoxygenic phototrophic bacteria, in: *Methods in Aquatic Bacteriology* (B. Austin, ed.), Wiley, Chichester, pp. 207–240.

Imhoff, J. F., 1989a, Genus *Ectothiorhodospira* in: *Bergey's Manual of Systematic Bacteriology,* Volume 3 (J. T. Staley, M. P. Bryant, N. Pfennig, and J. G. Holt, eds.), Williams and Wilkins, Baltimore, pp. 1654–1658.

Imhoff, J. F., and Trüper, H. G., 1981, *Ectothiorhodospira abdelmalekii* sp. nov., a new halophilic and alkaliphilic phototrophic bacterium, *Zentralbl. Bakteriol. Hyg. I. Abt. Orig.* **C2**:228–234.

Imhoff, J. F., and Trüper, H. G., 1989, The purple nonsulfur bacteria, in: *Bergey's Manual of Systematic Bacteriology,* Volume 3 (J. T. Staley, M. P. Bryant, N. Pfennig, and J. G. Holt, eds.), Williams and Wilkins, Baltimore, pp. 1658–1661.

Imhoff, J. F., Sahl, H. G., Soliman, G. S. H., and Trüper, H. G., 1979, The Wadi Natrun: Chemical composition and microbial mass developments in alkaline brines of eutrophic desert lakes, *Geomicrobiology* **1**:219–234.

Imhoff, J. F., Tindall, B., Grant, W. D., and Trüper, H. G., 1981, *Ectothiorhodospira vacuolata* sp. nov., a new phototrophic bacterium from soda lakes, *Arch. Microbiol.* **130**:238–242.

Imhoff, J. F., Kushner, D. J., Kushwaha, S. C., and Kates, M., 1982, Polar lipids in phototrophic bacteria of the Rhodospirillaceae and Chromatiaceae families, *J. Bacteriol.* **150**:1192–1201.

Imhoff, J. F., Trüper, H. G., and Pfennig, N., 1984, Rearrangement of the species and genera of the phototrophic "purple nonsulfur bacteria", *Int. J. Syst. Bacteriol.* **34**:340–343.

Ivanovsky, R. N., Sinton, N. V., and Kondratieva, E. N., 1980, ATP-linked citrate lyase activity in the green sulfur bacterium *Chlorobium limicola* forma *thiosulfatophilum, Arch. Microbiol.* **128:**239–241.

Jannasch, H. W., 1957, Die bakterielle Rotfärbung der Salzseen des Wadi Natrun, *Arch. Hydrobiol.* **53:**425–433.

Jannsen, P. H., and Harfoot, C. G., 1991, *Rhodopseudomonas rosea* sp. nov., a new purple nonsulfur bacterium, *Int. J. Syst. Bacteriol.* **41:**26–30.

Jörgensen, B. B., and Fenchel, T., 1974, The sulfur cycle of a marine sediment model system, *Mar. Biol.* **24:**189–201.

Kämpf, C., and Pfennig, N., 1980, Capacity of Chromatiaceae for chemotrophic growth. Specific respiration rates of *Thiocystis violacea* and *Chromatium vinosum, Arch. Microbiol.* **127:**125–135.

Kämpf, C., and Pfennig, N., 1986, Isolation and characterization of some chemoautotrophic Chromatiaceae, *J. Basic Microbiol.* **9:**507–515.

Kohler, H.-P., Ahring, B., Abella, C., Ingvorsen, K., Keweloh, H., Laczko, E., Stupperich, E., and Tomei, F., 1984, Bacteriological studies on the sulfur cycle in the anaerobic part of the hypolimnion and in the surface sediments of Rotsee Switzerland, *FEMS Microbiol. Lett.* **21:**279–289.

Kompantseva, E. J., 1985, *Rhodobacter euryhalinus* sp. nov., a new halophilic purple bacterial species, *Mikrobiologiya* **54:** 974–982 [in Russian].

Kompantseva, E. I., 1989, A new species of budding purple bacterium: *Rhodopseudomonas julia* sp. nov., *Microbiology* **58:**254–259.

Kompantseva, E. J., and Gorlenko, V. M., 1984, A new species of moderately halophilic purple bacterium *Rhodospirillum mediosalinum* sp. nov., *Mikrobiologiya* **53:**775–781.

Kondratieva, E. N., Zhukov, V. G., Ivanovsky, R. N., Petushkova, Y. P., and Monosov, E. Z., 1976, The capacity of phototrophic sulfur bacterium *Thiocapsa roseopersicina* for chemosynthesis, *Arch. Microbiol.* **108:**287–292.

Liaaen-Jensen, S., 1965, Bacterial carotenoids. XVIII. Arylcarotenes from *Phaeobium, Acta Chem. Scand.* **19:**1025–1030.

Liaaen-Jensen, S., 1978, Chemistry of carotenoid pigments, in: *The Photosynthetic Bacteria* (R. K. Clayton and W. R. Sistrom, eds.), Plenum Press, New York, pp. 233–247.

Madigan, M. T., 1986, *Chromatium tepidum* sp. nov., a thermophilic photosynthetic bacterium of the family Chromatiaceae, *Int. J. Syst. Bacteriol.* **36:**222–227.

Madigan, M. T., 1988, Microbiology, physiology, and ecology of phototrophic bacteria, in: *Biology of Anaerobic Microorganisms* (A. J. B. Zehnder, ed.), Wiley, Chichester, pp. 39–11.

Madigan, M. T., and Brock, T. D., 1975, Photosynthetic sulfide oxidation by *Chloroflexus aurantiacus*, a filamentous, photosynthetic, gliding bacterium, *J. Bacteriol.* **122:**782–784.

Matheron, R., and Baulaigue, R., 1976, Bacteries fermentatives, sulfato-reductrices et phototrophes sulfureuses en cultures mixtes, *Arch. Microbiol.* **109:**319–320.

Mayer, H., 1984, Significance of lipopolysaccharide structure for taxonomy and phylogenetical relatedness of Gram-negative bacteria, in: *The Cell Membrane* (E. Haber, ed.), Plenum Press, New York, pp. 71–83.

Meissner, J., Krauss, J. H., Jürgens, U. J., and Weckesser, J., 1988a, Absence of a characteristic cell wall lipopolysaccharide in the phototrophic bacterium *Chloroflexus aurantiacus, J. Bacteriol.* **170:**3213–3216.

Meissner, J., Pfennig, N., Krauss, J. H., Mayer, H., and Weckesser, J., 1988b, Lipopolysaccharides of *Thiocystis violacea, Thiocapsa pfennigii*, and *Chromatium tepidum*, species of the family Chromatiaceae, *J. Bacteriol.* **170:**3217–3222.

Molisch, H., 1907, *Die Purpurbakterien nach neuen Untersuchungen,* G. Fischer, Jena.

Montesinos, E., Guerrero, R., Abella, C., and Esteve, I., 1983, Ecology and physiology of the competition for light between *Chlorobium limicola* and *Chlorobium phaeobacteroides* in natural habitats, *Appl. Environ. Microbiol.* **46:**1007–1016.

Neutzling, O., Imhoff, J. F., and Trüper, H. G., 1984, *Rhodopseudomonas adriatica* sp. nov., a new species of the Rhodospirillaceae, dependent on reduced sulfur compounds, *Arch. Microbiol.* **137**:256–261.

Nishimura, Y., Shimizu, M., and Iizuka, H., 1981, Bacteriochlorophyll formation in radiation-resistent *Pseudomonas radiora, J. Gen. Appl. Microbiol.* **27**:427–430.

Nissen, H., and Dundas, I. D., 1984, *Rhodospirillum salinarum* sp. nov., a halophilic photosynthetic bacterium from a Portuguese saltern, *Arch. Microbiol.* **138**:251–256.

Oelze, J., and Drews, G., 1972, Membranes of photosynthetic bacteria, *Biochim. Biophys. Acta* **265**:209–239.

Oren, A., Kessel, M., and Stackebrandt, E., 1989, *Ectothiorhodospira marismortui* sp. nov., an obligatory anaerobic, moderately halophilic purple sulfur bacterium from a hypersaline sulfur spring on the shore of the Dead Sea, *Arch. Microbiol.* **151**:524–529.

Ormerod, J., Nesbakken, T., and Torgersen, Y., 1990, in: *Current Research in Photosynthesis,* Volume IV (M. Baltscheffsky, ed.), Kluwer Academic Publishers, Dordrecht, The Netherlands, pp. 935–938.

Overmann, J., and Pfennig, N., 1989, *Pelodictyon phaeoclathrathiforme*, sp. nov., a new brown-colored member of the Chlorobiaceae forming net-like colonies, *Arch. Microbiol.* **152**:401–406.

Oyaizu, H., Debrunner-Vossbrinck, B., Mandelco, L., Studier, J. A., and Woese, C. R., 1987, The green non-sulfur bacteria: A deep branching in the eubacterial line of descent, *Syst. Appl. Microbiol.* **9**:47–53.

Pfennig, N., 1967, Photosynthetic bacteria, *Annu. Rev. Microbiol.* **21**:285–324.

Pfennig, N., 1969, *Rhodopseudomonas acidophila*, sp. n., a new species of the budding purple nonsulfur bacteria, *J. Bacteriol.* **99**:597–602.

Pfennig, N., 1974, *Rhodopseudomonas globiformis*, sp. n., a new species of the Rhodospirillaceae, *Arch. Microbiol.* **100**:197–206.

Pfennig, N., 1977, Phototrophic green and purple bacteria: A comparative systematic survey, *Annu. Rev. Microbiol.* **31**:275–290.

Pfennig, N., 1978, General physiology and ecology of photosynthetic bacteria, in: *The Photosynthetic Bacteria* (R. E. Clayton, and W. R. Sistrom, eds.), Plenum Press, New York, pp. 3–18.

Pfennig, N., 1989a, Green sulfur bacteria, in: *Bergey's Manual of Systematic Bacteriology*, Volume 3 (J. T. Staley, M. P. Bryant, N. Pfennig, and J. C. Holt, eds.), Williams and Wilkins, Baltimore, pp. 1682–1683.

Pfennig, N., 1989b, Multicellular filamentous green bacteria, in: *Bergey's Manual of Systematic Bacteriology*, Volume 3 (J. T. Staley, M. P. Bryant, N. Pfennig, and J. C. Holt, eds.), Williams and Wilkins, Baltimore, p. 1697.

Pfennig, N., 1989c, Addendum to the green sulfur bacteria, in: *Bergey's Manual of Systematic Bacteriology*, Volume 3 (J. T. Staley, M. P. Bryant, N. Pfennig, and J. C. Holt, eds.), Williams and Wilkins, Baltimore, p. 1696–1697.

Pfennig, N., 1989d, Ecology of phototrophic purple and green sulfur bacteria, in: *Autotrophic Bacteria* (H. G. Schlegel and B. Bowien, eds.), Springer-Verlag, Heidelberg, pp. 97–116.

Pfennig, N., and Trüper, H. G., 1971, Higher taxa of the phototrophic bacteria, *Int. J. Syst. Bacteriol.* **21**:17–18.

Pfennig, N., and Trüper, H. G., 1974, The phototrophic bacteria. in: *Bergey's Manual of Determinative Bacteriology* (R. E. Buchanan and N. E. Gibbons, eds.), Williams and Wilkins, Baltimore, pp. 24–64.

Pierson, B. K., and Castenholz, R. W., 1974a, A phototrophic, gliding filamentous bacterium of hot springs, *Chloroflexus aurantiacus*, gen. and sp. nov., *Arch. Microbiol.* **100**:5–24.

Pierson, B. K., and Castenholz, R. W., 1974b, Studies of pigments and growth in *Chloroflexus aurantiacus*, a phototrophic, filamentous bacterium, *Arch. Microbiol.* **100**:283–301.

Pierson, B., K., Giovannoni, S. J., Stahl, D. A., and Castenholz, R. W., 1985, *Heliothrix oregonen-

sis gen. nov., spec. nov., a phototrophic filamentous gliding bacterium containing bacteriochlorophyll a, *Arch. Microbiol.* **142:**164–167.

Pivovarova, T. A., and Gorlenko, V. M., 1977, Fine structure of *Chloroflexus aurantiacus* var. *mesophilus* (nom. prof.) grown in the light under aerobic and anaerobic conditions. *Microbiology* **46:**276–282.

Puchkova, N. N., 1984, Green sulfur bacteria inhabiting shallow saline water bodies, *Mikrobiologiya* **53:**324–328 [in Russian].

Rodriguez-Valera, F., Ventosa, A., Juez, G., and Imhoff, J. F., 1985, Variation of environmental features and microbial populations with salt concentrations in a multi-pond saltern, *Microb. Ecol.* **11:**107–115.

Sato, K., 1978, Bacteriochlorophyll formation by facultative methylotrophs, *Protaminobacter ruber* and *Pseudomonas* AM 1, *FEBS Lett.* **85:**207–210.

Schmidt, K., 1978, Biosynthesis of carotenoids, in: *The Photosynthetic Bacteria* (R. K. Clayton, and W. R. Sistrom, eds.), Plenum Press, New York, pp. 729–750.

Schmidt, K., and Bowien, B., 1983, Notes on the description of *Rhodopseudomonas blastica*, *Arch. Microbiol.* **136:**242.

Seewaldt, E., Schleifer, K.-H., Bock, E., and Stackebrandt, E., 1982, The close phylogenetic relationship of *Nitrobacter* and *Rhodopseudomonas palustris*, *Arch. Microbiol.* **131:**287–290.

Shiba, T., 1984, Utilization of light energy by the strictly aerobic bacterium *Erythrobacter* sp. OCH 114, *J. Gen. Appl. Microbiol.* **30:**239–244.

Shiba, T., 1991, *Roseobacter litoralis* gen. nov., sp. nov., and *Roseobacter denitrificans* sp. nov., aerobic pink-pigmented bacteria which contain bacteriochlorophyll a, *Syst. Appl. Microbiol.* **14:**140–145.

Shiba, T., and Simidu, U., 1982, *Erythrobacter longus* gen. nov., spec. nov., an aerobic bacterium which contains bacteriochlorophyll a, *Int. J. Syst. Bacteriol.* **32:**211–217.

Shiba, T., Simidu, U., and Taga, N., 1979, Distribution of aerobic bacteria which contain bacteriochlorophyll a, *Appl. Environ. Microbiol.* **38:**43–45.

Shimada, K., Hayashi, H., and Tasumi, M., 1985, Bacteriochlorophyll–protein complexes of aerobic bacteria, *Erythrobacter longus* and *Erythrobacter* species OCH 114, *Arch. Microbiol.* **143:**244–247.

Stackebrandt, E., and Woese, C. R., 1981, The evolution of procaryotes, in: *Molecular and Cellular Aspects of Microbial Evolution* (M. J. Carlile, J. R. Collins, and B. E. B. Moseley, eds.), Cambridge University Press, Cambridge, pp. 1–31.

Stackebrandt, E., Fowler, V. J., Schubert, W., and Imhoff, J. F., 1984, Towards a phylogeny of phototrophic purple bacteria—The genus *Ectothiorhodospira*, *Arch. Microbiol.* **137:**366–370.

Stackebrandt, E., Murray, R. G. E., and Trüper, H. G., 1988, *Proteobacteria* classis nov., a name for the phylogenetic taxon that includes the "purple bacteria and their relatives", *Int. J. Syst. Bacteriol.* **38:**321–325.

Stadtwald-Demchick, R., Turner, F. R., and Gest, H., 1990, *Rhodopseudomonas cryptolactis*, sp. nov., a new thermotolerant species of budding phototrophic purple bacteria, *FEMS Microbiol. Lett.* **71:**117–122.

Staehelin, L. A., Fuller, R. C., and Drews, G., 1978, Visualization of the supramolecular architecture of chlorosomes (chlorobium vesicles) in freeze-fractured cells of *Chloroflexus aurantiacus*, *Arch. Microbiol.* **119:**269–277.

Staehelin, L. A., Golecki, J. R., and Drews, G., 1980, Supramolecular organization of chlorosomes (chlorobium vesicles) and of their membrane attachment sites in *Chlorobium limicola*, *Biochim. Biophys. Acta* **589:**30–45.

Stanier, R. Y., Pfennig, N., and Trüper, H. G., 1981, Introduction to the phototrophic prokaryotes, in: *The Prokaryotes* (M. P. Starr, H. Stolp, H. G. Trüper, A. Balows, and H. G. Schlegel, eds.), Springer-Verlag, New York, pp. 197–211.

Steiner, R., Schäfer, W., Blos, I., Wieschoff, H., and Scheer, H., 1981, 2,10-Phytadienol as esterifying alcohol of bacteriochlorophyll *b* from *Ectothiorhodospira halochloris, Z. Naturforsch.* **36c:**417–420.

Takahashi, M., and Ichimura, S., 1968, Vertical distribution and organic matter production of photosynthetic sulfur bacteria in Japanese lakes, *Limnol. Oceanogr.* **13:**644–655.

Trüper, H. G., 1989, Genus *Erythrobacter,* in: *Bergey's Manual of Systematic Bacteriology,* Volume 3 (J. T. Staley, M. P. Bryant, N. Pfennig, and J. G. Holt, eds.), Williams and Wilkins, Baltimore, pp. 1708–1709.

Trüper, H. G., and Pfennig, N., 1981, Characterization and identification of the anoxygenic phototrophic bacteria, in: *The Prokaryotes* (M. P. Starr, H. Stolp, H. G. Trüper, A. Balows, and H. G. Schlegel, eds.), Springer-Verlag, New York, pp. 299–312.

Van Gemerden, H., 1968, On the ATP generation by *Chromatium* in darkness, *Arch. Mikrobiol.* **64:**118–124.

Van Gemerden, H., 1974, Coexistence of organisms competing for the same substrate: An example among the purple sulfur bacteria, *Microb. Ecol.* **1:**104–119.

Van Gemerden, H., and Beeftink, H. H., Coexistence of *Chlorobium* and *Chromatium* in a sulfide-limited continuous culture, *Arch. Microbiol.* **129:**32–34.

Van Gemerden, H., and Beeftink, H. H., 1983, Ecology of phototrophic bacteria, in: *The Phototrophic Bacteria* (J. G. Ormerod, ed.), Blackwell, Oxford, pp. 146–185.

Veldhuis, M. J. W., and van Gemerden, H., 1986, Competition between purple and brown phototrophic bacteria: Sulfide, acetate, and light as limiting factors, *FEMS Microbiol. Ecol.* **38:**31–38.

Weckesser, J., Drews, G., Mayer, H., and Fromme, I., 1974, Lipopolysaccharide aus Rhodospirillaceae, Zusammensetzung und taxonomische Relevanz, *Zentralbl. Bakteriol. Hyg. I. Abt. Orig. A* **228:**193–198.

Weckesser, J., Drews, G., and Mayer, H., 1979, Lipopolysaccharides of photosynthetic prokaryotes, *Ann. Rev. Microbiol.* **33:**215–239.

Wetzel, R. G., 1973, Productivity investigations of interconnected lakes. I. The eight lakes of the Oliver and Walters chains, northeastern Indiana, *Hydrobiol. Stud.* **3:**91–143.

Widdel, F., 1988, Microbiology and ecology of sulfate- and sulfur-reducing bacteria, in: *Biology of Anaerobic Microorganisms* (A. J. B. Zehnder, ed.), Wiley, Chichester, pp. 469–585.

Wijbenga, D.-J., and van Gemerden, H., 1981, The influence of acetate on the oxidation of sulfide by *Rhodopseudomonas capsulata, Arch. Microbiol.* **129:**115–118.

Woese, C. R., 1987, Bacterial evolution, *Microbiol. Rev.* **51:**221–271.

Woese, C. R., Stackebrandt, E., Weisburg, W. G., Paster, B. J., Madigan, M. T., Fowler, V. J., Hahn, C. M., Blanz, P., Gupta, R., Nealson, K. H., and Fox, G. E., 1984a, The phylogeny of purple bacteria: The alpha subdivision, *Syst. Appl. Microbiol.* **5:**315–326.

Woese, C. R., Weisburg, W. G., Paster, B. J., Hahn, C. M., Tanner, R. S., Krieg, N. R., Koops, H.-P, Harms, H., and Stackebrandt, E., 1984b, The phylogeny of purple bacteria: The beta subdivision, *Syst. Appl. Microbiol.* **5:**327–336.

Woese, C. R., Stackebrandt, E., Macke, T. J., and Fox, G. E., 1985a, A phylogenetic definition of the major eubacterial taxa, *Sys. Appl. Microbiol.* **6:**143–151.

Woese, C. R., Weisburg, W. G., Hahn, C. M., Paster, B. J., Zablen, L. B., Lewis, B. J., Macke, T. J., Ludwig, W., and Stackebrandt, E., 1985b, The phylogeny of purple bacteria: The gamma subdivision, *Syst. Appl. Microbiol.* **6:**25–33.

Woese, C. R., Debrunner-Vossbrinck, B. A., Oyaizu, H., Stackebrandt, E., and Ludwig, W., 1985c, Gram-positive bacteria: Possible photosynthetic ancestry, *Science* **229:**762–765.

Yurkov, V. V., and Gorlenko, V. M., 1990, *Erythrobacter sibiricus* sp. nov., a new freshwater aerobic bacterial species containing bacteriochlorophyll *a, Microbiology,* **59:**85–89.

Zablen, L., and Woese, C. R., 1975, Procaryote phylogeny IV: Concerning the phylogenetic status of a photosynthetic bacterium, *J. Mol. Evol.* **5:**25–34.

Physiology of the Photosynthetic Prokaryotes 3

JOHN G. ORMEROD

1. INTRODUCTION

Living organisms grow by synthesizing in an ordered fashion the complex macromolecules of their own cells from simpler molecules. In general, the energy requirements for this can be met either by degrading part of the nutritional substrate for respiration (heterotrophic organisms) or by converting light energy into chemical energy as in the phototrophic organisms. The proportions of these two types of organisms on the earth are difficult to estimate, but their activities balance each other. In the long term both types are dependent on each other for major nutrients—heterotrophs must have the oxygen and organic molecules produced by photosynthesis; the phototrophs depend on the heterotrophs for keeping the oxygen content of the atmosphere at a tolerable level and for carbon dioxide, produced by respiration. The phototrophs also depend on sunlight, which is the driving force for the whole system. The two modes of life, heterotrophy and photototrophy, must have existed side by side on the surface of the earth for thousands of millions of years.

This picture of the living world is a general one which considers only the main participants. There are exceptions, such as heterotrophs that do not, indeed cannot, use oxygen in their energy metabolism. And among photosynthetic organisms, there are some that never produce oxygen. These can be either autotrophic (i.e., using CO_2 as sole carbon source) or heterotrophic. These exceptions are all prokaryotes and while they make up a limited proportion of the living organisms on the earth today (though not in the distant past), the sheer variety of their metabolic types demands the attention of the biotechnologist.

JOHN G. ORMEROD ● Biology Department, Division of Molecular Cell Biology, University of Oslo, Blindern, 0316 Oslo 3, Norway.
Photosynthetic Prokaryotes, edited by Nicholas H. Mann and Noel G. Carr. Plenum Press, New York, 1992.

This chapter will deal with the physiology of the various kinds of photosynthetic prokaryotes and will describe, explain, and discuss the process by which light energy is conserved in the form of the chemical bonds of organic material.

2. THE BASIC FEATURES OF PHOTOSYNTHESIS

The various kinds of prokaryotic photosynthesis are exemplified by the families of microorganisms known as cyanobacteria, purple bacteria (and the metabolically similar, but unrelated, *Chloroflexus*), green sulfur bacteria, and heliobacteria. Each of these groups is distinctive; yet their photosyntheses show a number of common, basic features: (i) in all cases, the process takes place in a lipid bilayer membrane; (ii) the first step, the gathering of light energy, is performed by arrays of pigment–protein complexes known as antennae, which are situated in or on the surface of the membrane, and (iii) the antennae deliver the energy by resonance transfer to reaction centers, which are very specialized microenvironments in the membrane where the first, energetically decisive photochemical events of photosynthesis take place.

Light energy can be considered as a flow of photons whose energy content is inversely related to the wavelength of the light. In general, the antenna must be capable of absorbing photons which have sufficient energy to drive the reaction center being served. Reaction centers are present in relatively small numbers and their absorption windows are specific, fixed, and narrow. On the other hand, antennae are much larger and their size and often their absorption spectrum can be regulated to suit growth conditions. Light is frequently the limiting factor for photosynthetic activity in nature, particularly under anaerobic conditions, and the various groups of photosynthetic prokaryotes each possess antennae that fit them for a particular spectral and ecological niche.

Photosynthetic reaction centers might be called one of the wonders of nature. These minute, pigment–protein complexes were first described and isolated by Clayton and co-workers (Reed and Clayton, 1968). There is plenty of evidence to indicate that reaction centers in all phototrophs function in the same way: certain chlorophyll molecules inside them, designated P, the primary donor, become activated by light energy (from the antenna) in such a way that they become powerful reducing agents. High-energy electrons are donated to a series of electron and hydrogen carriers in the membrane. In a short space of time a stable state is achieved, which can be characterized as a charge separation across the membrane with P^+ on the outside and a negatively charged carrier on the inside. This energy-rich state constitutes the basic, chemical energy source of photosynthesis.

Although the magnitude (energy content) of the charge separation is roughly similar in all known reaction centers, the actual range of redox potentials in which reaction centers operate is different in the various types of photosynthetic organisms. The operative redox range is determined by the redox potentials of the electron carriers in their bound state. Reaction centers that work in a low (electronegative) redox range are said to be of the photosystem I (PS I) type. Those that operate in a more electropositive redox range are called PS II-type reaction centers.

The designations PS I and PS II refer to the photosystems of oxygenic photosynthesis. All oxygenic phototrophs possess two kinds of reaction center, the one a strong reductant (PS I), the other a weak reductant (PS II). The two are redox-coupled in series. The anoxygenic phototrophs possess only one or the other of these types. The differences in redox range have important metabolic consequences and constitute one of the reasons for the variety of types of photosynthetic organisms in the biosphere.

3. VARIATIONS OF THE PHOTOSYNTHETIC PROCESS IN PROKARYOTES

We can distinguish three types of chlorophyll (Chl)-mediated photosynthesis: (i) the oxygenic process found in cyanobacteria, algae, and green plants, which involves both PS I- and PS II-type reaction centers; (ii) anoxygenic photosynthesis characteristic of purple bacteria and involving PS II-type reaction centers; and (iii) a strictly anaerobic kind of anoxygenic photosynthesis which is driven by PS I-type reaction centers and is present in the green sulfur bacteria and the heliobacteria. These characteristics are summarized in Table I.

3.1. Prokaryotes with PS II-Type Reaction Centers

This type of photosynthesis will be dealt with first because more is known about it than the other types.

Table I. Properties of Photosynthetic Prokaryotes

Group of organisms	Reaction center type	Usual electron donor or substrate	Special characteristics
Purple bacteria	II	H_2S, organic	Facultative
Chloroflexus	II	Organic, H_2S	Chlorosomes, facultative
Green sulfur bacteria	I	H_2S	Chlorosomes, anaerobic
Heliobacteria	I	Organic	Anaerobic
Cyanobacteria	I + II	H_2O	Phycobilisomes, aerobic

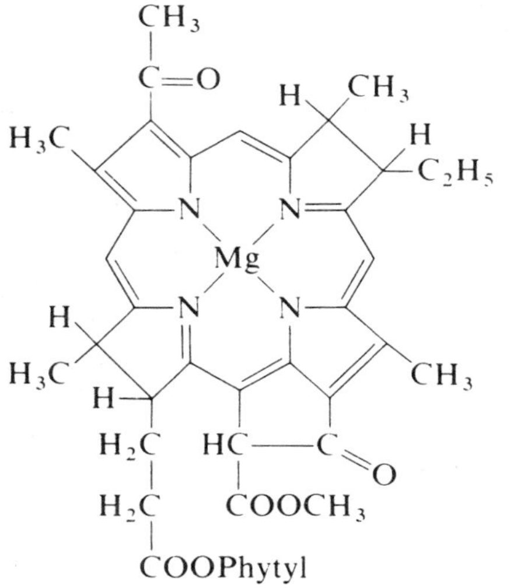

Figure 1. Structure of bacteriochlorophyll *a*.

3.1.1. Purple Bacteria

Apart from the fact that purple bacterial photosynthesis is simpler than the other types, the ease of handling of these organisms has been responsible for their frequent use as experimental organisms, and hence for the spectacular advances in knowledge of the process in recent years.

The bright colors of purple bacterial cultures are due mainly to carotenoids. The bluish green bacteriochlorophyll (Bchl) *a* is hidden, but can be extracted with methanol. A few species have Bchl *b* instead of *a*. The major *in vivo* absorption peaks of Bchl *a* and *b* are at 800–870 and about 1020 nm, respectively, and in methanol, 770 and 796 nm, respectively. The chemical structure of Bchl *a* is shown in Fig. 1; all the chlorophylls have the same basic structure. They are anchored by the hydrophobic tail on C7 as well as by ligands to side chains. Changes in the double bond pattern cause considerable differences in absorption spectrum. The Bchl serves as the antenna (together with carotenoids) as well as being the primary reaction center pigment. The antenna is localized in protein complexes in the cell membrane in close proximity to the reaction centers. The membrane is usually invaginated, as vesicles, lamellae, or tubes, depending on the species.

The synthesis of antennae and reaction centers is regulated by light intensity, temperature, and nutritional factors so as to ensure maximum photosynthetic efficiency under the prevailing conditions. In general, low light intensity and high temperatures stimulate the biosynthesis of the photosynthetic apparatus; oxygen inhibits Bchl synthesis. The carotenoids associated with the photosynthetic apparatus in all phototrophs act as antennae, but they also protect the cell from photooxidation in the presence of oxygen (Sistrom *et al.*, 1956).

In a recent major scientific breakthrough reaction centers from two purple bacteria were crystallized and their structures determined by X-ray diffraction (Deisenhofer *et al.*, 1985; J. P. Allen *et al.*, 1987). The precise positions of the electron carriers involved are now known in minute detail (Fig. 2). The structures of the two reaction centers are remarkably similar and can probably serve as a model for all photosynthetic reaction centers.

The primary electron donor is a specially positioned pair of Bchl molecules (P in Fig. 2) which within a few picoseconds after excitation react with the adjacent molecule of bacteriopheophytin (Bchl without its Mg atom, denoted F in Fig. 2). The redox potential of the bacteriopheophytin couple is about -0.5 V. Within 200 psec the nearby quinone molecule (Q in Fig. 2) becomes reduced. The quinone couple has a redox potential of around -0.1 V. Although this drop of 0.4 V in redox value represents a considerable loss of energy, it is essential in order to prevent back reaction with P^+ (redox potential $+0.5$ V) and allows the reaction center to enter the stable state.

There are three polypeptides in the purple bacterial reaction center: L, M, and H (Fig. 2). The L and M polypeptides (mol. wt. 30–35 kD) each have five transmembrane helices. These constitute the main scaffolding for the redox components, which are mostly liganded to amino acid residues in them. The H subunit has only one transmembrane helix, and is believed to have an accessory, anchoring function. Only the electron carriers on the right-hand side of the structure are involved in the electron transfer process.

The clue to the high energetic efficiency of photosynthesis lies in the redox potentials of the carrier molecules and their rigid positioning in relation to each other by the reaction center polypeptides. It is these factors which guarantee the forward reaction and prevent back reaction and loss of energy through fluorescence.

Contrary to what was earlier believed, the purple bacterial reaction center does not reduce NAD^+. There is instead a cycling of electrons from the reduced quinone via ubiquinol and a cytochrome $b–c_1$ complex to cytochrome c at the outside of the membrane (C in Fig. 2) and thence back to P^+, the oxidized primary donor. This results in movement of protons to the periplasmic space. The proton gradient so produced constitutes a proton motive force (pmf), and this is believed to be the sole energetic contribution of the reaction center to the metabolism of purple bacteria (Knaff, 1978). The pmf is utilized for ATP synthesis and transport, and to fuel reverse electron transfer involved in NAD^+ and ferredoxin reduction.

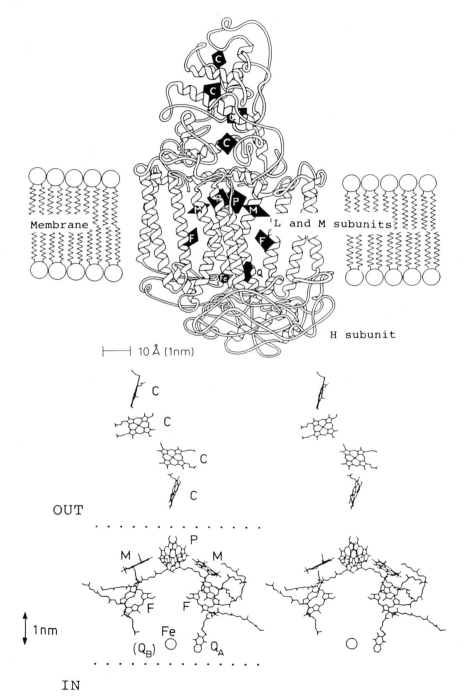

Membrane

L and M subunits

H subunit

⊢————⊣ 10 Å (1nm)

C

C

C

C

OUT

. . . .

M   P   M

F   F

1nm

Fe

(Q_B) ○   Q_A

. . . .

IN

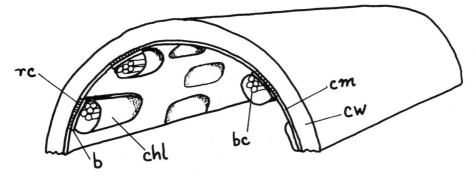

Figure 3. Diagram of a green bacterium cut apart to show chlorosomes: chl, chlorosome; cm, cytoplasmic membrane; cw, cell wall (simplified); b, baseplate; rc, reaction centers; bc, rods of antenna bacteriochlorophyll.

3.1.2. *Chloroflexus*

This thermophilic green bacterium contains two kinds of Bchl. The main antenna pigment, Bchl *c*, has its major *in vivo* absorption peak at about 750 nm and is localized in chlorosomes. These are cigar-shaped vesicles (Fig. 3), about 100×30 nm, bounded by a thin, single-layered membrane of unknown composition. They are filled with rods, 5 nm in diameter (Staehelin *et al.*, 1978), which are believed to contain the Bchl *c*. The chlorosomes are attached to the inside of the cell membrane by a protein baseplate which contains Bchl *a* and is believed to overlie the reaction centers. A membrane antenna containing Bchl *a* is also associated with the reaction centers. The synthesis of the photosynthetic apparatus, including the size and number of the chlorosomes, is regulated by light intensity, growth rate, and nutritional factors (Oelze and Fuller, 1987) and is inhibited by oxygen.

The reaction center of *Chloroflexus* has been isolated and its composition determined (Pierson and Thornber, 1983). A considerable amount is known about its photochemistry and the ensuing electron transport (Blankenship, 1985). Except for the lack of an H-subunit, it is structurally

---

Figure 2. Structure of the reaction center of a purple bacterium, *Rhodopseudomonas viridis*. The upper part of the figure shows the membrane containing the H peptide and the two main polypeptides, L and M, both with five transmembrane helices. To these are attached the pair of Bchl *b* molecules constituting the primary donor (P), two molecules each of Bchl *b* (M), bacteriopheophytin (F), and quinone (Q), and an Fe atom. Above this complex is a molecule of cytochrome *c* containing four heme groups (C). The lower part of the figure depicts the electron carriers as they would appear without their protein scaffolding (in stereo). [Andreasson and Vänngård, (1988).]

very similar to the purple bacterial reaction center, even to the extent of high sequence homology of the L and M polypeptides (Ovchinnikov *et al.,* 1988a,b). As with the purple bacteria, pmf is probably the main product also of this reaction center. The resemblance of the reaction centers of these two groups of organisms seems surprising in view of the evolutionary distance between the two (Woese, 1987).

3.2. Organisms with PS I-Type Reaction Centers

The two groups of bacteria to be considered here are both strictly anaerobic. While sensitivity to oxygen imposes limits on the types of environment in which such bacteria can live, it should be kept in mind that the anaerobic way of life is not without its advantages: a low external redox potential facilitates the operation of various favorable metabolic mechanisms and pathways which are not found in organisms that are associated with aerobic conditions.

3.2.1. Green Sulfur Bacteria

The green sulfur bacteria are probably the most energetically efficient of all the phototrophs and as such should be of special interest to the biotechnologist. They are nutritionally fastidious and more difficult to handle than purple bacteria. The requirement for anaerobiosis has hampered the application of genetic techniques, but a start has now been made (Ormerod, 1988).

The organization of the photosynthetic apparatus in green sulfur bacteria is similar to that of *Chloroflexus* (Fig. 3). The main antenna pigment, which is in the chlorosomes, is Bchl *c, d,* or *e*. These pigments all have their major *in vivo* absorption peak between 700 and 760 nm (and in methanol, from 640 to 670 nm). Light intensity has an inverse effect on the number, size, and Bchl content of the chlorosomes, but the Bchl *a* content of the cells is constant.

The structure of the green sulfur bacterial reaction center has not yet been established, but there is evidence for the presence of a single 65-kD polypeptide (Hurt and Hauska, 1984). The primary donor is Bchl *a*. The operative redox range is sufficiently low as to allow direct reduction of ferredoxin, which can couple directly to biosynthetic reactions in the cell, or reduce NAD^+ (Buchanan and Evans, 1969). Under these conditions, the oxidized primary donor, P_{840}^+, becomes reduced back to the ground state by an accessory electron donor, such as H_2S, thiosulfate, or H_2, via cytochromes. Inorganic electron donors are essential for the growth of green sulfur bacteria. An ATP synthase has been isolated which presumably is driven by pmf generated by electron transfer associated with the reaction center.

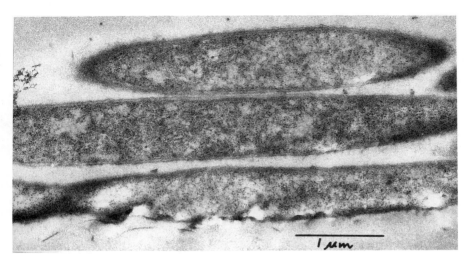

Figure 4. Thin section of *Heliobacterium fasciatum*. Note the absence of chlorosomes and membrane invaginations.

3.2.2. Heliobacteria

The organisms of this most recently discovered group of photosynthetic bacteria (Gest and Favinger, 1983) appear structurally to be among the simplest of the phototrophs, having neither invaginations of the cytoplasmic membrane, nor anything resembling chlorosomes (Fig. 4) (Gest and Favinger, 1983). This is in accord with these organisms' requirement for high light intensity. They contain Bchl g, which has its major absorption peak *in vivo* at about 790 nm and in methanol at 747 nm. The cellular content of Bchl g is inversely related to the light intensity (Torgersen, 1989).

The reaction center of *Heliobacterium chlorum* shows a functional resemblance to those of PS I and green sulfur bacteria (Fuller *et al.*, 1985; Amesz, 1990). There is a single polypeptide of molecular weight 94 kD (Amesz, 1990). The primary donor (P_{798}) is Bchl g (Fuller *et al.*, 1985). Like the reaction center of green sulfur bacteria, that of heliobacteria operates in a low redox range and it seems likely that ferredoxin, if present, would be reduced directly by it. However, the metabolic requirements for reducing power in heliobacteria are not known. A membrane-bound cytochrome [c-551 (Fuller *et al.*, 1985)] reduces P_{798}^+ to the ground state.

3.3. Prokaryotes with PS I- and PS II-Type Reaction Centers: The Cyanobacteria and Prochlorophytes

Among the phototrophs, the cyanobacteria and higher plants dominate overwhelmingly in terms of biomass. This is presumably because they

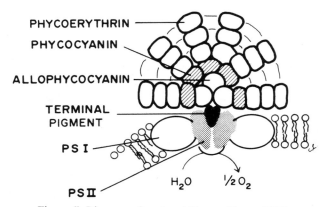

Figure 5. Diagram of a phycobilisome (Gantt, 1986).

are capable of continuous photosynthetic growth in the presence of oxygen and because their electron donor for CO_2 reduction, water, is available almost everywhere. In spite of relative uniformity as regards metabolic makeup, cyanobacteria occupy a wide range of ecological niches and show enormous differences in optimum light intensities for growth.

The membrane containing the photosynthetic apparatus (the thylakoid membrane) is usually laminated, and may be quite distinct from the cytoplasmic membrane. The PS I and PS II reaction centers are both served by membrane antennae containing Chl *a*, which has its major *in vivo* absorption peak around 700 nm, but in addition, the PS II units are associated with extramembraneous antennae known as phycobilisomes. These consist of discs of phycobiliprotein (Fig. 5) arranged on the thylakoid membrane surface so as to coincide with the PS II reaction centers. They contain water-soluble, reddish or bluish linear tetrapyrroles known as phycobilins which absorb light between the absorption peaks of Chl and carotenoids. The cyanobacteria are the only photosynthetic prokaryotes that can utilize this region of the spectrum effectively. Phycocyanin and allophycocyanin absorb at about 615 and 650 nm, respectively, and phycoerythrin at 500 to 570 nm. In some phycoerythrin-containing cyanobacteria, the proportions of the different phycobilins are regulated by light quality (chromatic adaptation).

The functional linkage between PS I and PS II means that by utilizing two photons, a strong reductant *and* a strong oxidant are formed (Fig. 6). The oxidant is powerful enough to extract electrons from water, under the influence of a manganese–protein complex. At the other end of the chain, the reductant is powerful enough to reduce ferredoxin, which can then reduce $NADP^+$ (Packham and Barber, 1987).

The redox carriers involved in PS II are P_{680} (primary donor), plastoquinol, a cytochrome *b/f* complex, and plastocyanin, which is a low mo-

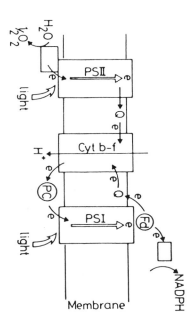

Figure 6. Diagram showing electron flow in PS I and PS II (Packham and Barber, 1987).

lecular weight (10-12 KD) water soluble copper protein. Plastocyanin reduces P_{700}^+ of PS I. Light-activated P_{700} (redox potential around -1 V) reduces ferrodoxin by way of an electron carrier chain comprising a molecule of Chl a and two iron–sulfur centers. The oxidized P_{680} (PS II) becomes reduced by the manganese–protein complex that mediates the dehydrogenation of water molecules, giving rise to molecular oxygen, the waste product of oxygenic photosynthesis. Proton motive force is generated by both of the photosystems.

Neither of the reaction centers from oxygenic organisms has been examined by X-ray crystallography, but they both appear to be more complex than the corresponding ones from anoxygenic bacteria. However, the core of the structures, where the photochemistry takes place, is believed to be similar in all cases.

Two of the PS II proteins, D1 and D2, show a high degree of sequence homology with the L and M subunits of the purple bacterial reaction center. The D1 (or Q_B) protein is the site of photoinhibition, caused by prolonged exposure to high light intensity. This protein has a high turnover rate, and its synthesis is inhibited during photoinhibition. It binds the herbicides atrazine and DCMU, both of which specifically inhibit this type of reaction center. The other proteins in PS II probably have peripheral functions. The oxygen-evolving manganese–protein complex contains four manganese ions and three polypeptides as well as Ca^{2+} and Cl^- ions.

In the PS I reaction center the primary donor, P_{700}, is associated with a polypeptide dimer (2×82 kD). A number of other polypeptides whose function is not certain are also present [see Scheller and Møller (1990) for review].

Under anaerobic conditions, some cyanobacteria reduce CO_2 with H_2S, using PS I only, PS II being inactivated [see Padan and Cohen (1982) for review].

Cyanobacteria differ from photosynthetic eukaryotes in lacking Chl *b*. Two oxygenic photosynthetic prokaryotes with both Chl *a* and *b* have recently been discovered. *Prochloron* is unicellular and grows in association with certain marine invertebrates (Lewin, 1976). *Prochlorothrix* is filamentous and was found growing in a shallow lake in Holland (Burger-Wiersma *et al.*, 1986). These organisms do not contain phycobilins; they have a Chl *a/b* antenna resembling that of green plants.

4. METABOLISM OF CARBON SOURCES AND ELECTRON DONORS

As far as is known, the mechanisms for macromolecule biosynthesis in photosynthetic prokaryotes are similar to those known from other bacteria and the reader is referred to textbooks on microbial biochemistry for this information (for example, Gottschalk, 1986). This section will deal primarily with the reactions involved in the photosynthetic assimilation of CO_2 and organic substrates into the central precursors of biosynthesis. In addition, the oxidation of inorganic electron donors and endogenous dark metabolism will be described briefly.

The normal prokaryotic cell contains, on a dry weight basis, about 55% C and 11% N. Protein accounts for about one-half and nucleic acids about one-tenth or more of the dry weight. The rest is polysaccharide, lipid, small molecules, ions, etc.

In order to produce cell material, the organism has first to make precursors such as acetyl CoA, pyruvate, phosphoenol pyruvate, oxaloacetate, succinyl CoA, and α-ketoglutarate. These are generally formed by reactions of, or associated with, the Krebs cycle. Triose, tetrose, pentose, and hexose phosphates are formed via the Calvin cycle (see Section 4.5 for a description of the cycle), or, in organisms which lack this mechanism, the oxidative pentose phosphate pathway. The way in which these biosynthetic reactions are driven depends largely on the type of reaction center(s) present (Ormerod and Sirevåg, 1983).

4.1. Purple Bacteria

On the basis of their relation to H_2S, the purple bacteria have traditionally been divided into two groups, the purple sulfur bacteria (γ group) and the purple non-sulfur bacteria (α and β groups).

All purple bacteria have simple growth requirements: usually one or two B vitamins, and simple nitrogen and carbon sources or CO_2 and an electron donor (see Chapter 2). The purple sulfur bacteria occur in nature in localities where H_2S is present. The H_2S is electron donor for the reduction of CO_2 by the Calvin cycle. The H_2S ($E_0' = -270$ mV) reduces a c-type cytochrome in the membrane and sulfur ($S°$) is formed. The cytochrome reduces NAD^+ with the help of pmf generated by the illuminated reaction centers. The sulfur formed is stored intracellularly (except in *Ectothiorhodospira*) and eventually oxidized to sulfate via sulfite and adenosine 5'-phosphosulfate. Other reduced sulfur compounds or H_2 can also be used as electron donors. The hydrogenase in purple bacteria is in the membrane, and cannot reduce NAD^+ without the consumption of pmf. For a more detailed review of electron donor metabolism in purple bacteria, see Hansen (1983).

The non-sulfur and some sulfur purple bacteria grow rapidly in the light on a wide range of organic substrates, including mono- and dicarboxylic acids, alcohols, and sugars. Some can even use aromatic compounds (Dutton and Evans, 1978). Since the energy source for assimilation is pmf from the reaction center, the organic substrate is not used for energy purposes and the cell yield per unit of substrate utilized is very high.

The Krebs cycle plays a prominent role in the metabolism of purple bacteria growing on organic substrates. Parts of the cycle can function in the reverse direction, because enzymes such as α-ketoglutarate synthase and pyruvate synthase may be present. These enzymes are dealt with more fully in Section 4.3.

Any excess electrons arising from assimilation of an organic substrate are used for CO_2 fixation by the Calvin cycle, the enzymes of which become derepressed under these conditions. This occurs with organic substrates that are more reduced than cell material [the elementary composition of dried purple bacteria is $C_5H_8O_2N$ (Van Gemerden, 1968)]. Photoautotrophic growth of purple bacteria with H_2 also involves CO_2 fixation by the Calvin cycle. Organic substrates which are more oxidized than cell material cause CO_2 production.

Although the growth of some purple bacteria may be inhibited by O_2, they are not killed by it. Indeed, many of them can grow under aerobic and microaerophilic conditions in the dark, using O_2 as electron acceptor for the oxidation of organic substrates (the non-sulfur types), or H_2S (purple sulfur bacteria).

It has also been shown (De Wit and Van Gemerden, 1990) that the purple sulfur bacterium *Thiocapsa roseopersicina* growing in the light in an anaerobic chemostat continues to grow phototrophically at an undiminished rate when air is admitted; the oxygen present is not utilized by the cells until the Bchl becomes diluted out (its synthesis is completely inhibited by O_2). This aspect of the physiology of purple bacteria is important, because in nature, these organisms are often found living under conditions where illumination and oxygen concentration fluctuate. Their ability to utilize two energy sources permits purple bacteria to occupy such ecological niches effectively.

As with other organisms which use the Calvin cycle, the oxygenase function of ribulose bisphosphate carboxylase/oxygenase in purple bacteria comes into function in the light when the O_2 concentration is high and the CO_2 concentration low, and glycollate is formed and presumably photorespired (Takabe and Akazawa, 1977).

The metabolism of organic substrates by purple bacteria during aerobic growth in the dark involves the Krebs cycle and cytochrome-containing electron transport chains which create pmf.

Some purple non-sulfur bacteria can grow slowly in the dark by anaerobic fermentation of added pyruvate to acetate, formate, hydrogen, and CO_2 [see Uffen (1978) for review]. Faster growth under these conditions can be obtained with sugars, provided that dimethyl sulfoxide or trimethylamine oxide is added as electron acceptor (Yen and Marrs, 1976; Madigan and Gest, 1978). In contrast to aerobic, dark-grown cells, which contain no Bchl, the anaerobically grown ones are fully pigmented.

Two distinct carbonaceous reserve materials are formed by purple bacteria: glycogen and poly-β-hydroxy butyrate. The latter is formed mostly from organic substrates which are metabolized via acetyl CoA. Reserve material may constitute over half of the dry weight of the cell under nitrogen-limiting conditions. The reserves are utilized for biosynthesis when the cells become replete with nitrogen, or as energy source under anaerobic conditions in the dark; glycogen is then fermented to a mixture of lower fatty acids. Poly-β-hydroxy butyrate is of commercial interest in the plastics industry [also from phototrophic bacteria (Brandl *et al.*, 1989)].

The biosynthesis of chlorophyll is of course of great importance in photosynthetic prokaryotes, but our knowledge of the process is incomplete. Few of the enzymes have been measured, let alone characterized. The first committed intermediate in the pathway, 5-aminolevulinate, was until recently thought to be formed universally from glycine and succinyl CoA (the aminolevulinate synthase reaction). It is now known that only purple bacteria of the α group (and their descendents) have this enzyme. All other phototrophs use the glutamate pathway (Avissar *et al.*, 1989), which involves substrate activation by tRNA. Two molecules of aminolevulinate then react to form a substituted pyrrole ring, porphobilinogen. Four of the pyrroles join together to form uroporphyrinogen. This is eventually converted into protoporphyrin, at which stage Mg (for Chl) or Fe (for heme) is incorporated. The final products result after a further series of reactions. Vitamin B_{12} is formed from uroporphyrinogen.

The regulation of Bchl synthesis is also poorly understood. It appears to occur at the aminolevulinate stage and at the Mg protoporphyrin stage (Oelze, 1988).

4.2. *Chloroflexus*

This facultative anaerobe grows in close association with cyanobacteria and is believed to utilize carbon compounds secreted by them. The carbon metabolism of *Chloroflexus* has been looked at only to a limited extent, but it appears to resemble that of purple bacteria in involving the Krebs cycle, both anaerobically in the light and aerobically in the dark (Sirevåg and Castenholz, 1979). In acetate-grown cells, the glyoxylate cycle is present (Løken and Sirevåg, 1979). However, the Calvin cycle is absent from *Chloroflexus*, and autotrophic CO_2 fixation (with H_2 as electron donor) occurs by a novel mechanism involving 3-hydroxypropionate and acetyl CoA as intermediates (Holo, 1989). *Chloroflexus* can also use H_2S for reduction of CO_2 (Madigan and Brock, 1977). Like the purple bacteria, *Chloroflexus* makes poly-β-hydorxybutyrate (Sirevåg and Castenholz, 1979) and glycogen (Holo, 1989).

4.3. Green Sulfur Bacteria

The green sulfur bacteria are often found in rather forbidding surroundings, like the gloomy depths of lakes, where H_2S and CO_2 abound, but where the light intensity is so low as to be unmeasurable. They can wring the last drops of energy out of such environments.

These bacteria have rather strict chemical requirements for growth and survival. They are obligately anaerobic and phototrophic, and fix CO_2 by the reductive tricarboxylic acid cycle [Fig. 7 (Evans *et al.*, 1966); see Ormerod and Sirevåg (1983) for review], using reduced sulfur compounds or H_2 as electron donors; in essence this is a reversed Krebs cycle. In keeping with the ability of their reaction centers to reduce ferredoxin (Section 3.2.1), two enzymes which use this coenzyme occupy a central place in the carbon metabolism. These are α-ketoglutarate-ferredoxin oxidoreductase and pyruvate-ferredoxin oxidoreductase. The reductive tricarboxylic acid cycle supplies the cell with acetyl CoA, pyruvate, oxaloacetate, succinyl CoA, and α-ketoglutarate.

A few simple organic compounds can be assimilated at low concentrations (5 mM), provided that CO_2 and a suitable inorganic electron donor are present. These requirements can be ascribed to the apparent inability of the green sulfur bacteria to extract electrons and CO_2 from the organic substrates for the purpose of reductive carboxylations. It will be recalled, on the other hand, that many purple bacteria can use organic substrates without supplements.

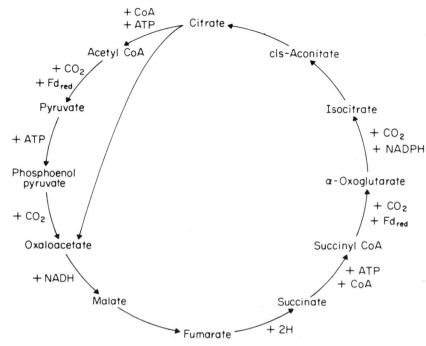

Figure 7. The reductive tricarboxylic acid cycle (Evans *et al.*, 1966).

With green sulfur bacteria, the utilization of organic substrates (e.g., acetate) can halve the generation time (to about 3.5 hr or less) and double the cell yield. Propionate is metabolized via succinate by a vitamin B_{12}-requiring reaction. Access to suitable organic substrates is undoubtedly common in nature. The green sulfur bacteria seem to have poorly developed regulatory systems, and branched-chain ketoacids are excreted when they are incubated in the light with CO_2 and an electron donor, in the absence of a nitrogen source (Sirevåg and Ormerod, 1970).

Glycogen, but not poly-β-hydroxybutyrate, is stored as reserve material in green sulfur bacteria. In the dark, the glycogen is fermented to fatty acids (mainly acetate) and succinate.

4.4. Heliobacteria

The growth requirements of these spore-forming phototrophs are similar to those of non-sulfur purple bacteria: simple organic compounds and nitrogen sources, plus biotin. An important difference is the requirement for strict anaerobiosis and a reduced sulfur source. Virtually nothing is known about their carbon metabolism and it will be of particular interest to elucidate the connection between the products of their PS I-type reaction center (Section 3.2.2) and biosynthesis. The physiology of spore formation and germination should also prove to be an exciting field of research, since these are the first endospore-forming phototrophs to be discovered.

4.5. Cyanobacteria

In spite of a large variety of morphological types which live in an enormous range of ecological situations, the cyanobacteria appear to be relatively uniform from the point of view of carbon metabolism. In common with purple bacteria, they fix CO_2 by the Calvin cycle (Fig. 8). This cycle has some of the same reactions as the universal oxidative pentose phosphate mechanism as well as some unique ones, like ribulose bisphosphate carboxylase/oxygenase (RuBisCo). Large amounts of RuBisCo are required because of its low affinity for CO_2 and the low pCO_2 in the atmosphere. RuBisCo is probably the most abundant protein in the biosphere.

In most cyanobacteria, RuBisCo consists of eight large and eight small subunits (L_8S_8), with a total molecular weight of 500–600 kD. The L subunits contain the catalytic site, but the function of the small subunits is not clear. (That the small subunit is not strictly necessary is shown by the fact

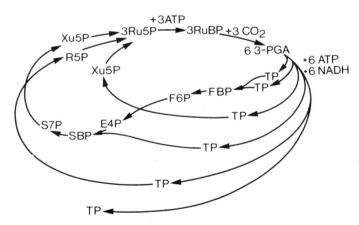

Figure 8. The Calvin cycle. E4P, Erythrose-4-phosphate; F6P, fructose-6-phosphate; FBP, fructose-1,6-bisphosphate; 3-PGA, 3-phosphoglycerate; R5P, ribose-5-phosphate; Ru5P, ribulose-5-phosphate; RuBP, ribulose-1,5-bisphosphate; S7P, sedoheptulose-7-phosphate; SBP, sedoheptulose-1,7-bisphosphate; TP, triose phosphate; Xu5P, xylulose-5-phosphate (Ormerod and Sirevåg, 1983).

that it is absent from the enzyme in *Rhodospirillum rubrum*.) The K_m for CO_2 is in the millimolar range and the enzyme has to be activated by CO_2 and Mg^{2+}. Some organisms contain polygonal crystalline masses of functional RuBisCo enclosed within a single-layered membrane and known as carboxysomes. Cyanobacteria are capable of concentrating CO_2 inside the cell by a mechanism involving carbonic anhydrase (Aizawa and Miyachi, 1986).

The Calvin cycle is driven by NADPH and ATP, both produced by the thylakoids. Regulation of cycle activity is complicated and is mediated in part by thioredoxin, a small redox protein whose reduction rate is affected by the photosynthetic apparatus.

The Calvin cycle supplies important intermediates for biosynthesis: hexose, pentose, and triose phosphates. Pyruvate and acetyl CoA are formed from phosphoglycerate, and the main amino acid precursor, α-ketoglutarate, is formed from acetyl CoA and oxaloacetate, via citrate. The Krebs cycle stops here though, because α-ketoglutarate dehydrogenase is lacking in cyanobacteria, and acetyl CoA cannot be oxidized. The remaining reactions of the Krebs cycle are present and presumably serve (in the reverse direction) to form succinyl CoA, required for amino acid synthesis.

In bright light, with low pCO_2 and high pO_2, the oxygenase activity of RuBisCo takes over and instead of phosphoglycerate, phosphoglycollate is formed, which is converted to glycollate. This may be excreted or photorespired, forming CO_2 and phosphoglycerate. Because it consumes O_2, NADH, and ATP, photorespiration may be regarded as a way of getting rid of the oxygen-dependent products of excess light energy (Ormerod, 1983). Such mechanisms are important in photosynthetic organisms because of the irregularity of the potentially harmful energy source. For much of the time, cyanobacteria in nature are probably under conditions of stress due primarily to the inconstancy of the energy source.

In the dark, cyanobacteria degrade reserve glycogen by the oxidative pentose phosphate mechanism. The reduced pyridine nucleotide formed enters the aerobic redox chain, believed to be in the cytoplasmic membrane, yielding pmf. This dark pathway is inhibited in the light putatively through the action of reduced thioredoxin on glucose-6-phosphate dehydrogenase (Pelroy *et al.*, 1972).

Some cyanobacteria can use organic substrates, some even in the dark. Of course, these are limited to such as can enter the rather restricted assimilatory system (e.g., sugars).

A rather unusual mechanism is involved in the synthesis of the linear tetrapyrroles (phycobilins); they are formed from heme by the action of a monooxygenase; thus their synthesis is dependent on aerobic conditions. This is true also for one of the steps in Chl synthesis in cyanobacteria. It will be recalled that oxygen inhibits formation of Bchl in purple and green bacteria, again emphasizing the fundamental difference between the oxygenic and the anoxygenic phototrophs.

5. NITROGEN METABOLISM

In nature, fermentation of organic material by heterotrophic anaerobes often results in an excess of ammonia which is available for the phototrophs. In terms of energy, this is the best nitrogen source for all prokaryotic phototrophs. However, under aerobic conditions, NH_3 becomes oxidized to NO_3 by nitrifying bacteria, and cyanobacteria can also use this nitrogen source readily. Under conditions where there is no fixed nitrogen available, many phototrophs synthesize the enzymes for nitrogen fixation.

Reduction of NO_3^- (via NO_2^-) and N_2 to amino groups involves NH_3 as an intermediate. Ammonia is assimilated by the so called GS-GOGAT mechanism (glutamine synthetase, glutamine-oxoglutarate aminotransferase). The affinity of glutamine synthetase for NH_3 is very high, but the cost is one ATP per ammonium ion assimilated. Glutamine synthetase is a very large enzyme and is subject to a regulatory system which is of a complexity commensurable with its importance. The regulation of the enzyme in many organisms involves adenylylation of a tyrosine residue on each of the 12 subunits. The assimilatory product of the GS-GOGAT system, glutamate, is the main source of amino groups for the other amino acids.

The cyanobacteria are among nature's most important diazotrophs and the ability to fix N_2 is found in a number of genera, including a few single-celled types and the heterocyst-forming, filamentous genera. Apart from one or two known exceptions, all anoxygenic phototrophs can fix N_2. The enzymes involved seem to be very similar in all organisms that have been investigated. The nitrogenase complex consists of two main proteins: the Mo—Fe protein (dinitrogenase) and the Fe protein (dinitrogenase reductase). The latter, after activation by Mg^{2+} and ATP, is reduced by ferredoxin and then acts as reductant for dinitrogenase in a reaction in which a molecule of N_2 is reduced to two molecules of NH_3. The process consumes a lot of ATP and reducing power.

The sensitivity of nitrogenase to O_2 limits the conditions under which N_2 fixation can occur. Oxygen creates special problems for cyanobacteria in this sense, because it is produced right inside the cell. Some of the filamentous genera such as *Anabaena* form special, thick-walled cells at more or less regular intervals in the filament, called heterocysts. These lack PS II and the key enzymes of the Calvin cycle and their cell walls are less permeable to oxygen than the walls of the other cells. The heterocysts are the sites of N_2 fixation in these genera. They receive organic substrate (carbohydrate) from adjacent vegetative cells and export back glutamine (Neuer *et al.*, 1983). The heterocysts possess PS I, and can create a pmf for the production of ATP as well as use electrons from the oxidation of the imported organic substrate for reduction of ferredoxin, which is required for N_2 fixation.

Other filamentous genera, such as *Trichodesmium,* can fix N_2 in cells situated in the central region of their filaments, when these are bunched together to reduce the light intensity (and so the production of O_2). Some single-celled genera (e.g., *Gloeothece*) fix nitrogen at night, using stored carbon and energy sources (Mullineaux *et al.,* 1981; Kallas *et al.,* 1983).

Purple bacteria and green sulfur bacteria can fix N_2, but apparently *Chloroflexus* cannot (Heda and Madigan, 1988). The heliobacteria are avid N_2 fixers (Heda and Madigan, 1988), an attribute which may be significant for the fertility of paddy fields, apparently one of their natural habitats.

The enzymes for N_2 fixation are not inducible; rather, they are derepressible, lack of fixed nitrogen being the signal. When nitrogenase is present in the absence of N_2, the system reduces protons and forms gaseous H_2. In nitrogen-starved purple bacteria this process results in the photoconversion of organic substrates completely to H_2 and CO_2 (Gest *et al.,* 1962). Hydrogen production by nitrogenase probably occurs to some extent in all diazotrophs even during N_2 fixation. It is considered to be a waste of energy since it is driven by ATP and high-energy electrons. An uptake hydrogenase present in the diazotroph can refix some of the H_2. Nitrogen fixation is inhibited by NH_3, the "switch-off" effect, and the synthesis of nitrogenase is repressed by NH_3.

While carbon and energy reserves are universal, the formation of nitrogenous reserves is uncommon, although most bacteria can probably utilize ribosomes as a nitrogen reserve under starvation conditions. Cyanobacteria are unusual, therefore, in making a nitrogenous material which is specifically a reserve. This is cyanophycin, a polyaspartic acid with an arginine residue attached by a peptide bond to each β-carboxyl. Cyanophycin is formed as intracellular granules at the onset of the stationary phase or when growth is severely limited by one or other essential factor such as light or phosphate or presumably CO_2 [see Allen (1984) for review]. The polymer is utilized when conditions permit resumption of growth. The formation of cyanophycin also appears to be associated with N_2 fixation, because heterocysts form it (Carr, 1988), probably when the rate of flow of acceptor carbon from the vegetative cells is slower than that of N_2 fixation.

Phycobiliproteins have also been shown to function as a nitrogen reserve, and in some cases, this function appears to be specific (Wyman *et al.,* 1985), in the sense that some of the phycoerythrin synthesized as reserve material is not active as antenna. It has been known for many years that nitrogen-starved cultures of cyanobacteria become greenish, and this is probably caused by the breakdown of phycobilins.

It is interesting to reflect on the interplay of energy metabolism with metabolic processes like carbon assimilation, nitrogen fixation, and the formation of the various reserve materials in cyanobacteria. Most organisms increase the formation of carbonaceous reserves when growth is limited by lack of a nutrient other than carbon; cyanobacteria appear to be the only organisms that form nitrogenous reserves when starved of a nutrient other than nitrogen. Under these conditions, the cell material produced has a higher nitrogen content than that of any other known organism (Ciferri, 1983). This illustrates the remarkable flexibility of these organisms in the face of environmental variations and it is not surprising that they occupy such a dominating ecological position in nature.

6. ENRICHMENT AND CULTIVATION OF PHOTOSYNTHETIC PROKARYOTES

The merits and methods of enrichment will be discussed first, then some of the methods and media used for growing pure cultures of phototrophic prokaryotes will be described.

6.1. Enrichment Cultures

Time and time again, scientific investigation combined with serendipity has shown the existence in nature of microorganisms possessing new or unprecedented combinations of characteristics. Materials such as soil, mud, etc., frequently contain many types of photosynthetic prokaryotes. Simply suspending these materials in nutrient media and incubating under specific conditions can lead to growth of novel strains. By intelligent manipulation of enrichment conditions it is possible to select for organisms with particular properties. A good example is the bacterium now known as *Rhodopseudomonas viridis*, which was first isolated from an enrichment culture illuminated with light of wavelength 900–1200 nm, obtained by means of an infrared filter (Eimhjellen *et al.*, 1963, 1967). Before this, no phototroph was known that could utilize such radiation. The experiment led to the discovery of a previously unknown kind of Bchl (Bchl *b*) which absorbs at about 1020 nm *in vivo*.

Photosynthetic prokaryotes often occur as blooms in nature, either in water or on land surfaces, and the dominant organism can frequently be isolated from such blooms without prior enrichment. Many phototrophs, particularly cyanobacteria, have been isolated in this way. Indeed, the enrichment culture technique does not seem to have been employed for cyanobacteria to the same extent as for anoxygenic phototrophs, and its more widespread use could well uncover the existence of novel cyanobacteria, perhaps suited to particular biotechnological applications. The attractive feature of enrichment cultures is that organisms can be isolated which are present in very small numbers in the inoculum.

Conditions of Enrichment

Inoculum materials from temperate climatic areas have been most commonly used and some materials are known to be fairly specific for particular types of phototrophs. For example, anaerobic marine muds usually contain phototrophic sulfur bacteria. Materials from the tropics have been given much less attention. Recently, paddy field soils were shown to contain endospore-forming heliobacteria [giving the added possibility of selection by pasteurization (Ormerod *et al.*, 1990)]. It is quite possible that there are other novel phototrophs in such environments.

When designing enrichment media, attention should be paid to pH, electron donors, and carbon, nitrogen, and sulfur sources, as well as the provision of B vitamins and mineral salts. Cyanobacteria would usually be enriched for under aerobic conditions, whereas closed bottles either full of medium or with a nitrogen gas phase are used for anoxygenic phototrophs.

In general, when searching for new organisms, it is wise to set up a number of enrichment cultures with different combinations of variables. Certain organisms can be discouraged by using filtered light or by adding inhibitors; for example the herbicide atrazine (see Section 3.3) prevents the growth of organisms with PS II-type reaction centers. Enrichment can also be performed in continuous culture, particularly if organisms with high affinity for the limiting substrate are desired; the interested reader should consult other publications for details (e.g., Harder *et al.*, 1977).

There are a large number of tried and tested media for growing the various kinds of photosynthetic prokaryotes and variations of these can also be used for enrichment cultures; see Section 6.2. There is room for a lot of simple experimentation in this field. It should be kept in mind that the specific conditions for enrichment are often not those which give the best or fastest growth of the bacterium which becomes dominant; rather, the conditions should be those which give the desired organism a competitive edge over others present in the inoculum.

Omission of fixed nitrogen (with air or a nitrogen gas phase) leads to dominance of diazotrophs. Attention should also be paid to sulfur source, since many anoxygenic phototrophs cannot reduce sulfate.

Most purple bacteria grow at a pH of 7 or above, and the non-sulfur types can be enriched for with a variety of organic substrates. Usually, nonfermentable substrates such as succinate or acetate are used, but in some cases the use of a fermentable organic substrate may, through prior heterotrophic fermentation, pave the way for subsequent growth of particular phototrophs.

Phototrophic sulfur bacteria can be enriched for in a Winogradsky column, made by mixing approximately equal volumes of anaerobic mud, $CaSO_4$, and cellulose powder in a tall glass cylinder and filling up with mineral medium or brackish or sea water containing a B vitamin mixture.

After 1 week or so of incubation in the light (which may be filtered as desired), colonies of purple and green sulfur bacteria appear on the inside of the glass. As a general rule, purple sulfur bacteria need a lower sulfide concentration (<4 mM) than green sulfur bacteria; the latter require a lower pH (5–6.8). When the growth of colored organisms is observed, microscopic examination and determination of absorption spectrum of the culture (and a methanol extract) should be performed. A pure culture can usually be obtained by streaking out on agar medium similar to that used for the enrichment. With anaerobic organisms, this can be done in an anaerobic chamber, or the plates can be incubated in an anaerobic jar. Alternatively a small volume of the anaerobic enrichment culture can be serially diluted in molten agar medium in tubes and a pure culture of the anaerobe isolated in this way. The isolation of pure cultures of cyanobacteria (especially filamentous types) has proved more difficult than that of anoxygenic phototrophs (Rippka *et al.*, 1981). It may be necessary to use other techniques than plating out, for example, washing, serial dilution in liquid medium, or addition of antibiotics.

Sometimes a stable, mixed culture will suit a particular biotechnological or other purpose. "Chloropseudomonas ethylicum" is such a culture. It consists of a green sulfur bacterium living symbiotically with a sulfur-reducing bacterium (Pfennig and Biebl, 1976).

By imaginative use of enrichment techniques, the skillful biotechnologist should be able to obtain a flying start in the acquisition of organisms suited to particular tasks, perhaps to be improved by genetic manipulation.

6.2. Conditions of Cultivation

Difficulty is frequently encountered in getting a newly acquired organism to grow, be it one from an enrichment or from a culture collection. In general the medium used for starting up should resemble that which the organism has been growing on before transfer. Changing the medium, even to one reported as being suitable for the organism, may result in a long lag period or no growth at all. The reason for this is probably that after regular transfer on a particular medium, strains are selected which grow best on that medium. The best way to encourage growth of an unfamiliar organism is to use a large inoculum of a fresh, exponentially growing culture, thereby easing the selection of mutant strains.

With media of pH 7 or above, particular attention should be paid to the composition of the mineral salt mixture, so as to avoid the formation of a permanent precipitate on autoclaving, thereby removing trace elements from solution. A suitable medium for purple sulfur bacteria was described by Pfennig and Lippert (1966). A number of media for non-sulfur purple bacteria have been described; see, for example, Cohen-Bazire *et al.* (1957)

or Ormerod *et al.* (1961). A growth temperature of 30–35°C and a relatively high light intensity, say 1000 lux (except for the large-celled species of purple sulfur bacteria) are usually suitable.

Chloroflexus has an optimum pH of about 8 and growth temperature of 55°C and likes a fairly high light intensity. See Pierson and Castenholz (1974) and Løken and Sirevåg (1979) for suitable media with organic substrates. Autotrophic growth of this bacterium on sulfide was reported by Madigan and Brock (1977), but in our laboratory *Chloroflexus* has proved difficult in this respect, best results being obtained with H_2 as electron donor (Holo and Sirevåg, 1986).

Green sulfur bacteria do not grow at a pH above 7, and are usually cultivated at pH 6.0–6.8 in a mineral medium and with a light intensity of 100–1000 lux. The pH tends to fall during growth owing to sulfuric acid formation. Fastest growth and highest yields are obtained if acetate or propionate is added (up to 5 mM). Reductants such as thioglycollate (0.05%) or sodium ascorbate (0.1%) protect the cells from traces of oxygen. Resazurin (0.1 mg liter^{-1}) is useful as a redox indicator. A suitable medium for green sulfur bacteria was designed by Pfennig (1961), and has been modified to ease the preparation and improve keeping properties (Rieble *et al.*, 1989).

Heliobacteria are, as far as is known, all obligately anaerobic and photoheterotrophic, and grow rapidly in 1% (w/v) yeast extract. A defined medium has been described by Gest and Favinger (1983). The heliobacteria are able to use a few organic carbon sources, such as lactate, and require biotin and a reduced sulfur source. They are avid nitrogen-fixers and at least some of them form heat-resistant endospores.

A useful medium for cyanobacteria was described by Stanier *et al.* (1971). Sterile air enriched with 5% (v/v) CO_2 may be passed through the culture during growth, in which case the desired pH can be obtained by adding Na_2CO_3. The light intensity and temperature employed depend on the particular strain.

7. CONCLUDING REMARKS

The physiology of photosynthetic prokaryotes is inevitably tied up with the type of reaction center(s) present in their cells.

It would seem as though the full potential of PS I-type reaction centers (in green sulfur bacteria for example) is best realized in a low-redox environment, as usually accompanies anaerobic conditions. By utilizing reduced ferredoxin for CO_2 fixation, more of the energy made available by the reaction center is conserved. This, together with the massive chlorosome antenna, should make the green sulfur bacteria first choice for biotechnological applications where light energy is at a premium. Also, their poor carbon-regulatory systems lead to accumulation of potentially interesting intermediates. These organisms have already been used for scrubbing gases of H_2S and producing sulfur and organic carbon (Cork *et al.,* 1983). Cyanobacteria do not utilize reduced ferredoxin for CO_2 fixation, and therefore lose some of the energy which green sulfur bacteria manage to conserve. This may be related to their aerobic growth mode. Indeed, considering the reactivity of oxygen, it is remarkable that the cyanobacteria are able to house in one and the same cell a reaction center which reduces ferredoxin and one which produces O_2. The biotechnological potential of cyanobacteria must take account of their ability to grow in the presence of O_2 on water, CO_2, N_2 and salts and to make cell material of high nitrogen content.

It is frequently the case that an organism which is adapted to a particular type of environment is also by its metabolism restricted to that environment. This is true for most photosynthetic prokaryotes, but non-sulfur purple bacteria are an exception: they are the most versatile of all phototrophs—the jacks of all trades, so to speak. However, what they have gained in versatility, they seem to have lost in ability to dominate as blooms in nature. As objects of utility in biotechnology they are admirably suited for special metabolic purposes.

Heliobacteria, because of their active nitrogenase, might fertilize tropical soils, and the heat-resistant endospores would ease the task of administering them.

REFERENCES

Aizawa, K., and Miyachi, S., 1986, Carbonic anhydrase and CO_2 concentrating mechanisms in microalgae and cyanobacteria, *FEMS Microbiol. Rev.* **39**:215–233.

Allen, J. P., Feher, G., Yeates, T. O., Komiya, H., and Rees, D. C., 1987, Structure of the reaction center from *Rhodobacter sphaeroides* R-26: The protein subunits, *Proc. Natl. Acad. Sci. USA* **84**:6162–6166.

Allen, M. M., 1984, Cyanobacterial cell inclusions, *Annu. Rev. Microbiol.* **38**:1–25.

Amesz, J., 1990, Antenna systems of green bacteria and heliobacteria, in: *Current Research in Photosynthesis*, Volume II (M. Baltscheffsky, ed.), Kluwer, Dordrecht, pp. 25–31.

Andreasson, L. E., and Vänngård, T., 1988, Nytt ljus over fotosyntesen, *Kem. Tidskr.* **1988**(12):41–46.

Avissar, Y. J., Ormerod, J. G., and Beale, S. I., 1989, Distribution of δ-aminolevulinic acid biosynthetic pathways among phototrophic bacterial groups, *Arch. Microbiol.* **151**:513–519.

Blankenship, R. E., 1985, Electron transport in green photosynthetic bacteria, *Photosynth. Res.* **6**:317–333.

Brandl, H., Knee, E. J., Fuller, R. C., Gross, R. A., and Lenz, R. W., 1989, Ability of the phototrophic bacterium *Rhodospirillum rubrum* to produce various poly (β-hydroxy-alkanoates): Potential sources for biodegardable polyesters, *Int. J. Biol. Macromol.* **11**:49–55.

Buchanan, B. B., and Evans, M. C. W., 1969, Photoreduction of ferredoxin and its use in $NAD(P)^+$ reduction by a subcellular preparation from the photosynthetic bacterium *Chlorobium thiosulfatophilum*, *Biochim. Biophys. Acta* **180**:123–129.

Burger-Wiersma, T., Veenhuis, M., Korthals, H. J., Van De Wiel, C. C. M., and Mur, L. R., 1986, A new prokaryote containing chlorophylls *a* and *b*, *Nature* **320:**262–264.

Carr, N. G., 1988, Nitrogen reserves and dynamic reservoirs in cyanobacteria, in: *Biochemistry of the Algae and Cyanobacteria* (L. J. Rogers and J. R. Gallon, eds.), Oxford University Press, Oxford, pp. 13–21.

Ciferri, O., 1983, *Spirulina*, the edible microorganism, *Microbiol. Rev.* **47:**551–578.

Cohen-Bazire, G., Sistrom, W. R., and Stanier, R. Y., 1957, Kinetic studies of pigment synthesis by non-sulfur purple bacteria, *J. Cell. Comp. Physiol.* **49:**25–68.

Cork, D. J., Garunas, R., and Sajjad, A., 1983, *Chlorobium limicola* forma *thiosulfatophilum:* Biocatalyst in the production of sulfur and organic carbon from a gas stream containing H_2S and CO_2, *J. Bacteriol.* **45:**913–918.

Deisenhofer, J., Epp, O., Miki, K., Huber, R., and Michel, H., 1985, Structure of the protein subunits in the photosynthetic reaction centre of *Rhodopseudomonas viridis* at 3 Å resolution, *Nature* **318:**618–624.

De Wit, R., and Van Gemerden, H., 1990, Growth of the phototrophic purple sulfur bacterium *Thiocapsa roseopersicina* under oxic/anoxic regimes in the light, *FEMS Microbiol. Ecol.* **73:**69–76.

Dutton, P. L., and Evans, W. C., 1978, Metabolism of aromatic compounds by Rhodospirillaceae, in: *The Photosynthetic Bacteria* (R. K. Clayton and W. R. Sistrom, eds.), Plenum Press, New York, pp. 719–726.

Eimhjellen, K. E., Aasmundrud, O., and Jensen, A., 1963, A new bacteriochlorophyll, *Biochem. Biophys. Res. Commun.* **10:**232–236.

Eimhjellen, K. E., Steensland, H., and Trätteberg, J., 1967, A *Thiococcus* sp. nov. gen., its pigments and internal membrane system, *Arch. Mikrobiol.* **59:**82–92.

Evans, M. C. W., Buchanan, B. B., and Arnon, D. I., 1966, A new ferredoxin dependent reduction cycle in a photosynthetic bacterium, *Proc. Natl. Acad. Sci. USA* **55:**928–934.

Fuller, R. C., Sprague, S. G., Gest, H., and Blankenship, R. E., 1985, Unique photosynthetic reaction center from *Heliobacterium chlorum*, *FEBS Lett.* **182:**345–349.

Gantt, E., 1986, Phycobilisomes, in: *Photosynthesis III* (L. A. Staehelin and C. J. Arntzen, eds.), Springer, Berlin, pp. 260–268.

Gest, H., and Favinger, J. L., 1983, *Heliobacterium chlorum*, an anoxygenic brownish-green photosynthetic bacterium containing a "new" form of bacteriochlorophyll, *Arch. Microbiol.* **136:**11–16.

Gest, H., Ormerod, J. G., and Ormerod, K. S., 1962, Photometabolism of *Rhodospirillum rubrum:* Light-dependent dissimilation of organic compounds to carbon dioxide and molecular hydrogen by an anaerobic citric acid cycle, *Arch. Biochem. Biophys.* **97:**21–33.

Gottschalk, G., 1986, *Bacterial Metabolism*, 2nd ed., Springer, New York.

Hansen, T. A., 1983, Electron donor metabolism in phototrophic bacteria, in: *The Phototrophic Bacteria* (J. G. Ormerod, ed.), Blackwell, Oxford, pp. 76–99.

Harder, W., Kuenen, J. G., and Matin, A., 1977, A review, microbial selection in continuous culture, *J. Appl. Bacteriol.* **43:**1–24.

Heda, G. D., and Madigan, M. T., 1988, Nitrogen metabolism and N_2 fixation in phototrophic green bacteria, in: *Green Photosynthetic Bacteria* (J. M. Olson, J. G. Ormerod, J. Amesz, E. Stackebrandt, and H. G. Trüper, eds.), Plenum Press, New York, pp. 175–187.

Holo, H., 1989, *Chloroflexus auranctiacus* secretes 3-hydroxypropionate, a possible intermediate in the assimilation of CO_2 and acetate, *Arch. Microbiol.* **151:**252–256.

Holo, H., and Sirevåg, R., 1986, Autotrophic growth and CO_2 fixation of *Chloroflexus auranctiacus*, *Arch. Microbiol.* **145:**173–180.

Hurt, E. C., and Hauska, G., 1984, Purification of membrane-bound cytochromes and a photoactive P840 protein complex of the green sulfur bacterium *Chlorobium* f. *thiosulfatophilum*, *FEBS Lett.* **168:**149–154.

Kallas, T., Rippka, R., Coursin, T., Rebiere, M. C., Tandeau de Marsac, N., and Cohen-Bazire,

G., 1983, Aerobic nitrogen fixation by non-heterocystous cyanobacteria, in: *Photosynthetic Prokaryotes* (C. G. Papageorgiou and L. Packer, eds.), Elsevier, Amsterdam, pp. 281–302.

Knaff, D. B., 1978, Reducing potentials and the pathway of NAD$^+$ reduction, in: *The Photosynthetic Bacteria* (R. K. Clayton and W. R. Sistrom, Eds.), Plenum Press, New York, pp. 629–640.

Lewin, R. A., 1976, Prochlorophyta as a proposed new division of algae, *Nature* **261**:697–698.

Løken, Ø., and Sirevåg, R., 1979, Evidence for the presence of the glyoxylate cycle in *Chloroflexus*, *Arch. Microbiol.* **132**:276–279.

Madigan, M. T., and Brock, T. D., 1977, CO_2 fixation in photosynthetically grown *Chloroflexus aurantiacus*, *FEMS Microbiol. Lett.* **1**:301–304.

Madigan, M. T., and Gest, H., 1978, Growth of a photosynthetic bacterium in darkness, supported by oxidant-dependent sugar fermentation, *Arch. Microbiol.* **117**:119–122.

Mullineaux, P. M., Gallon, J. R., and Chaplin, A. E., 1981, Acetylene reduction in cyanobacteria grown under alternating light–dark cycles, *FEMS Microbiol. Lett.* **10**:245–247.

Neuer, G., Papen, H., and Bothe, H., 1983, Heterocyst biochemistry and differentiation, in: *Photosynthetic Prokaryotes* (G. C. Papageorgiou and L. Packer, eds.), Elsevier, Amsterdam, pp. 219–242.

Oelze, J., 1988, Regulation of tetrapyrrol synthesis by light in chemostat cultures of *Rhodobacter sphaeroides*, *J. Bacteriol.* **170**:4652–4657.

Oelze, J., and Fuller, R. C., 1987, Growth and control of development of the photosynthetic apparatus in *Chloroflexus auranctiacus*, *Arch. Microbiol.* **148**:132–136.

Ormerod, J. G., 1983, The carbon cycle in aquatic ecosystems, in: *Microbes in Their Natural Environments* (J. H. Slater, R. Whittenbury, and J. M. Wimpenny, eds.), Cambridge University Press, Cambridge, pp. 463–482.

Ormerod, J. G., 1988, Natural genetic transformation in *Chlorobium*, in: *Green Photosynthetic Bacteria* (J. M. Olsen, J. G. Ormerod, J. Amesz, E. Stackebrandt, and H. G. Trüper, eds.), Plenum Press, New York, pp. 315–319.

Ormerod, J. G., and Sirevåg, R., 1983, Essential aspects of carbon metabolism, in: *The Phototrophic Bacteria* (J. G. Ormerod, ed.), Blackwell, Oxford, pp. 100–119.

Ormerod, J. G., Ormerod, K. S., and Gest, H., 1961, Light dependent utilisation of organic compounds and photoproduction of molecular hydrogen by photosynthetic bacteria: Relationships with nitrogen metabolism, *Arch. Biochem. Biophys.* **94**:449–463.

Ormerod, J. G., Nesbakken, T., and Torgersen, Y., 1990, Phototrophic bacteria that form heat resistant endospores, in: *Current Research in Photosynthesis*, Volume IV (M. Baltscheffsky, ed.), Kluwer, Dordrecht, pp. 935–938.

Ovchinnikov, Y. A., Abdulev, A. S., Zolotarev, A. S., Shmukler, B. E., Zargarov, A. A., Kutuzov, M. A., Telezhinskaya, I. N., and Levina, N. B., 1988a, Photosynthetic reaction centre of *Chloroflexus auranctiacus* I. Primary structure of L-subunit, *FEBS Lett.* **231**:237–242.

Ovchinnikov, Y. A., Abdulev, A. S., Schmuckler, B. E., Zargarov, A. A., Kutuzov, M. A., Telezhinskaya, I. N., Levina, N. B., and Zolotarev, A. S., 1988b, Photosynthetic reaction centre of *Chloroflexus auranctiacus*. Primary structure of M-subunit, *FEBS Lett.* **232**:364–368.

Packham, N. K., and Barber, J., 1987, Structural and functional comparison of anoxygenic and oxygenic organisms, in: *The Light Reactions* (J. Barber, ed.), Elsevier, Amsterdam, pp. 1–30.

Padan, E., and Cohen, Y., 1982, Anoxygenic photosynthesis, in: *The Biology of Cyanobacteria* (N. G. Carr and B. A. Whitton, eds.), Blackwell, Oxford, pp. 215–235.

Pelroy, R. A., Rippka, R., and Stanier, R. Y., 1972, The metabolism of glucose by unicellular blue-green algae, *Arch. Mikrobiol.* **87**:303–322.

Pfennig, N., 1961, Eine vollsynthetische Nährlösung zur . . . , *Naturwissenschaften* **48**:136.

Pfennig, N., and Biebl, H., 1976, *Desufuromonas acetoxidans gen. nov.* and *sp. nov.*, a new anaerobic, sulfur reducing, acetate-oxidising bacterium, *Arch. Microbiol.* **110**:3–12.

Pfennig, N., and Lippert, D. T., 1966, Uber das vitamin B_{12}-bedurfnis phototropher schwefelbakterien, *Arch. Mikrobiol.* **55:**245–246.

Pierson, B. K., and Castenholz, R. W., 1974, Studies of pigments and growth in *Chloroflexus auranctiacus*, a phototrophic filamentous bacterium, *Arch. Microbiol.* **100:**283–305.

Pierson, B. K., and Thornber, J. P., 1983, Isolation and spectral characteristics of photochemical reaction centers from the thermophilic green bacterium *Chloroflexus auranctiacus* strain J-10-fl, *Proc. Natl. Acad. Sci. USA* **80:** 80–84.

Reed, D. W., and Clayton, R. K., 1968, Isolation of a reaction center fraction from *Rhodopseudomonas sphaeroides, Biochem. Biophys. Res. Commun.* **30:**471–475.

Rieble, S., Ormerod, J. G., and Beale, S. I., 1989, Transformation of glutamate to δ-aminolevulinic acid by soluble extracts of *Chlorobium vibrioforme, J. Bacteriol.* **171:**3782–3787.

Rippka, R., Waterbury, J. B., and Stanier, R. Y., 1981, Isolation and purification of cyanobacteria: Some principles, in: *The Prokaryotes,* Volume I (M. P. Starr, H. Stolp, H. G. Truper, A. Balows, and H. G. Schlegel, eds.), Springer, Berlin, pp. 212–220.

Scheller, H. V., and Møller, B. L., 1990, Photosystem I polypeptides, *Physiol Plant.* **78:**484–494.

Sirevåg, R., and Castenholz, R. C., 1979, Aspects of carbon metabolism in *Chloroflexus, Arch. Microbiol.* **120:**151–153.

Sirevåg, R., and Ormerod, J. G., 1970, Carbon dioxide fixation in green sulphur bacteria, *Biochem. J.* **120:**399–408.

Sistrom, W. R., Griffiths, M., and Stanier, R. Y., 1956, The biology of a photosynthetic bacterium which lacks colored carotenoids, *J. Cell. Comp. Physiol.* **48:**473–515.

Staehelin, L. A., Golecki, R., Fuller, R. C., and Drews, G., 1978, Visualization of the supramolecular architecture of chlorosomes (Chlorobium type vesicles) in freeze-fractured cells of *Chloroflexus auranctiacus, Arch. Microbiol.* **119:**269–277.

Stanier, R. Y., Kunisawa, R., Mandel, M., and Cohen-Bazire, G., 1971, Purification and properties of unicellular blue green algae (order Chroococcales), *Bacteriol. Revs.* **35:**171–205.

Takabe, T., and Akazawa, T., 1977, A comparative study on the effect of O_2 on photosynthetic carbon metabolism by *Chlorobium thiosulfatophilum* and *Chromatium vinosum, Plant Cell Physiol.* **18:**753–765.

Torgersen, Y. A., 1989, Characterization of an obligately phototrophic bacterium that contains bacteriochlorophyll *g*, Cand. Scient. thesis, Oslo University, Oslo, Norway [in Norwegian].

Uffen, R. L., 1978, Fermentative metabolism and growth of photosynthetic bacteria, in: *The Photosynthetic Bacteria* (R. K. Clayton and R. W. Sistrom, eds.), Plenum Press, New York, pp. 857–872.

Van Gemerden, H., 1968, Utilization of reducing power in growing cultures of *Chromatium, Arch. Mikrobiol.* **64:**111–117.

Woese, C. R., 1987, Bacterial evolution, *Microbiol. Rev.* **51:**221–271.

Wyman, M., Gregory, R. P. F., and Carr, N. G., 1985, Novel role for phycoerythrin in a marine cyanobacterium, *Synechococcus* strain DC2, *Science* **230:**818–820.

Yen, H.-C., and Marrs, B. L., 1976, Growth of *Rhodopseudomonas capsulata* under anaerobic dark conditions with dimethyl sulfoxide, *Arch. Biochem. Biophys.* **181:**411–418.

Genetics of the Photosynthetic Prokaryotes

4

VENETIA A. SAUNDERS

1. INTRODUCTION

Photosynthetic prokaryotes exhibit a diversity of morphological and metabolic capabilities that deserve intensive genetic analysis. Members of the group are able to grow phototrophically and chemotrophically and can fix nitrogen. The availability of a repertoire of genetic systems now permits application of genetic technology to identifying the mechanisms governing photosynthesis, nitrogen fixation, and membrane biogenesis within a group that includes multicellular as well as unicellular organisms.

While the photosynthetic prokaryotes comprise the photosynthetic bacteria, cyanobacteria, and Prochlorophyta, this chapter will focus on purple non-sulfur photosynthetic bacteria (*Rhodospirillaceae*), principally of the genus *Rhodobacter,* and on selected cyanobacteria, reflecting the direction of genetic studies. The development of genetic tools, notably gene transfer systems and transposable genetic elements, will be considered as a prelude to their use in the genetic manipulation of these photosynthetic organisms.

2. GENOME ORGANIZATION

2.1. Genomes

The deoxyribonucleic acid base composition of photosynthetic prokaryotes spans a range from 35 to 73 mole % guanine and cytosine (G + C), which is almost as wide as that for the entire bacterial kingdom (Mandel *et al.,* 1971; Herdman *et al.,* 1979a; de Bont *et al.,* 1981). Greatest genetic

VENETIA A. SAUNDERS ● School of Natural Sciences, Liverpool Polytechnic, Liverpool L3 3AF, United Kingdom.

Photosynthetic Prokaryotes, edited by Nicholas H. Mann and Noel G. Carr. Plenum Press, New York, 1992.

heterogeneity is displayed by the cyanobacteria, although the majority of DNA base compositions fall within the lower half of the range. By contrast, base compositions of the Rhodospirillaceae are confined to the upper half of the range, varying between 62 and 73% G + C. Interspecies hybridizations with total DNA of *Rhodobacter capsulatus* and *Rhodobacter sphaeroides* produced 11–28% of the homology observed in self-hybridizations (de Bont *et al.*, 1981). Further hybridization studies with the cloned *R. capsulatus* photosynthesis gene cluster (see Section 3.2.1) and DNA from other members of the Rhodospirillaceae have revealed only a small degree of sequence relatedness for such genes, despite the apparent phenotypic similarities of these organisms (Beatty and Cohen, 1983).

Genome sizes for photosynthetic bacteria of the genus *Rhodobacter* are 1.6×10^9 daltons (Gibson and Niederman, 1970; Yen *et al.*, 1979), values similar to those of many other bacteria. Recently, it has been demonstrated by Suwanto and Kaplan (1989) that *R. sphaeroides* contains two unique circular genomes of approximately 3050 and 910 kilobases (kb). Both genomes contain rRNA cistrons, and the genes for the two distinct forms of ribulose 1,5-bisphosphate carboxylase oxygenase are on separate genomes. In addition, there are copies of genes for such enzymes as glyceraldehyde-3-phosphate dehydrogenase on both genomes, but the extent of DNA homology is low. It remains to be seen whether this fascinating observation of the occurrence of multiple discrete genomes is a widespread feature of the photosynthetic bacteria, and indeed of other groups of bacteria. Cyanobacterial genomes vary from about 1.6×10^9 to 8.6×10^9 daltons (Herdman *et al.*, 1979b), the largest being among the most complex reported for prokaryotes. Some of these genomes contain multiple copies of certain genes, such as the *psbA* genes (encoding Q_B protein of photosystem II) in *Anacystis nidulans* R2 (Golden *et al.*, 1986) and *nif* genes (for nitrogen fixation) in *Anabaena* (Rice *et al.*, 1982), *Calothrix* (Kallas *et al.*, 1983), and the photosynthetic bacterium *R. capsulatus* (Scolnik and Haselkorn, 1984; Klipp *et al.*, 1988).

Extensive modification of genomic DNA has been reported for a number of cyanobacterial species, as evidenced by resistance to cleavage by various restriction endonucleases (Lambert and Carr, 1984; Adams, 1988; Padhy *et al.*, 1988). DNA modification, particularly in the filamentous strains, is vastly in excess of that required to protect against resident endonucleases and does not appear to vary with the complement of restriction enzymes found in different hosts. Such conservation of methylation patterns tends to rule out host-controlled modification as the sole source of methylation. Indeed, it has been suggested that adenine and cytosine methylases, analogous to the *Escherichia coli dam* and *dcm* enzymes, contribute to the presence of methylated bases in these cyanobacteria (Lambert and Carr, 1984; Padhy *et al.*, 1988) (see Chapter 5).

DNA methylation may act in mismatch repair processes and have a role in DNA replication. Alternatively, or additionally, methylation may be important in regulating gene expression in cyanobacteria, especially during differentiation of specialized cells such as the nitrogen-fixing heterocysts. However, Adams (1988) could not detect any gross changes in methylation patterns of DNA from heterocysts and vegetative cells of *Anabaena* spp. Any changes in methylation during such cellular differentiation may thus be restricted to small, specific regions of the genome and/or to short periods in the cell cycle.

Rearrangement of some of the *nif* genes accompanies heterocyst differentiation in *Anabaena* (Golden *et al.*, 1985, 1987a; Haselkorn, 1986). The rearrangements involve deletion, operon fusion, and translocation events and are environmentally triggered by the deprivation of combined nitrogen. Transposable genetic elements are responsible for effecting such types of genome rearrangement in various microorganisms (Kingsman *et al.*, 1988). Analogous elements may mediate the *nif* gene rearrangement and may thus function as developmental regulators in cyanobacteria.

2.2. Plasmids

Plasmids are commonly found in photosynthetic prokaryotes where the majority remain phenotypically cryptic. Many strains harbor multiple plasmid species ranging in size from about 2 to 140 kb (for example, Roberts and Koths, 1976; Simon, 1978; Hu and Marrs, 1979; Lau *et al.*, 1980; Fornari *et al.*, 1984; Anderson and Eiserling, 1985). Homologous plasmids or plasmids sharing defined regions of homology have been described for strains of various unicellular cyanobacteria (Lau and Doolittle, 1979; van den Hondel *et al.*, 1979; Lau *et al.*, 1980). Such regions of homology may correspond to transposable genetic elements and may have been disseminated among the different strains and species by repeated intra- and interspecies transfer events. Extensive sequence homology is also displayed by a number of plasmids isolated from strains of the photosynthetic bacterium *R. sphaeroides* (Fornari *et al.*, 1984). Moreover, homologous sequences have been identified between plasmid and chromosome in *R. capsulatus* (Willison *et al.*, 1987).

In *Rhodospirillum rubrum* plasmid loss has resulted in photosynthetic incompetence (Kuhl *et al.*, 1983, 1984). Furthermore, plasmid rearrangements found in *R. sphaeroides* show a high correlation with an inability to grow photosynthetically (Saunders *et al.*, 1976; Nano and Kaplan, 1984). Rearranged plasmid sequences could be generated through the formation and subsequent resolution of cointegrates involving those plasmids that share homology. It has been proposed (Nano and Kaplan, 1984) that a genetic element, which may be a recombination system or may be akin to an insertion sequence, could mediate the rearrangements. The existence of such an element capable of altering the photosynthetic phenotype could be an asset in ecological competition, where metabolic switching is under the control of environmental stimuli.

There is some evidence to associate certain antibiotic-resistance determinants with plasmids in photosynthetic bacteria. Tucker and Pemberton (1978, 1980) described an *R. sphaeroides* viral R-plasmid, Rφ6P, capable of transfer to various strains of the species (see Section 3). Rφ6P is an unusual entity combining properties of an R-plasmid, encoding a β-lactamase (*bla*) gene conferring resistance to penicillin on the host, and of a virus. Rφ6P appears to behave as a plasmid in the prophage state. In *R. capsulatus*, hybridization studies indicate that a penicillin-resistance determinant is plasmid-encoded (Saunders, 1984) and Kuhl and Yoch (1981) propose an association between streptomycin resistance and plasmid DNA in *R. rubrum*. Genes involved in the regulation of CO_2 fixation may be plasmid-borne in *R. sphaeroides* (Rainey and Tabita, 1989). Plasmid sequences that complement mutants deficient in the capacity for CO_2-dependent growth have been identified in this organism (but see Section 2.1).

Cyanobacterial plasmids do not appear to encode any of the functions, such as antibiotic resistance or nitrogen fixation, that are commonly associated with plasmid DNA (Mazur *et al.*, 1980). Several of these plasmids have, however, been subjected to detailed structural analysis (for example, Weisbeek *et al.*, 1985) and much effort devoted to their use in the design of cloning vectors. Many of these vectors have been constructed by splicing together a ColE1-like replicon and a cyanobacterial plasmid replicon and tagging with transposons where appropriate (Kuhlemeier and van Arkel, 1987) (see also Chapter 5). Such vector constructions have not been routinely used with photosynthetic bacteria. There has, instead, been a tendency to conscript, for gene cloning, broad-host-range plasmids, of which several have been shown to transfer to and be maintained in photosynthetic bacteria (see Section 3 and Chapter 5).

2.3. Naturally-Occurring Bacteriophages

Various temperate and virulent phages of photosynthetic prokaryotes have been isolated. Temperate phages have been particularly widespread among strains of *R. sphaeroides* and typically fall into one of three morphological classes (for example, Mural and Friedman, 1974; Pemberton *et al.*, 1983; Duchrow and Giffhorn, 1987; Heitefuss and Giffhorn, 1989). By contrast, few temperate phages have been detected for the related bacterium *R. capsulatus*. However, Wall *et al.* (1975) isolated various virulent phages specific for *R. capsulatus*. These were classified into 16 phage types based on host range. Virulent phage RC1 of *R. capsulatus* could be propagated equally well in photosynthetically- or aerobically-grown host cells, while RS1, a virulent phage active only against *R. sphaeroides*, formed plaques less efficiently on photosynthetically-grown cells than on aerobically-grown cells, implying some degree of physiological specificity (Abeliovich and Kaplan, 1974). A more complex *R. sphaeroides* virulent phage with a genome of about 160 kb is φRsV. This phage is also highly specific to strains of *R. sphaeroides*, but forms plaques with the same efficiency under phototrophic or aerobic conditions (Duchrow *et al.*, 1988).

The first reports of viruses (cyanophages) infecting cyanobacteria were those of Safferman and Morris (1963) and Safferman *et al.* (1969) describing phages of the LPP group. Such phages attack three genera of filamentous cyanobacteria: *Lyngbya*, *Plectonema*, and *Phormidium*. Subsequently, a variety of cyanophages has been isolated and characterized (for example, Sherman and Connelly, 1976; Sherman and Brown, 1978; Hu *et al.*, 1981, Mendzhul *et al.*, 1985). Requirements for cyanophage replication vary. For instance, illumination of photosynthetically-grown *Plectonema boryanum* is required throughout the eclipse period for phage LPP1-G (Padan *et al.*, 1970), while illumination throughout the entire latent period is required for phage N1 in *Nostoc muscorum* (Adolph and Haselkorn, 1972).

Despite the discovery of a variety of phages for both cyanobacteria and photosynthetic bacteria, transduction *per se* has not proved a particularly useful tool in the genetic analysis of photosynthetic prokaryotes, with the exception of *R. capsulatus*. A transduction-like process, termed capsduction, available for *R. capsulatus* (Section 3.3.1) has made a crucial contribution to the molecular genetics of this species. The gene transfer agent (GTA) mediating the process resembles a small phage with an icosahedral head and a short tail of variable length (Marrs, 1978, 1983). However, no plaque-forming activity appears to be associated with this gene transfer system. Furthermore, there is no transfer of the ability to produce the GTA to recipients. In these and other respects the GTA represents a novel biological agent.

3. GENE TRANSFER MECHANISMS

3.1. Transformation

The ability to introduce purified DNA into bacteria by transformation has been described for a number of photosynthetic prokaryotes (Porter, 1986; Tandeau de Marsac and Houmard, 1987; Saunders and Saunders, 1988) (Table I).

The first report of transformation in the photosynthetic bacterium *R. sphaeroides* described uptake of DNA from the temperate phage Rφ6P (Tucker and Pemberton, 1980). Simultaneous infection with a helper phage Rφ9 was required for entry of the Rφ6P DNA into recipients. A more generalized transformation system has since been developed for *R.*

Table I. Selected Transformable Strains of Photosynthetic Prokaryotes

Strain	Ref.
Anacystic nidulans R2 (*Synechococcus* sp. PCC 7942)	Chauvat *et al.* (1983), Gendel *et al.* (1983), Golden *et al.* (1978b), Kuhlemeier *et al.* (1981), Kuhlemeier and van Arkel (1987)
Agmenellum quadruplicatum PR-6 (*Synechococcus* sp. PCC 7002)	Buzby *et al.* (1983, 1985), Stevens and Porter (1980)
Synechococcus sp. PCC 6301	Herdman (1973), Lightfoot *et al.* (1988), Orkwiszewski and Kaney (1974)
Synechocystis sp. PCC 6803	Chauvat *et al.* (1986), Dzelzkalns and Bogorad (1986), Grigorieva and Shestakov (1982), Williams (1988)
Aphanocapsa sp. 6714 (*Synechocystis* sp. 6714)	Astier and Espardellier (1976)
Rhodobacter sphaeroides	Tucker and Pemberton (1980), Fornari and Kaplan (1982)

sphaeroides, in which recipient cells were treated with high concentrations (500 mM) of Tris prior to exposure to the transforming DNA (Fornari and Kaplan, 1982). Transformation frequencies were dependent upon the concentrations of $CaCl_2$ and PEG 6000 used in the transformation mixture and could be as high as 10^{-5} per viable cell. In contrast, the transformation system thus far developed for *R. capsulatus* gives frequencies of only 10^{-9} per viable cell (Jasper *et al.,* 1978).

There are several highly transformable strains among the unicellular cyanobacteria. *A. nidulans* R2 was initially shown to be naturally competent to take up DNA by Shestakov and Khyen (1970). Subsequently, further reports describing transformation of this and other unicellular cyanobacteria have appeared (for example, Herdman, 1973; Mitronova *et al.,* 1973; Orkwiszewski and Kaney, 1974; Astier and Espardellier, 1976; Devilly and Houghton, 1977; Stevens and Porter, 1980; Grigorieva and Shestakov, 1982; Buzby *et al.,*1983; Golden and Sherman, 1984; Kuhlemeier *et al.* 1981; Kuhlemeier and van Arkel, 1987; Lightfoot *et al.,* 1988).

In these cyanobacteria the transforming DNA may be covalently closed circular (CCC) DNA molecules comprising endogenous plasmids with antibiotic resistance transposons (van den Hondel *et al.,* 1980) and/or *E. coli* plasmid replicons (Buzby *et al.,* 1983, 1985; Golden and Sherman, 1983). Such transforming DNA may remain extrachromosomal or may integrate into the recipient genome.

Transformation of *A. nidulans* R2 with biphasic shuttle vectors, such as pUC303 (comprising the cyanobacterial plasmid pUH24 and the *E. coli* plasmid pACYC184), results in recombination between homologous sequences on the incoming vector, pUC303, and the resident plasmid replicon, pUH24 (Kuhlemeier *et al.*, 1983). Deletion of vector sequences generally attends the recombination event, leading to instability. Such vector instability could be overcome by using pUH24-cured recipients. These strains may be advantageous for gene cloning in *A. nidulans*, since they would circumvent the possible loss of cloned DNA effected by recombination between a recombinant vector plasmid and homologous resident plasmid. However, this would be at the expense of transformation efficiency, which is reduced some 30-fold in spontaneous pUH24$^-$ variants of *A. nidulans* R2, when transformed by vectors derived from pUH24 (Chauvat *et al.*, 1983). Homologous recombination between shared sequences on incoming and resident replicons has also been observed during transformation of *Synechocystis* sp. PCC 6803 (Chauvat *et al.*, 1986; Kuhlemeier and van Arkel, 1987). Provision of homologous sequences enhanced the transformation efficiency and enabled the construction of a highly transformable strain of *Synechocystis*.

Biphasic vectors consisting of cyanobacterial plasmids (for example, pAQ1) from *Agmenellum quadruplicatum* PR-6 and *E. coli* plasmids (such as pBR322 or pBR325) have been used to transform strain PR-6 (Buzby *et al.*, 1983). Elimination of the *Aqu*I (*Ava*I) restriction enzyme recognition site from the hybrid plasmids greatly enhanced their ability to transform strain PR-6. Furthermore, multimeric forms of the hybrid plasmids [with or without the *Aqu*I (*Ava*I) recognition site] proved more effective than analogous monomeric forms at transforming the organism. Such higher transformation efficiency with multimers may reflect the mode of processing of the DNA during its entry and establishment in recipients. In this respect the *A. quadruplicatum* system is reminiscent of that operating in naturally competent *Bacillus subtilis*, where transforming DNA is converted to single strands during uptake and a double-stranded replicon is subsequently reconstituted from partially homologous single strands in the transformants (de Vos *et al.*, 1981).

Chromosomal DNA carrying selectable markers can also be used to transform cyanobacterial strains. The incoming DNA recombines with homologous sequences on the resident chromosome (Fig. 1). This property can be exploited to insert heterologous DNA into the chromosome. Bracketing the heterologous DNA by homologous DNA directs it to a specific chromosomal locus. Transformation by such chromosomal recombination can be effected by homologous DNA presented as a linear fragment or as a cloned insert in a recombinant vector. Applications of this system include gene inactivation by recombination with a cloned altered allele (Golden *et al.*, 1986), gene duplication (Williams and Szalay, 1983; Golden *et al.*, 1987b), and insertional mutagenesis (Buzby *et al.*, 1985; Labarre *et al.*, 1989) (see Chapter 5).

Integrative transformation of *Synechocystis* sp. PCC 6803 can be ef-

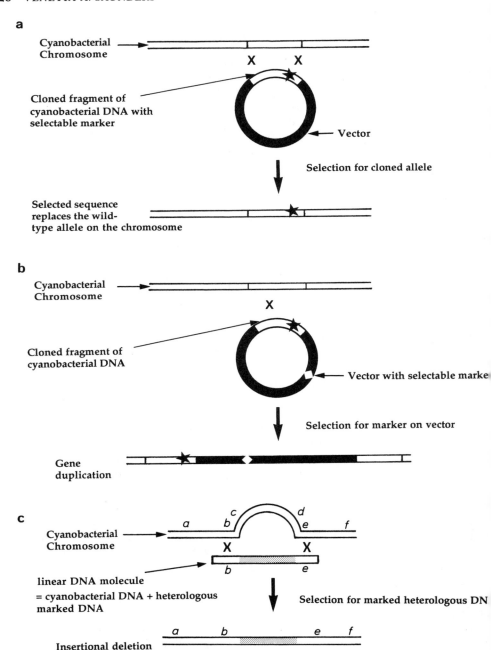

Figure 1. Possible mechanisms for integrative transformation in cyanobacteria. (a) Replacement of wild-type chromosomal allele with a marked homologous allele cloned in a vector. Selection for the marker (*) on the cloned allele results in gene conversion or reciprocal recombination (involving double crossover), which replaces the chromosomal allele with the

fected through the use of transforming DNA comprising restriction fragments bearing antibiotic-resistance determinants ligated randomly to a pool of genomic restriction fragments with compatible termini. Stable integration of the resistance markers is presumed to occur by homologous recombination in a process involving precise deletion of the target DNA (see Fig. 1). Random disruption of the host DNA by insertional transformation generates various mutants among the antibiotic-resistant transformants (Labarre *et al.*, 1989).

Transformation of *Synechococcus* sp. PCC 6301 using *Synechococcus* chromosomal DNA inserted into ColE1-derived vectors has been reported by Lightfoot *et al.* (1988). The size of the insert DNA and the presence of internal deletions, which interrupt the linear sequence of this DNA, affected transformation efficiencies. Integrative transformation involving homologous recombination in which a single crossover event effects integration of circular plasmid DNA into the recipient chromosome is probably operating in this cyanobacterium.

A transformation system that is independent of homologous recombination and of autonomously replicating plasmids has been described for *Synechocystis* sp. PCC 6803 (Dzelzkalns and Bogorad, 1986). Segments of transforming *E. coli* plasmids inserted randomly and stably into the *Synechocystis* genome following exposure of the recipients to low doses of UV irradiation. Such transformation events may have been promoted by the availability of increased amounts of proteins associated with recombination and/or by transient alleviation of host-controlled restriction in the UV-treated cells. Such phenomena are known to be induced in *E. coli* and other prokaryotes in response to UV irradiation (Walker, 1984). This UV-induced transformation system should enable any foreign DNA that happens to be present in the cells to become illegitimately incorporated into the host genome. The system may be utilized as a mutagenic agent where the genetic lesion is tagged with a selectable marker. Additionally, heterologous DNA sequences may be introduced directly into *Synechocystis*, without the need for recloning into a shuttle or integrative cloning vector.

There is a dearth of transformation systems for filamentous cyanobacteria. Factors such as the physiological state of the cells and the conditions

---

selected sequence. (b) Gene duplication by integration of the entire recombinant vector. Selection for a marker on the vector results in cells with the entire vector integrated into the chromosome. A single crossover event between homologous sequences on the chromosome and the cloned cyanobacterial allele results in integration of the vector and duplication of the cloned DNA at the chromosomal locus. (c) Insertional deletion by random cloning into the chromosome. The transforming DNA comprises heterologous marked DNA randomly ligated to restriction fragments of the host genome. The heterologous DNA may be ligated between cyanobacterial fragments b and e that are not contiguous. Gene conversion or a double crossover event on either side of the heterologous DNA results in its insertion with attendant deletion of c–d between b and e on the chromosome. ×, position of recombination; shaded area, heterologous DNA.

for selecting the transferred markers have a crucial influence on transformation (for example, Chauvat *et al.*, 1983; Golden and Sherman, 1984; Daniell and McFadden, 1986; Lightfoot *et al.*, 1988). The use of strains with altered restriction-modification systems that have enhanced the transformation efficiency in certain unicellular cyanobacteria (Buzby *et al.*, 1985) may facilitate transformation in filamentous forms. Recently electroporation, which has been used to introduce DNA into various bacteria (Calvin and Hanawalt, 1988), has been successfully applied to the filamentous cyanobacterium *Anabaena* sp. strain M131 (Thiel and Poo, 1989). Moreover, there are indications that other cyanobacteria that are refractory to existing transformation regimes may be amenable to electroporation.

3.2. Conjugation

Conjugation involves the transfer of genetic material in a process requiring cellular contact between donors and recipients and is almost invariably mediated by plasmids. Broad-host-range conjugative plasmids belonging principally to incompatibility (Inc) group P have proved effective in promoting conjugation in photosynthetic prokaryotes. Not only are the conjugative plasmids themselves transferred, but they can, in some cases, serve as helper plasmids in the mobilization of nonconjugative plasmids and the chromosome. IncP-mediated conjugation is generally more efficient on a solid substratum than in liquid medium, due to the rigidity of IncP sex pili.

3.2.1. Conjugative Plasmids and Chromosome Mobilization

Many of the reports of conjugal transfer within photosynthetic prokaryotes have featured IncPα plasmids, RP1, RP4, RK2, R18, and R68 (and their derivatives). Such plasmids are very similar as judged by gross restriction map and heteroduplex analysis (Thomas and Smith, 1987).

Olsen and Shipley (1973) first described conjugative plasmid transfer in the Rhodospirillaceae. However, the plasmid R1822 was not stably maintained in these photosynthetic bacteria in the absence of appropriate selection. Subsequently, stable transfer of RP4 from *Pseudomonas aeruginosa* or *E. coli* to *R. sphaeroides* (Miller and Kaplan, 1978) and to *R. capsulatus* (Yu *et al.*, 1981) was reported. Recipients acquired resistance to kanamycin and tetracycline, but not to ampicillin or carbenicillin. This implied a lack of expression of the RP4 *bla* (β-lactamase) gene in these organisms. A derivative of the broad-host-range plasmid R68, designated R68.45, with an enhanced ability to mobilize the chromosome of *P. aeruginosa* and other bacteria (Haas and Holloway, 1978; Haas and Reimmann, 1989) is also capable of transferring to the *Rhodobacter* (Marrs *et al.*, 1977, Sistrom; 1977, Yu *et al.*, 1981) and to the unicellular cyanobacterium *Synechococcus* sp. PCC 6301

(Delaney and Reichelt, 1983). Furthermore, R68.45 can promote chromosome mobilization and has proved useful in the genetic analysis of *R. sphaeroides* (Sistrom *et al.*, 1984). Chromosome-mobilizing ability (CMA) of R68.45 has been attributed to its integration into the chromosome via transposition-mediated events involving insertion sequence IS*21* (Willetts *et al.*, 1981). IS*21*, which occurs as a single copy on R68, is tandemly repeated on R68.45 and is essential for CMA (Haas and Riess, 1983). However, stable Hfr (high frequency of recombination) strains have not been isolated from R68.45 donors in different bacterial species. The formation of such strains appears to require replication-defective variants of R68.45 (Reimmann *et al.*, 1988). Temperature-sensitive replication defectives, such as pME487, have been used to isolate Hfr strains of *P. aeruginosa*. At the restrictive temperature the plasmid can integrate into the chromosome to suppress replication defects. Such Hfr strains transferred larger tracts of the genome than did R68.45 donors (Reimmann *et al.*, 1988). A similar strategy exploiting integrative suppression to create Hfr donors of *R. capsulatus* was intended in the experiments of Willison *et al.* (1985). pTH10, an RP1 derivative that is temperature sensitive for replication in *E. coli* and could transfer to *R. capsulatus* at a frequency of 2×10^{-2} per recipient, was conscripted for chromosome mobilization. However, in contrast to its behavior in *E. coli*, pTH10 did not exhibit temperature sensitivity in *R. capsulatus*, nor did it integrate into the host chromosome at the "restrictive" temperature. Nevertheless, pTH10 was capable of mediating chromosomal gene transfer at a frequency that was 10^4 times higher than that of the parental plasmid, RP1. While the precise mechanism underlying pTH10-mediated chromosome mobilization is unknown, it appears that transfer occurs from multiple origins. The CMA of pTH10 has enabled a genetic map of *R. capsulatus* to be constructed and the position of *nif* genes to be assigned (Willison *et al.*, 1985) (Section 5).

A range of IncP and IncW resistance plasmids has been found to transfer to and be maintained in the Rhodospirillaceae (Tucker and Pemberton, 1979a; Quivey *et al.*, 1981; Pemberton *et al.*, 1983) (Table II). Furthermore, the plasmids could transfer not only between strains of such photosynthetic bacteria, but also to a number of other organisms, including *E. coli*, *Alcaligenes eutrophus*, *Pseudomonas putida*, *P. aeruginosa*, *Agrobacterium tumefaciens*, and *Cellvibrio* sp. (Pemberton *et al.*, 1983). Such traffic of plasmids between a diversity of hosts can be exploited to carry genes that have been manipulated in one organism, into another. In addition, the plasmids provide vehicles for the introduction of transposable genetic elements into the Rhodospirillaceae.

RP4::Mu*cts*62 has been transferred to *R. sphaeroides*, where the Mu phage genome is expressed (Tucker and Pemberton, 1979b). This opens up avenues for Mu-mediated genetic analysis of photosynthetic bacteria. Mu can integrate efficiently at many sites on the *E. coli* chromosome, in turn

Table II. Selected R-Plasmid Transfers to Photosynthetic Bacteria

Plasmid	Recipient	Ref.
Inc P		
R1822	*Rhodobacter sphaeroides,*	Olsen and Shipley (1973)
	Rhodospirillum rubrum	
RP4	*R. sphaeroides*	Miller and Kaplan (1978),
		Tucker and Pemberton (1979a)
RP4::Mu*cts*61	*Rhodobacter capsulatus*	Yu *et al.* (1981)
RP4::Mu*cts*62	*R. sphaeroides*	Tucker and Pemberton (1979b)
RP1	*R. capsulatus*	Yu *et al.* (1981)
pBLM2	*R. capsulatus*	Marrs (1981)
RP1::Tn*501*	*R. sphaeroides*	Pemberton and Bowen (1981)
R68.45	*R. sphaeroides*	Tucker and Pemberton (1979a),
		Sistrom (1977)
	Rhodocyclus gelatinosus	Sistrom (1977)
	R. capsulatus	Yu *et al.* (1981), Marrs *et al.*
		(1977)
R751	*R. sphaeroides*	Tucker and Pemberton (1979a)
	Rsp. rubrum	Quivey *et al.* (1981)
R751::Tn*5*	*Rsp. rubrum*	Quivey *et al.* (1981)
R702	*R. sphaeroides*	Tucker and Pemberton (1979a)
RK2	*Rsp. rubrum*	Quivey *et al.* (1981)
pPH1JI	*R. sphaeroides*	Saunders (1984)
IncW		
R388	*R. sphaeroides*	Tucker and Pemberton (1979a)
S-a	*R. sphaeroides*	Tucker and Pemberton (1979a)

providing a portable region of homology with which plasmids carrying additional copies of the prophage can recombine. Furthermore, Mu-mediated chromosome mobilization and prime plasmid formation have been described in various organisms (Symonds *et al.,* 1987).

RP1::Tn*501* has also been transferred to *R. sphaeroides,* and high-frequency transfer of the chromosome was achieved (Pemberton and Bowen, 1981). Recombinants were recovered at frequencies between 10^{-3} and 10^{-7} per donor using RP1::Tn*501,* compared with less than 10^{-8} per donor for the wild-type plasmid. The hybrid plasmid, which appeared to promote polarized transfer from one or possibly two origins on the chromosome, enabled compilation of the first genetic map for *R. sphaeroides.* A possible mechanism for chromosome mobilization by RP1::Tn*501* involves cointegrate formation during transposition of Tn*501* to the *R. sphaeroides* chromosome. Tn*501* transposes at high frequency, but due to the weak resolvase (*tnpR* gene product) activity of the transposon, the cointegrates that form are only poorly resolved. Thus, the RP1::Tn*501* plasmid is likely to remain stably inserted into the *R. sphaeroides* chromosome in the course of cointegrate formation and thereby promote chromosome transfer (Pemberton *et al.*, 1983). Assuming this to be the mechanism accounting for Tn*501*-enhanced CMA, it may be possible to modulate expression of the *tnpR* gene of this and related transposons in order to influence the stability of cointegrate formation and, in turn, the mobilizing ability of conjugative plasmids carrying such transposable elements. Indeed, cosmid mobilization from *E. coli* to *R. sphaeroides* has been accomplished by adopting such a rationale (see Section 3.2.2).

pBLM2, a derivative of RP1 exhibiting enhanced CMA in *R. capsulatus*, has been isolated by Marrs (1981). The frequency of pBLM2-mediated transfer of chromosomal genes was up to 6×10^{-4} per donor. Genes for photosynthesis, rifampicin and streptomycin resistance, and tryptophan and cytochrome biosynthesis could all be mobilized. pBLM2-mediated chromosomal transfer appeared to originate at more than one site. A collection of R-primes (including pRPS4, pRPS223, and pRPS404) incorporating genes for photosynthesis was generated using pBLM2 (Marrs, 1981). These R-prime plasmids were unstable in *R. capsulatus,* but stable in the *E. coli* and *Pseudomonas fluorescens* strains tested. The plasmid-borne genes for photosynthesis functioned in *R. capsulatus,* but not in these alternative hosts. In one of the R-primes, pRPS404, the 46 kb tract of *R. capsulatus* chromosomal DNA is bracketed by direct repeats of IS*21*, which in a recombination proficient (Rec$^+$) background can recombine, effecting deletion of the intervening chromosomal DNA. This can in turn provide a means of gene replacement (homogenotization) (Youvan *et al.*, 1982; Taylor *et al.*, 1983) (see Section 4.2).

Other R-prime plasmids have been isolated, including derivatives of R68.45 (for example, pWS1 and pWS2) containing chromosomal genes of *R. sphaeroides* (Sistrom *et al.*, 1984). Recombination-deficient strains of *R. sphaeroides* facilitated the isolation of such R-primes. R-primes are useful for various genetic manipulations. For example, they can be used for complementation analysis (Marrs, 1981; Kaufmann *et al.*, 1984), to introduce mutations into specific genes through gene replacement (Youvan *et al.*, 1982; Biel and Marrs, 1985), and for physical and genetic mapping of photosynthetic and other genes (Taylor *et al.*, 1983; Zsebo *et al.*, 1984; Sistrom *et al.*, 1984). Furthermore, R-prime plasmids provide a rich source of DNA for cloning genes, such as those for photosynthesis and nitrogen fixation (Marrs, 1983; Taylor *et al.*, 1983; Klug and Drews, 1984).

3.2.2. Transfer of Nonconjugative Plasmids

Certain conjugative IncP plasmids have proved useful in promoting the transfer of nonconjugative plasmids from *E. coli* to *Rhodobacter* and to cyanobacteria (for example, Saunders, 1984; Wolk *et al.*, 1985; Zinchenko *et al.*, 1984; Thiel and Wolk, 1987; Elhai and Wolk, 1988; Hunter and Turner, 1988). The conjugative plasmid provides transfer (*tra*) functions and the nonconjugative plasmid may be mobilized either autonomously or by cointegrate formation with the mobilizing plasmid.

The choice of mobilization system depends upon the experimental objectives. For instance, where gene cloning is the priority, broad-host-range nonconjugative plasmids, particularly of IncQ group (such as RSF1010 and derivatives), are especially useful. They can be mobilized efficiently, have high copy number, and can be maintained in the autonomous state in organisms such as *R. sphaeroides* (Ashby *et al.*, 1987; Hunter and Turner, 1988). pNH2, a derivative of RSF1010, can be transferred efficiently (with a frequency of 5×10^{-1} per recipient, at best) from *E. coli* to *R. sphaeroides* utilizing the *tra* functions of either pRK2073 or RP4. The latter plasmid may be present in the *E. coli* donor either extrachromosomally or integrated into the chromosome to prevent self-transfer. This highly efficient mobilization system has been employed to clone genes for photosynthesis by complementation of photosynthetically-incompetent *R. sphaeroides* mutants (Hunter and Turner, 1988). The cosmid cloning vector pHC79 has been mobilized from *E. coli* to *R. sphaeroides* by cointegration with the IncP conjugative plasmid R751 carrying the transposon Tn*813* (*tnpR⁻*, *tnpA⁺*). The lack of resolvase activity of Tn*813* enabled the cointegrate R751::Tn*813*::pHC79 to be maintained stably in *R. sphaeroides*. This system provided the basis for characterizing cloned carotenoid biosynthesis genes from *R. sphaeroides* (Pemberton and Harding, 1986).

Mobilization using a binary-vehicle broad-host-range cloning system based on plasmid RK2 (Ditta *et al.*, 1980) has been exploited to construct and analyze gene libraries of *R. capsulatus* (Klug and Drews, 1984; Colbeau *et al.*, 1986). The transfer and replication regions of RK2 were separated onto two plasmids; pRK290 carried the replication system and served as a broad-host-range cloning vector, while pRK2013 contained RK2 *tra* functions and a kanamycin-resistance determinant fused to the ColE1 replication system. pRK2013 was required to serve as a helper plasmid for mobilization of pRK290. To force for chromosomal integration of *R. capsulatus* DNA that was transferred in the pRK290 vector, an incompatible plasmid pPH1JI was subsequently introduced into the recipient by conjugation. Following appropriate selection for pPH1JI, the vector plasmid pRK290 was displaced from *R. capsulatus*, while the transferred DNA presumably recombined with the host chromosome (Klug and Drews, 1984). The same vector strategy was employed to mobilize the cosmid pLAFR1 (constructed from pRK290), carrying a library of *R. capsulatus* chromosomal DNA into mutants unable to fix nitrogen (Nif⁻). This permitted the isolation of *nif* genes of *R. capsulatus* by complementation (Avtges *et al.*, 1985). Taylor *et al.* (1983) utilized pBR322 derivatives pDPT42 and pDPT44 (both carrying a kanamycin-resistance marker) to mobilize photosynthetic genes from their repository strains of *E. coli* to *R. capsulatus* mutants for marker rescue analysis. Transfer was mediated by a plasmid construct comprising plasmid R751, providing *tra* functions, and a ColE1 derivative, supplying requisite mobility functions *in trans*, for mobilization of pBR322 vectors. Intergeneric mobilization was facilitated by using putative restriction-defective *R. capsulatus* mutants as recipients. Nonconjugative plasmids with a limited host range, such as those derived from pBR322, are applicable as transposon carriers, for insertion mutagenesis and for homogenotization experiments, in strains into which they can be mobilized, but are not stably maintained. Suicide vectors based on pBR325 (for example, pSUP201 and pSUP202) and on pACYC184 (for example, pSUP101) have been employed for transposition mutagenesis in the *Rhodobacter* (see Section 4). The vector design strategy is based on high-frequency mobilization from donor strains that carry RP4 *tra* genes incorporated into the chromosome. The mobilizable vector plasmids contain the IncP-type specific recognition site for mobilization (Mob site) believed to contain the origin of transfer, *oriT* (Simon *et al.*, 1983). Loading the vectors with transposons permits introduction of such elements into hosts for random transposon mutagenesis, with attendant elimination of the transposon carrier (Fig. 2). In addition, the system is exploitable for site-directed mutagenesis and for site-specific gene transfer.

pBR322 derivatives have also been used in the construction of shuttle vectors capable of transfer from *E. coli* to filamentous cyanobacteria *Ana-*

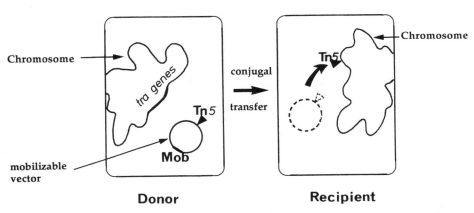

Figure 2. Transposon mutagenesis with a mobilizable Tn5 carrier vector. The mobilizing donor strain carries IncP broad-host-range transfer (*tra*) genes integrated into the chromosome. The Tn5 carrier is a mobilizable vector carrying the IncP mobilization site (Mob). The vector is mobilized into the recipient by *trans*-acting mobilization functions. Transposition of Tn5 to the recipient chromosome may then occur. The vector is not stably maintained. Neomycin selects for stable Tn5 insertion into the recipient genome. [From Simon *et al.* (1983).]

baena and *Nostoc* spp. and to the unicellular cyanobacterium *Anacystis* (Wolk *et al.*, 1984; Flores and Wolk, 1985) (see Chapter 5).

3.3. Transduction

A collection of both temperate and virulent phages has been isolated for photosynthetic prokaryotes (see Section 1.3). However, the role of these phages as mediators of genetic exchange within this biological group remains limited. *Bona fide* transduction does not appear to have been reported for any cyanobacterium. More encouraging results have been obtained with photosynthetic bacteria. Phage-mediated transfer of a variety of genetic markers, albeit at low frequency, has been demonstrated by Kaplan and co-workers using the temperate phage RS-2 in *R. sphaeroides* (Saunders, 1978). Furthermore, Pemberton and Tucker (1977) reported high-frequency transduction of a penicillin-resistance determinant between *R. sphaeroides* strains, promoted by the viral R plasmid RΦ6P. The RΦ6P prophage probably carries the *bla* (β-lactamase) gene as a transposon (Tucker and Pemberton, 1978).

However, the only transduction-like system of practical importance to date is that involving the GTA of *R. capsulatus* (Marrs, 1974). The GTA-mediated process, capsduction, resembles generalized transduction, but there are some obvious differences (Marrs, 1978, 1983). These include a failure to transmit to recipients the capacity to produce the GTA or to amplify GTA-encoding DNA during production of the agent and a general lack of overt viral activities. Such features make capsduction a unique genetic process.

3.3.1. Capsduction

Capsduction occupies a historic position in the genetics of photosynthetic bacteria. It was the first operational genetic exchange system to be described and is still a popular tool with *R. capsulatus* geneticists. The GTA mediating capsduction randomly packages fragments (some 5 kb in length) of the donor genome. Successful establishment of the GTA-borne markers in recipients demands integration into the resident chromosome by homologous recombination. Recombination frequencies of 10^{-3} per recipient have been attained and markers from any region of the genome can be transferred (Yen *et al.*, 1979). Such GTA-mediated genetic exchange occurs exclusively between strains of *R. capsulatus*. No analogous system has been discovered for other photosynthetic bacteria. The requirement for chromosomal integration of markers carried by the GTA has been instrumental in the exploitation of capsduction in interposon mutagenesis (Section 4.3) and in genetic mapping (Section 5).

4. TRANSPOSONS, INTERPOSONS, AND THEIR APPLICATIONS

4.1. Properties of Transposons

Transposable genetic elements are discrete DNA sequences that are capable of moving from one genetic locus to another on the same or different replicon (Berg and Howe, 1989). Such elements are responsible for a range of genome rearrangements and can even be used to engineer organisms where no genetic system was previously available. Transposons (Tns) which carry accessory marker genes, such as antibiotic-resistance determinants, provide useful tools for a variety of genetic manipulations in photosynthetic prokaryotes. For example, resistance transposons have been used in transposition (insertion) mutagenesis (see Section 4.2) and to mediate cointegrate formation, either between conjugative and nonconjugative plasmids for plasmid mobilization (see Section 3.2.2), or between conjugative plasmids and the chromosome to generate transient or stable Hfr strains for chromosome mobilization (see Section 3.2.1). In addition, antibiotic-resistance transposons provide selectable markers to tag the chromosome (Tandeau de Marsac *et al.*, 1982) in turn facilitating gene cloning. Physical and genetical mapping can be simplified through the use of Tn-induced mutants, where the Tn is, of necessity, linked to the altered genotype, and site-specific Tn mutagenesis can be used to tailor strains for particular purposes.

4.2. Transposon Mutagenesis

Antibiotic-resistance Tns are particularly useful agents in mutagenesis, since presence of the resistance determinant facilitates monitoring of the transposition event (Bennett *et al.*, 1988). Use of such Tns as mutagenic agents requires a vehicle to carry the Tn into the host. In addition, a means of selecting against the Tn carrier is needed. Various strategies have been adopted to provide suitable Tn carriers for mutagenesis of photosynthetic prokaryotes. For example, promiscuous IncP plasmids that can be modified (by, for example, carriage of Mu or of temperature-sensitive mutations affecting replication) to render them unstable in the host have been used. One such suicide plasmid, pJB4JI, a derivative of pPH1JI that carries Mu*cts* and Tn5 [specifying resistance to kanamycin (Km^r) and streptomycin (Sm^r)] has been transferred to *R. sphaeroides*. Maintenance of pJB4JI was severely reduced compared with the parental plasmid in *R. sphaeroides*, while the frequency of pigment mutants was about 2% of the transconjugants (Saunders, 1984).

Site-directed Tn7 mutagenesis of genes for the photosynthetic apparatus of *R. capsulatus* was accomplished using the IncP R-prime pRPS404, which bears most of the photosynthesis genes (see Chapter 5).

Homologous recombination in *R. capsulatus* traded the Tn-mutagenized DNA on the R-prime for the wild-type copy on the chromosome. In turn, the wild-type DNA, transiently on the R-prime, was subsequently deleted by an intramolecular recombination event involving the flanking copies of IS*21* on the plasmid (see Fig. 3). Such spontaneous deletion of this wild-type DNA alleviates the need for a second incompatible plasmid to exclude the recombined plasmid (bearing the wild-type alleles) and reveal the lesion-bearing genes. Zsebo and Hearst (1984) have also utilized this system to insert Tn*5.7* into the *R. capsulatus* chromosome (see Chapter 5).

Alternative Tn carriers utilize modified *E. coli*-specific vectors, such as pSUP101, pSUP201, and pSUP202, that incorporate the IncP Mob site and can therefore be mobilized by coexisting IncP *tra* genes (see Section 3.2.2). Such a mobilization system, which enables carriage of Tns into recipients at high frequency in the absence of vector maintenance, has proved popular for Tn mutagenesis in the *Rhodobacter*.

pSUP201 carrying Tn*5* has been used in *R. capsulatus*, where the vector is not maintained, but Tn*5* is well expressed (for example, Kaufmann *et al.*, 1984; Hudig *et al.*, 1986; Klipp *et al.*, 1988; Daniels *et al.*, 1988). Kaufmann and colleagues (1984), using pSUP201::Tn*5*, obtained up to 3×10^{-5} Km^r transconjugants per donor and isolated Tn*5* mutants of *R. capsulatus* defective in the photosynthetic apparatus. Complementation analysis with the R-prime pRPS404 revealed that some genes for the B800–B850 light-harvesting complex (LH2) lie outside the main photosynthesis gene cluster and this has been corroborated by the work of Youvan and Ismail (1985). Site-specific Tn*5* mutagenesis was also carried out. Tn*5*-containing restriction fragments from specific photosynthetic mutants of *R. capsulatus* were cloned into the broad-host-range vector pRK290 and subsequently transferred to a wild-type strain. The Tn*5*-mutagenized DNA could insert into the wild-type chromosome by homologous recombination involving two crossover events, as proposed by Ruvkun and Ausubel (1981) for *Rhizobium*. Introduction of a second IncP plasmid pPH1JI could displace pRK290, now carrying the wild-type chromosomal DNA, from the cells (Fig. 4), in turn revealing the lesions of the photosynthetic apparatus. (Kaufmann *et al.*, 1984). This approach has also been applied to the generation of Nif$^-$ mutants of *R. capsulatus* (Avtges *et al.*, 1985).

In addition, Nif$^-$ mutants of *R. capsulatus* mutants have been obtained by random Tn*5* mutagenesis, again using pSUP202 as Tn carrier (Klipp *et al.*, 1988). About 0.5% of the Tn mutants were Nif$^-$. pJB3JI, a kanamycin-sensitive derivative of R68.45 (Section 3.2.1), was transferred to the Nif$^-$ mutants for the purpose of generating R-primes carrying Tn*5*-mutated *nif* genes. Subsequent cloning enabled Tn*5* insertions to be mapped and *nif* genes to be located by hybridization to *Klebsiella pneumoniae nif* probes specific for individual *nif* determinants. Hunter (1988) has also used pSUP202 for Tn*5*-induced mutation in *R. sphaeroides* (see Chapter 5).

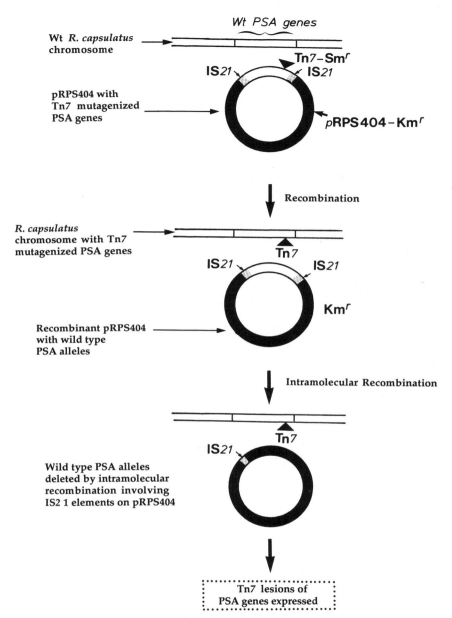

Figure 3. A model for site-directed Tn7 mutagenesis in *R. capsulatus* The R-prime pRPS404 is mutagenized with Tn7 in *E. coli* and transferred to wild-type *R. capsulatus*. Homologous recombination exchanges the Tn7-mutagenized photosynthetic apparatus (PSA) genes for the wild-type chromosomal alleles. Copies of IS*21* that flank the wild-type PSA alleles now on pRPS404 mediate an intramolecular recombination event to delete these wild-type alleles from the R-prime. In the absence of streptomycin selection all copies of the R-prime will delete to the parental plasmid. Sm^r, Streptomycin resistance; Km^r, kanamycin resistance. [Modified from Youvan *et al.* (1982).]

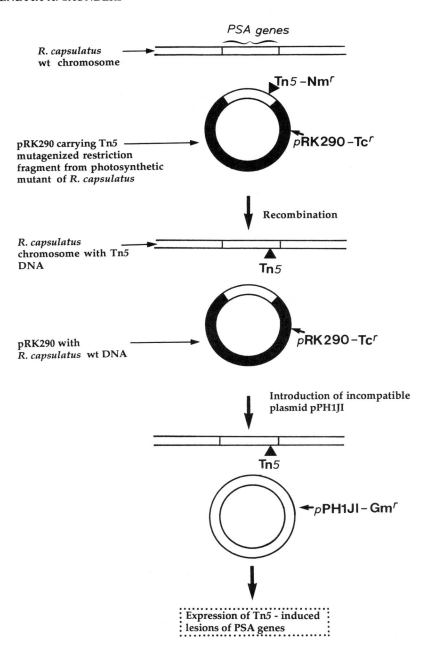

Figure 4. A strategy for site-specific Tn5 mutagenesis in *R. capsulatus*. pRK290 containing Tn5-mutagenized restriction fragments from photosynthetic mutants of *R. capsulatus* was transferred to wild-type *R. capsulatus*. Homologous recombination involving two crossover events inserts the Tn5 containing DNA into the chromosome. A second IncP plasmid incom-

Transposon mutagenesis has been used in the cyanobacterium *A. nidulans* R2 (Tandeau de Marsac *et al.*, 1982; Kuhlemeier *et al.*, 1984; Kuhlemeier and van Arkel, 1987). Mutants have been isolated by using cells harboring pCH1 (pUH24::Tn*901*). The transposon Tn*901* (specifying resistance to ampicillin) subsequently provided a marker for cloning those cyanobacterial genes into which the element had inserted. In turn, such Tn-inactivated genes have served as probes to identify corresponding wild-type genes from gene libraries.

In filamentous cyanobacteria restriction appears to have hampered the use of Tns for mutant induction. Tn*5* derivatives lacking appropriate restriction sites have been constructed, but do not apparently transpose stably in cyanobacteria (Wolk *et al.*, 1985). Insertional mutagenesis via integrative transformation (see Section 3.1) may provide a useful alternative to transposition mutagenesis in cyanobacteria, particularly where multiple copies of the genome are found (Labarre *et al.*, 1989).

4.3. Interposon Mutagenesis

Site-specific interposon (cartridge) mutagenesis permits the generation of insertion and deletion mutations in target genes. Interposons for this purpose are segments of DNA (generally 4.0 kb or less) that carry a selectable marker (for example, antibiotic resistance) bracketed by useful restriction sites and ideally transcription-translation terminators in inverted repeats (Prentki and Krisch, 1984). Interposon insertions are generated at points in the target sequence, by utilizing a single restriction site, while interposon deletions involve replacement of a deleted restriction fragment with an interposon at the target site and require at least two restriction sites. In either case, the interposon is flanked on both sides by DNA that permits homologous recombination with the recipient genome in order to incorporate the interposon at a predetermined site. Interposons may be used for insertion mutagenesis under conditions where Tns are not appropriate. For example, Tns may exhibit an unacceptable level of insertional specificity or provide transcriptional activity into adjacent genes.

In photosynthetic bacteria interposon mutagenesis was first used by Scolnik and Haselkorn (1984) to study the nitrogenase genes of *R. capsulatus*. Plasmids carrying *R. capsulatus* genes were mutagenized *in vitro* by using interposons containing antibiotic-resistance markers. Such plasmids were introduced into *E. coli* and thence mobilized to a GTA-overproducing mutant of *R. capsulatus*. The GTA was used to deliver the inactivated

---

patible with pRK290 is introduced with simultaneous selection for neomycin resistance (for retention of Tn*5*) and gentamicin resistance (conferred by pPH1JI). All neomycin- and gentamicin-resistant clones were sensitive to tetracycline, indicating loss of pRK290. Nm^r, Tc^r, Gm^r: resistance to neomycin, tetracycline, and gentamicin, respectively; PSA, photosynthetic apparatus.

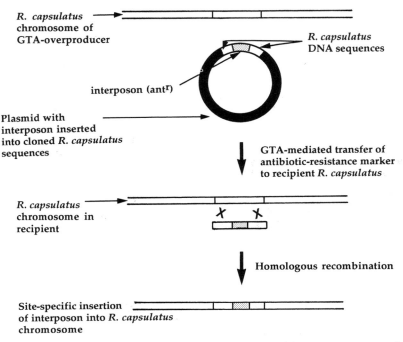

Figure 5. Interposon mutagenesis. *R. capsulatus* DNA is cloned into a vector. An antibiotic resistance (ant^r) interposon is inserted into the cloned DNA so that it is flanked on each side by *R. capsulatus* DNA. The recombinant vector is transferred to an *R. capsulatus* strain that overproduces GTA. The interposon is introduced into an *R. capsulatus* recipient using the GTA. Homologous recombination between the transferred *R. capsulatus* DNA and the resident DNA results in insertion of the interposon into the recipient chromosome. ×, Position of recombination.

marked genes to a suitable *R. capsulatus* recipient. Recombinational replacement of the wild-type DNA by the corresponding mutagenized DNA occurred in the transductants (Fig. 5). Provision of vehicles other than the GTA for introducing the mutagenized genes into the strain of interest has extended the applicability of this method to other photosynthetic bacteria. In *R. sphaeroides* the suicide vector system of Simon *et al.* (1983) (see Section 3.2.2) has been used. Mutagenesis relies on the fact that the vector is not maintained in the species.

Interposon mutagenesis has been used to generate mutations in an array of photosynthetic genes. For example, mutants of *R. capsulatus* and *R. sphaeroides* that carry deletions in genes encoding photosynthetic apparatus structural polypeptides have been constructed and subjected to complementation analysis (Youvan *et al.*, 1985; Burgess *et al.*, 1989). In addition, *R. capsulatus* mutants that specifically lack cytochrome c_2 have been obtained

(Daldal *et al.*, 1986). Such mutants are capable of growing both photo-synthetically and by respiration. Since cytochrome c_2 is the primary electron donor to the reaction center during photosynthesis, the ability of a cytochrome c_2-deficient mutant to carry out photosynthesis indicates that an alternative route must exist for recycling electrons to the reaction center in *R. capsulatus*. Thus, although cytochrome c_2 is required for efficient photosynthetic growth, it does not appear to be essential for that growth. Deletion and insertion mutagenesis of the *puf* operon of both *R. capsulatus* (Bauer *et al.*, 1988) and *R. sphaeroides* (Davis *et al.*, 1988) has also been undertaken. This affords opportunities for a functional analysis of the operon in order to elucidate the mechanisms controlling its expression.

5. GENETIC MAPPING

The genetic techniques that now exist for analyzing photosynthetic prokaryotes have permitted the construction of genetic maps, principally for *Rhodobacter* spp. There is a dearth of such maps for cyanobacteria, due to the limited availability of appropriate genetic systems. Instead, physical maps have been compiled through the use of recombinant DNA technology (for example, Herrero and Wolk, 1986).

5.1. Genes for Photosynthesis

Significant advances have been made in the molecular genetics of photosynthesis in the *Rhodobacter*. Techniques such as conjugation, capsduction, and Tn and interposon mutagenesis along with classical mutagenesis have been employed in the construction of genetic maps for *R. capsulatus* and *R. sphaeroides*. In *R. capsulatus* most of the genes for photosynthesis are found in a 46-kb segment carried on the R-prime plasmid pRPS404 (Taylor *et al.*, 1983; Zsebo and Hearst, 1984). However, some genes for the B800–B850 light-harvesting complex lie outside this segment (Kaufmann *et al.*, 1984). The photosynthesis gene cluster is located between the genes for histidine and nitrogen fixation on the map of Willison *et al.* (1985). In *R. sphaeroides* photosynthesis genes are clustered on the chromosome close to genes involved in aromatic amino acid biosynthesis and a gene for adenine biosynthesis (Pemberton *et al.*, 1983; Bowen and Pemberton, 1985; Pemberton and Harding, 1986). A number of operons encoding genes for reaction center and light-harvesting polypeptides (for example, *puf, puc,* and *puh*) and for cytochromes (for example, *pet*) have been identified. Tight linkage of the genes governing photosynthesis is thus apparent in both *R. capsulatus* and *R. sphaeroides;* however, physical maps indicate certain dissimilarities between these two related species (Taylor *et al.*, 1983; Pemberton and Harding, 1986).

5.2. Genes for Nitrogen Fixation

Much of the information about the molecular genetics of nitrogen fixation in photosynthetic bacteria comes from studies with *R. capsulatus.* Analysis of Nif⁻ mutants of *R. capsulatus* with the GTA established six *nif* linkage groups (Wall and Braddock, 1984; Wall *et al.*, 1984). Unlike the contiguous arrangement found in *Klebsiella pneumoniae*, the *nif* genes of *R. capsulatus* are arranged in dispersed groups on the genome. Verification of this arrangement comes from the use of a combination of techniques, including hybridization (using *K. pneumoniae nifH, D, K* genes as probe), Tn mutagenesis, complementation, marker rescue, and conjugation (Avtges *et al.*, 1983, 1985; Willison *et al.*, 1985). The *R. capsulatus* structural genes *nifH, D,* and *K,* coding for the nitrogenase complex, appear to be organized into an operon, as is the case for *K. pneumoniae,* and are transcribed from *nifH → nifK.* A regulatory gene that may be equivalent to *K. pneumoniae nifA* is linked to this operon, with *nifB* adjacent to it. Furthermore, the *nifA–nifB* region appears to be duplicated in *R. capsulatus* (Klipp *et al.*, 1988).

In cyanobacteria a variable organization of the genes for nitrogenase is exhibited. In the nonheterocystous cyanobacteria, such as *Cyanothece* and *Synechococcus, nifK, D,* and *H* are found in a contiguous cluster (Kallas *et al.*, 1985), while in most heterocystous forms, including *Anabaena* (Mazur *et al.*, 1980) and *Nostoc* (Kallas *et al.*, 1983), *nifD* and *nifH* are adjacent with *nifK* 11 Kb from *nifDH.* Excision of the 11 Kb sequence, which occurs during heterocyst differentiation, leads to the formation of the *nifKDH* operon. However, in *Fischerella* sp., *nifKDH* is in a contiguous array (Saville *et al.*, 1987). A notable feature of the *nif* gene arrangements in photosynthetic prokaryotes is gene duplication. However, the role of such duplications remains to be clarified.

6. CONCLUSION

There is an expanding armory of genetic tools for working with photosynthetic prokaryotes. Efficient transformation systems, available for some unicellular cyanobacteria, permit the direct introduction of cloned genes into these organisms. Such systems, however, remain wanting for filamentous cyanobacteria and for the photosynthetic bacterium *R. capsulatus.* Yet this has not unduly hampered genetic analysis of *R. capsulatus,* where capsduction in conjunction with the recently developed technique of interposon mutagenesis has played a significant part in elevating the genetics of this species to its current status. However, R-plasmid-mediated conjugation is proving the most generally applicable system, providing scope for genetic manipulations across the biological group. A number of powerful techniques, including vector mobilization, Hfr-like genetic mapping, and site-directed mutagenesis, effected through the agency of conjugation, can be applied to the *Rhodobacter* and to some cyanobacteria. Such transfer systems need to be developed for other photosynthetic prokaryotes in order to extend the arena for genetic interplay.

The availability of a range of genetic tools has encouraged the cloning, mapping, and sequencing of many of the cardinal genes for photosynthesis, N_2 fixation, and CO_2 fixation (Hallenbeck, 1987; Scolnik and Marrs, 1987; Houmard and Tandeau de Marsac, 1988).

By implanting specific cloned genes or their mutant alleles into photosynthetic prokaryotes, precise and extensive genomic modifications can now be effected. This paves the way to unraveling the genetic and biochemical intricacies of the processes for which these organisms are renowned.

REFERENCES

Abeliovich, A., and Kaplan, S., 1974, Bacteriophages of *Rhodopseudomonas sphaeroides:* Isolation and characterisation of a *Rhodopseudomonas* bacteriophage, *J. Virol.* **13:**1392–1399.

Adams, D. G., 1988, Isolation and restriction analysis of DNA from heterocysts and vegetative cells of cyanobacteria, *J. Gen. Microbiol.* **134:**2943–2949.

Adolph, K. W., and Haselkorn, R., 1972, Photosynthesis and the development of blue-green algal virus N-1, *Virology* **47:**370–374.

Anderson, L. K., and Eiserling, F. A., 1985, Plasmids of the cyanobacterium *Synechocystis* 6701, *FEMS Microbiol. Lett.* **29:**193–195.

Ashby, M. K., Coomber, S. A., and Hunter, C. N., 1987, Cloning, nucleotide sequence and transfer of genes for the B800-850 light harvesting complex of *Rhodobacter sphaeroides,* *FEBS Lett.* **213:**245–248.

Astier, C., and Espardellier, F., 1976, Mise en évidence d'un système de transfert genétique chez une cyanophycée du genre *Aphanocapsa, C. R. Hebd. Séances Acad. Sci. D* **282:**795–797.

Avtges, P., Scolnik, P., and Haselkorn, R., 1983, Genetic and physical map of the structural genes (*nifH, D, K*) coding for the nitrogenase complex of *Rhodopseudomonas capsulata, J. Bacteriol.* **156:**251–256.

Avtges, P., Kranz, R., and Haselkorn, R., 1985, Isolation and organization of genes for nitrogen fixation in *Rhodopseudomonas capsulata, Mol. Gen. Genet.* **201:**363–369.

Bauer, C. E., Young, D. A., and Marrs, B. L., 1988, Analysis of the *Rhodobacter capsulatus puf* operon, *J. Biol. Chem.* **263:**4820–4827.

Beatty, J. T., and Cohen, S. N., 1983, Hybridization of cloned *Rhodopseudomonas capsulata* photosynthesis genes with DNA from other photosynthetic bacteria, *J. Bacteriol.* **154:**1440–1445.

Bennett, P. M., Grinsted, J., and Foster, T. J., 1988, Detection and use of transposons, in: *Plasmid Technology* (J. Grinsted and P. M. Bennett, eds.), Academic Press, New York, pp. 205–231.

Berg, D. E., and Howe, M. M. (eds.), 1989, *Mobile DNA,* American Society for Microbiology, Washington, D.C.

Biel, A. J., and Marrs, B. L., 1985, Oxygen does not directly regulate carotenoid biosynthesis in *Rhodopseudomonas capsulata, J. Bacteriol.* **162:**1320–1321.

Bowen, A. R. S. G., and Pemberton, J. M., 1985, Mercury resistance transposon Tn *813* mediates chromosome transfer in *Rhodopseudomonas sphaeroides* and intergeneric transfer of pBR322, in: *Plasmids in Bacteria* (D. R. Helinski, S. N. Cohen, D. B. Clewell, D. A. Jackson, and A. Hollaender, eds.), Plenum Press, New York, pp. 105–115.

Burgess, J. G., Ashby, M. K., and Hunter, C. N., 1989, Characterization and complementation of a mutant of *Rhodobacter sphaeroides* with a chromosomal deletion in the light harvesting (LHZ) genes, *J. Gen. Microbiol.* **135**:1809–1816.

Buzby, J. S., Porter, R. D., and Stevens, S. E., Jr., 1983, Plasmid transformation in *Agmenellum quadruplicatum* PR-6: Construction of biphasic plasmids and characterization of their transformation properties, *J. Bacteriol.* **154**:1446–1450.

Buzby, J. S., Porter, R. D., and Stevens, S. E., Jr., 1985, Expression of the *Escherichia coli lacZ* gene on a plasmid vector in a cyanobacterium, *Science* **230**:805–807.

Calvin, N. M., and Hanawalt, P. C., 1988, High-efficiency transformation of bacterial cells by electroporation, *J. Bacteriol.* **170**:2796–2801.

Chauvat, F., Astier, C., Vedel, F., and Joset-Espardellier, F., 1983, Transformation in the cyanobacterium *Synechococcus* R2: Improvement of efficiency: role of the pUH24 plasmid, *Mol. Gen. Genet.* **191**:39–45.

Chauvat, F., De Vries, L., Van der Ende, A., and van Arkel, G. A., 1986, A host-vector system for gene cloning in the cyanobacterium *Synechocystis* PCC 6803, *Mol. Gen. Genet.* **204**:185–191.

Colbeau, A., Godfroy, A., and Vignais, P. M., 1986, Cloning of DNA fragments carrying hydrogenase genes of *Rhodopseudomonas capsulata*, *Biochimie* **68**:147–155.

Daldal, F., Cheng, S., Applebaum, J., Davidson, E., and Prince, R. C., 1986, Cytochrome c_2 is not essential for photosynthetic growth of *Rhodopseudomonas capsulata*, *Proc. Natl. Acad. Sci. USA* **83**:2012–2016.

Daniell, H., and McFadden, B. A., 1986, Characterization of DNA uptake by the cyanobacterium *Anacystis nidulans*, *Mol. Gen. Genet.* **204**:243–248.

Daniels, G. A., Drews, G., and Saier, M. H., Jr., 1988, Properties of a Tn5 insertion mutant defective in the structural gene (*fruA*) of the fructose-specific phosphotransferase system of *Rhodobacter capsulatus* and cloning of the *fru* regulon, *J. Bacteriol.* **170**:1698–1703.

Davis, J., Donohue, T. J., and Kaplan, S., 1988, Construction, characterization, and complementation of a Puf⁻ mutant of *Rhodobacter sphaeroides*, *J. Bacteriol.* **170**:320–329.

De Bont, J. A. M., Scholten, A., and Hansen, T. A., 1981, DNA–DNA hybridization of *Rhodopseudomonas capsulata*, *Rhodopseudomonas sphaeroides* and *Rhodopseudomonas sulfidophila* strains, *Arch. Microbiol.* **128**:271–274.

Delaney, S. F., and Reichelt, B. Y., 1983, Integration of R68.45 into the genome of a cyanobacterium results in genome mobilization, *Heredity* **51**:525–526.

Devilly, C. I., and Houghton, J. A., 1977, A study of genetic transformation in *Gloeocapsa alpicola*, *J. Gen. Microbiol.* **98**:277–280.

De Vos, W. M., Venema, G., Canosi, U., and Trautner, T. A., 1981, Plasmid transformation in *Bacillus subtilis:* Fate of plasmid DNA, *Mol. Gen. Genet.* **181**:424–433.

Ditta, G., Stanfield, S., Corbin, D., and Helinski, D. R., 1980, Broad host range DNA cloning system for Gram-negative bacteria: Construction of a gene bank of *Rhizobium meliloti*, *Proc. Natl. Acad. Sci. USA* **77**:7347–7351.

Duchrow, M., and Giffhorn, F., 1987, Physical map of the *Rhodobacter sphaeroides* bacteriophage φRsG1 genome and location of the prophage on the host chromosome, *J. Bacteriol.* **169**:4410–4414.

Duchrow, M., Heitefuss, S., Kalkus, J., Hoppert, M., and Giffhorn, F., 1988, Isolation and characterization of a virulent phage for *Rhodobacter sphaeroides*, *Arch. Microbiol.* **149**:476–479.

Dzelzkalns, V. A., and Bogorad, L., 1986, Stable transformation of the cyanobacterium *Synechocystis* sp. PCC6803 induced by UV irradiation, *J. Bacteriol.* **165**:964–971.

Elhai, J., and Wolk, C. P., 1988, Conjugal transfer of DNA to cyanobacteria, *Meth. Enzymol.* **167**:747–754.

Flores, E., and Wolk, C. P., 1985, Identification of facultatively heterotrophic N_2 fixing cyanobacteria able to receive plasmid vectors from *Escherichia coli* by conjugation, *J. Bacteriol.* **162**:1339–1341.

Fornari, C. S., and Kaplan, S., 1982, Genetic transformation of *Rhodopseudomonas sphaeroides* by plasmid DNA, *J. Bacteriol.* **152**:89–97.

Fornari, C. S., Watkins, M., and Kaplan, S., 1984, Plasmid distribution and analyses in *Rhodopseudomonas sphaeroides*, *Plasmid* **11**:39–47.

Friedberg, D., and Seijffers, J., 1979, Plasmids in two cyanobacterial strains, *FEBS Lett.* **107**:165–168.

Gendel, S., Straus, N., Pulleyblank, D., and Williams, J., 1983, Shuttle cloning vectors for the cyanobacterium *Anacystis nidulans*, *J. Bacteriol.* **156**:148–154.

Gibson, K. D., and Niederman, R. A., 1970, Characterization of two circular satellite species of deoxyribonucleic acid in *Rhodopseudomonas sphaeroides*, *Arch. Biochem. Biophys.* **141**:694–704.

Golden, S. S., and Sherman, L. A., 1983, A hybrid plasmid is a stable cloning vector for the cyanobacterium *Anacystis nidulans* R2, *J. Bacteriol.* **155**:966–972.

Golden, S. S., and Sherman, L. A., 1984, Optimal conditions for genetic transformation of the cyanobacterium *Anacystis nidulans* R2, *J. Bacteriol.* **158**:36–42.

Golden, J. W., Robinson, S. J., and Haselkorn, R., 1985, Rearrangement of nitrogen fixation genes during heterocyst differentiation in the cyanobacterium *Anabaena*, *Nature* **314**:419–423.

Golden, S. S., Brusslan, J., and Haselkorn, R., 1986, Expression of a family of *psbA* genes encoding a photosystem II polypeptide in the cyanobacterium *Anacystis nidulans* R2, *EMBO J.* **5**:2789–2798.

Golden, J. W., Mulligan, M. E., and Haselkorn, R., 1987a, Different recombination site specificity of two developmentally regulated genome rearrangements, *Nature* **327**:526–529.

Golden, S. S., Brusslan, J., and Haselkorn, R., 1987b, Genetic engineering of the cyanobacterial chromosome, *Meth. Enzymol.* **153**:215–231.

Grigorieva, G. R., and Shestakov, S. V., 1982, Transformation in the cyanobacterium *Synechocystis* sp6803, *FEMS Microbiol. Lett.* **13**:367–370.

Haas, D., and Holloway, B. W., 1978, Chromosome mobilization by the R plasmid R68.45; a tool in *Pseudomonas* genetics, *Mol. Gen. Genet.* **158**:229–237.

Haas, D., and Reimmann, C., 1989, Use of IncP plasmids in chromosomal genetics of Gram-negative bacteria, in: *Promiscuous Plasmids of Gram-Negative Bacteria* (C. M. Thomas, ed.), Academic Press, London, pp. 185–206.

Haas, D., and Riess, G., 1983, Spontaneous deletions of the chromosome-mobilizing plasmid R68.45 in *Pseudomonas aeruginosa* PAO, *Plasmid* **9**:42–52.

Hallenbeck, P. C., 1987, Molecular aspects of nitrogen fixation by photosynthetic prokaryotes, *CRC Crit. Rev. Microbiol.* **14**:1–48.

Haselkorn, R., 1986, Organization of the genes for nitrogen fixation in photosynthetic bacteria and cyanobacteria, *Annu. Rev. Microbiol.* **40**:525–547.

Heitefuss, S., and Giffhorn, F., 1989, Isolation and characterization of two endogenous phages of *Rhodobacter sphaeroides*, *J. Gen. Microbiol.* **135**:911–919.

Herdman, M., 1973, Transformation in the blue-green alga *Anacystis nidulans* and the associated phenomena of mutation, in: *Bacterial Transformation* (L. J. Archer, ed.), Academic Press, London, pp. 369–386.

Herdman, M., Janvier, M., Waterbury, J. B., Rippka, R., Stanier, R. Y., and Mandel, M., 1979a, Deoxyribonucleic acid and base composition of cyanobacteria, *J. Gen. Microbiol.* **111**:63–71.

Herdman, M., Janvier, M., Rippka, R., and Stanier, R. Y., 1979b, Genome size of cyanobacteria, *J. Gen. Microbiol.* **111**:73–85.

Herrero, A., and Wolk, C. P., 1986, Genetic mapping of the chromosome of the cyanobacterium *Anabaena variabilis, J. Biol. Chem.* **261**:7748–7754.

Houmard, J., and Tandeau de Marsac, N., 1988, Cyanobacterial genetic tools: Current status, *Meth. Enzymol.* **167**:808–847.

Hu, N. T., and Marrs, B. L., 1979, Characterization of the plasmid DNAs of *Rhodopseudomonas capsulata, Arch. Mikrobiol.* **121**:61–69.

Hu, N.-T., Thiel, T., Giddings, T. H., Jr., and Wolk, C. P., 1981, New *Anabaena* and *Nostoc* cyanophages from sewage settling ponds, *Virology* **114**:236–246.

Hudig, H., Kaufmann, N., and Drews, G., 1986, Respiratory deficient mutants of *Rhodopseudomonas capsulata, Arch. Microbiol.* **145**:378–385.

Hunter, C. N., 1988, Transposon Tn5 mutagenesis of genes encoding reaction centre and light-harvesting LH1 polypeptides of *Rhodobacter sphaeroides, J. Gen. Microbiol.* **134**:1481–1489.

Hunter, C. N., and Turner, G., 1988, Transfer of genes coding for apoproteins of reaction centre and light-harvesting LH1 complexes to *Rhodobacter sphaeroides, J. Gen. Microbiol.* **134**:1471–1480.

Jasper, P., Hu, N. T., and Marrs, B., 1978, Transfer of plasmid-borne Kanamycin-resistance genes to *Rhodopseudomonas capsulata* by transformation and conjugation, *Abstr. Annu. Mtg. Am. Soc. Microbiol.* **1978**:114.

Kallas, T., Rebière, M.-C., Rippka, R., and Tandeau de Marsac, N., 1983, The structural *nif* genes of the cyanobacteria *Gloeothece* sp. and *Calothrix* sp. share homology with those of *Anabaena* sp., but the *Gloeothece* genes have a different arrangement, *J. Bacteriol.* **155**:427–431.

Kallas, T., Coursin, T., and Rippka, R., 1985, Different organization of *nif* genes in nonheterocystous and heterocystous cyanobacteria, *Plant Mol. Biol.* **5**:321–329.

Kaufmann, N., Hüdig, H., and Drews, G., 1984, Transposon Tn5 mutagenesis of genes for the photosynthetic apparatus in *Rhodopseudomonas capsulata, Mol. Gen. Genet.* **198**:153–158.

Kingsman, A. J., Chater, K. F., and Kingsman, S. M. (eds.), 1988, *Transposition, Soc. Gen. Microbiol. Symp. 43,* Cambridge University Press, Cambridge.

Klipp, W., Masepohl, B., and Pühler, A., 1988, Identification and mapping of nitrogen fixation genes of *Rhodobacter capsulatus:* Duplication of a *nifA–nifB* region, *J. Bacteriol.* **170**:693–699.

Klug, G., and Drews, G., 1984, Construction of a gene bank of *Rhodopseudomonas capsulata* using a broad host range DNA cloning system, *Arch. Microbiol.* **139**:319–325.

Kuhl, S. A., and Yoch, D. C., 1981, Loss of photosynthetic growth of *Rhodospirillum rubrum* associated with loss of a plasmid, *Abstr. 81 Annu. Mtg. Am. Soc. Microbiol.* **1981**(H126):134.

Kuhl, S. A., Nix, D. W., and Yoch, D. C., 1983, Characterization of a *Rhodospirillum rubrum* plasmid: Loss of photosynthetic growth in plasmidless strains, *J. Bacteriol.* **156**:737–742.

Kuhl, S. A., Wimer, L. T., and Yoch, D. C., 1984, Plasmidless photosynthetically-incompetent mutants of *Rhodospirillum rubrum, J. Bacteriol.* **159**:913–918.

Kuhlemeier, C. J., and van Arkel, G. A., 1987, Host-vector systems for gene cloning in cyanobacteria, *Meth. Enzymol.* **153**:199–215.

Kuhlemeier, C. J., Borrias, W. E., van den Hondel, C. A. M. J. J., and van Arkel, G. A., 1981, Vectors for cloning in cyanobacteria: Construction and characterization of two recombinant plasmids capable of transformation in *Escherichia coli* K12 and *Anacystis nidulans* R2, *Mol. Gen. Genet.* **184**:249–254.

Kuhlemeier, C. J., Thomas, A. A. M., van der Ende, A., van Leen, R. W., Borrias, W. E., van den Hondel, C. A. M. J. J., and van Arkel, G. A., 1983, A host-vector system for gene cloning in the cyanobacterium *Anacystis nidulans* R2, *Plasmid* **10**:156–163.

Kuhlemeier, C. J., Logtenberg, T., Stoorvogel, W., van Heugten, H. A. A., Borrias, W. E., and van Arkel, G. A., 1984, Cloning of nitrate reductase genes from the cyanobacterium *Anacystis nidulans, J. Bacteriol.* **159:**36–41.

Labarre, J., Chauvat, F., and Thuriaux, P., 1989, Insertional mutagenesis by random cloning of antibiotic resistance genes into the genome of the cyanobacteria *Synechocystis* strain PCC 6803, *J. Bacteriol.* **171:**3449–3457.

Lambert, G. R., and Carr, N. G., 1984, Resistance of DNA from filamentous and unicellular cyanobacteria to restriction endonuclease cleavage, *Biochim. Biophys. Acta* **781:**45–55.

Lau, R. H., and Doolittle, W. F., 1979, Covalently closed circular DNAs in closely related unicellular bacteria, *J. Bacteriol.* **137:**648–652.

Lau, R. H., Sapienza, C., and Doolittle, W. F., 1980, Cyanobacterial plasmids: Their widespread occurrence and the existence of regions of homology between plasmids in the same and different species, *Mol. Gen. Genet.* **178:**203–211.

Lightfoot, D. A., Walters, D. E., and Wootton, J. C., 1988, Transformation of the cyanobacterium *Synechococcus* PCC 6301 using cloned DNA, *J. Gen. Microbiol.* **134:**1509–1514.

Mandel, M., Leadbetter, E. R., Pfennig, N., and Trüper, H. G., 1971, Deoxyribonucleic acid base compositions of phototrophic bacteria, *Int. J. Syst. Bacteriol.* **21:**222–230.

Marrs, B., 1974, Genetic recombination in *Rhodopseudomonas capsulata, Proc. Natl. Acad. Sci. USA* **71:**971–973.

Marrs, B. L., 1978, Genetics and bacteriophage, in: *The Photosynthetic Bacteria* (R. K. Clayton and W. R. Sistrom, eds.), Plenum Press, New York, pp. 873–883.

Marrs, B., 1981, Mobilization of the genes for photosynthesis from *Rhodopseudomonas capsulata* by a promiscuous plasmid, *J. Bacteriol.* **146:**1003–1012.

Marrs, B. L., 1983, Genetics and molecular biology, in: *Phototrophic Bacteria: Anaerobic Life in the Light* (J. G. Ormerod, ed.), Blackwell, Oxford, pp. 186–214.

Marrs, B., Wall, J. D., and Gest, H., 1977, Emergence of the biochemical genetics and molecular biology of photosynthetic bacteria, *Trends Biochem. Sci.* **2:**105–108.

Mazur, B. J., Rice, D., and Haselkorn, R., 1980, Identification of blue-green algal nitrogen fixation genes by using heterologous DNA hybridization probes, *Proc. Natl. Acad. Sci. USA* **77:**186–190.

Mendzhul, M. I., Nesterova, N. V., Goryushin, V. A., and Lysenko, T. G., 1985, *Tsianofagi-Virusi Tsianobakteri*, Naukova Dumka, Kiev.

Miller, L., and Kaplan, S., 1978, Plasmid transfer and expression in *Rhodopseudomonas sphaeroides, Arch. Biochem. Biophys.* **187:**229–234.

Mitronova, T. N., Shestakov, S. V., and Zhevner, V. D., 1973, Properties of a radiosensitive filamentous mutant of the blue-green alga *Anacystis nidulans, Mikrobiologiya* **42:**519–524.

Mural, R. J., and Friedman, D. I., 1974, Isolation and characterization of a temperate bacteriophage specific for *Rhodopseudomonas sphaeroides, J. Virol.* **14:**1288–1292.

Nano, F. E., and Kaplan, S., 1984, Plasmid rearrangements in the photosynthetic bacterium *Rhodopseudomonas sphaeroides, J. Bacteriol.* **158:**1094–1103.

Olsen, R. H., and Shipley, P., 1973, Host range and properties of the *Pseudomonas aeruginosa* R Factor R1822, *J. Bacteriol.* **113:**772–780.

Orkwiszewski, K. G., and Kaney, A. R., 1974, Genetic transformation of the blue-green bacterium *Anacystis nidulans, Arch. Microbiol.* **98:**31–37.

Padan, E., Ginsburg, D., and Shilo, M., 1970, The reproductive cycle of cyanophage LPP1-G in *Plectonema boryanum* and its dependence on photosynthetic and respiratory systems, *Virology* **40:**514–521.

Padhy, R. N., Hottat, F. G., Coene, M. M., and Hoet, P. P., 1988, Restriction analysis and quantitative estimation of methylated bases of filamentous and unicellular cyanobacterial DNAs, *J. Bacteriol.* **170:**1934–1939.

Pemberton, J. M., and Bowen, A. R. St. G., 1981, High frequency chromosome transfer in

Rhodopseudomonas sphaeroides promoted by broad-host-range plasmid RP1 carrying mercury transposon Tn*501*, *J. Bacteriol.* **147**:110–117.

Pemberton, J. M., and Harding, C. M., 1986, Cloning of carotenoid biosynthesis genes from *Rhodopseudomonas sphaeroides, Curr. Microbiol.* **14**:25–29.

Pemberton, J. M., and Tucker, W. T., 1977, Naturally occurring viral R. plasmid with a circular supercoiled genome in the extracellular state, *Nature* **266**:50–51.

Pemberton, J. M., Cooke, S., and Bowen, A. R. St. G., 1983, Gene transfer mechanisms among members of the genus *Rhodopseudomonas, Ann. Microbiol. (Inst. Pasteur)* **134B**:195–204.

Porter, R. D., 1986, Transformation in cyanobacteria, *CRC Crit. Rev. Microbiol.* **13**:111–132.

Prentki, P., and Krisch, H. M., 1984, *In vitro* insertional mutagenesis with a selectable DNA fragment, *Gene* **29**:303–313.

Quivey, R., Jr., Meyer, R. J., and Tabita, F. R., 1981, Plasmid transfer into *Rhodospirillum rubrum, Abstr. 81 Annu. Mtg. Am. Soc. Microbiol.* **1981**(H113):132.

Rainey, A. M., and Tabita, F. R., 1989, Isolation of plasmid DNA sequences that complement *Rhodobacter sphaeroides* mutants deficient in the capacity for CO_2-dependent growth, *J. Gen. Microbiol.* **135**:1699–1713.

Reimmann, C., Rella, M., and Haas, D., 1988, Integration of replication-defective R68.45-like plasmids into the *Pseudomonas aeruginosa* chromosome, *J. Gen. Microbiol.* **134**:1515–1523.

Rice, D., Mazur, B. J., and Haselkorn, R., 1982, Isolation and physical mapping of nitrogen fixation genes from the cyanobacterium *Anabaena* 7120, *J. Biol. Chem.* **257**:13157–13163.

Roberts, T. M., and Koths, K. E., 1976, The blue-green alga *Agmenellum quadruplicatum* contains covalently closed DNA circles, *Cell* **9**:551–557.

Ruvkun, G. B., and Ausubel, F. M., 1981, A general method for site-directed mutagenesis in prokaryotes, *Nature* **289**:85–88.

Safferman, R. S., and Morris, M. E., 1963, Algal virus: isolation, *Science* **140**:679–680.

Safferman, R. S., Morris, M. E., Sherman, L. A., and Haselkorn, R., 1969, Serological and electron microscopic characterization of a new group of blue-green algal viruses (LPP-2), *Virology* **39**:775–780.

Saunders, V. A., 1978, Genetics of *Rhodospirillaceae, Microbiol. Rev.* **42**:357–384.

Saunders, V. A., 1984, Genetics, metabolic versatility and differentiation in photosynthetic prokaryotes, in: *Aspects of Microbial Metabolism and Ecology* (G. A. Codd, ed.), Academic Press, New York, pp. 241–276.

Saunders, J. R., and Saunders, V. A., 1988, Bacterial transformation with plasmid DNA, in: *Plasmid Technology*, 2nd ed. (J. Grinsted and P. M. Bennett, eds.), Academic Press, New York, pp. 79–128.

Saunders, V. A., Saunders, J. R., and Bennett, P. M., 1976, Extrachromosomal DNA in wild-type and photosynthetically-incompetent strains of *Rhodopseudomonas sphaeroides, J. Bacteriol.* **125**:1180–1187.

Saville, B., Straus, N., and Coleman, J. R., 1987, Contiguous organization of nitrogenase genes in a heterocystous cyanobacterium, *Plant Physiol.* **85**:26–29.

Scolnik, P. A., and Haselkorn, R., 1984, Activation of extra copies of genes coding for nitrogenase in *Rhodopseudomonas capsulata, Nature* **307**:289–292.

Scolnik, P. A., and Marrs, B. L., 1987, Genetic research with photosynthetic bacteria, *Annu. Rev. Microbiol.* **41**:703–726.

Sherman, L. A., and Brown, R. M., Jr., 1978, Cyanophages and viruses of eukaryotic algae, **12**:145–234.

Sherman, L. A., and Connelly, M., 1976, Isolation and and characterization of a cyanophage infecting the unicellular blue-green algae *A. nidulans* and *S. cedrorum, Virology* **72**:540–544.

Shestakov, S. V., and Khyen, N. T., 1970, Evidence for genetic transformation in blue-green alga *Anacystis nidulans, Mol. Gen. Genet.* **107**:372–375.

Simon, R. D., 1978, Survey of extrachromosomal DNA found in filamentous cyanobacteria, *J. Bacteriol.* **136**:414–418.

Simon, R., Priefer, U., and Pühler, A., 1983, A broad host range mobilization system for *in vivo* genetic engineering: Transposon mutagenesis in Gram negative bacteria, *Biotechnology* **1**:784–791.

Sistrom, W. R., 1977, Transfer of chromosomal genes mediated by plasmid R68.45 in *Rhodopseudomonas sphaeroides*, *J. Bacteriol.* **131**:526–532.

Sistrom, W. R., Macaluso, A., and Pledger, R., 1984, Mutants of *Rhodopseudomonas sphaeroides* useful in genetic analysis, *J. Bacteriol.* **138**:161–165.

Stevens, S. E., and Porter, R. D., 1980, Transformation of *Agmenellum quadruplicatum*, *Proc. Natl. Acad. Sci. USA* **77**:6052–6056.

Suwanto, A., and Kaplan, S., 1989, Physical and genetic mapping of the *Rhodobacter sphaeroides* 2.4.1. genome: Presence of two unique circular chromosomes, *J. Bacteriol.* **171**:5850–5859.

Symonds, N., Toussaint, A., van de Putte, P., and Howe, M. M., 1987, *Phage Mu*, Cold Spring Harbor Laboratory, Cold Spring Harbor, New York.

Tandeau de Marsac, N., and Houmard, J., 1987, Advances in cyanobacterial molecular genetics, in: *The Cyanobacteria* (P. Fay and C. Van Baalen, eds.), Elsevier Biomedical Press, Amsterdam, pp. 251–302.

Tandeau de Marsac, N., Borrias, W. E., Kuhlemeier, C. J., Castets, A. M., van Arkel, G. A., and van den Hondel, C. A. M. J. J., 1982, A new approach for molecular cloning in cyanobacteria: Cloning of an *Anacystis nidulans met* gene using a Tn*901*-induced mutant, *Gene* **20**:111–119.

Taylor, D. P., Cohen, S. N., Clark, G. W., and Marrs, B. L., 1983, Alignment of genetic and restriction maps of the photosynthesis region of the *Rhodopseudomonas capsulata* chromosome by a conjugation-mediated marker rescue technique, *J. Bacteriol.* **154**:580–590.

Thiel, T., and Poo, H., 1989, Transformation of a filamentous cyanobacterium by electroporation, *J. Bacteriol.* **171**:5743–5746.

Thiel, T., and Wolk, C. P., 1987, Conjugal transfer of plasmids to cyanobacteria, *Meth. Enzymol.* **153**:232–243.

Thomas, C. M., and Smith, C. A., 1987, Incompatibility group P plasmids: Genetics, evolution and use in genetic manipulation, *Annu. Rev. Microbiol.* **41**:77–101.

Tucker, W. T., and Pemberton, J. M., 1978, Viral R. plasmid RΦ6P: Properties of the penicillinase plasmid prophage and the supercoiled, circular encapsidated genome, *J. Bacteriol.* **135**:207–214.

Tucker, W. T., and Pemberton, J. M., 1979a, Conjugation and chromosome transfer in *Rhodopseudomonas sphaeroides* mediated by W and P group plasmids, *FEMS Microbiol. Lett.* **5**:173–176.

Tucker, W. T., and Pemberton, J. M., 1979b, The introduction of RP4::Mu*cts*62 into *Rhodopseudomonas sphaeroides*, *FEMS Microbiol. Lett.* **5**:215–217.

Tucker, W. T., and Pemberton, J. M., 1980, Transformation of *Rhodopseudomonas sphaeroides* with deoxyribonucleic acid isolated from bacteriophage RΦ6P, *J. Bacteriol.* **143**:43–49.

Van den Hondel, C. A. M. J. J., Keegstra, W., Borrias, W. E., and van Arkel, G. A., 1979, Homology of plasmids in strains of unicellular cyanobacteria, *Plasmid* **2**:323–333.

Van den Hondel, C. A. M. J. J., Verbeek, S., van der Ende, A., Weisbeek, P. J., Borrias, W. E., and van Arkel, G. A., 1980, Introduction of transposon Tn*901* into a plasmid of *Anacystis nidulans:* Preparation for cloning in cyanobacteria, *Proc. Natl. Acad. Sci. USA* **77**:1570–1574.

Walker, G. C., 1984, Mutagenesis and inducible responses to deoxyribonucleic acid damage in *Escherichia coli*, *Microbiol. Rev.* **48**:60–93.

Wall, J. D., and Braddock, K., 1984, Mapping of *Rhodopseudomonas capsulata nif* genes, *J. Bacteriol.* **158**:404–410.

Wall, J. D., Weaver, P. F., and Gest, H., 1975, Gene transfer agents, bacteriophages, and bacteriocins of *Rhodopseudomonas capsulata*, *Arch. Microbiol.* **105**:217–224.

Wall, J. D., Love, J., and Quinn, S. P., 1984, Spontaneous Nif⁻ mutants of *Rhodopseudomonas capsulata*, *J. Bacteriol.* **159**:652–657.

Weisbeek, P. J., Teertstra, R., van Dijk, M., Bloemheuvel, G., de Boer, D., van der Plas, J., Borrias, W. E., and van Arkel, G. A., 1985, Modification and sequence analysis of the small plasmid pUH24 of *Anacystis nidulans* R2, *Abstr. 5 Int. Symp. Photosynthetic Prokaryotes* (Grindelwald, Switzerland) **1985**:328.

Willetts, N. S., Crowther, C., and Holloway, B. W., 1981, The insertion sequence IS*21* of R68.45 and the molecular basis for the mobilization of the bacterial chromosome, *Plasmid* **6**:30–52.

Williams, J. G. K., 1988, Construction of specific mutations in photosystem II photosynthetic reaction centre by genetic engineering methods in *Synechocystis* 6803, *Meth. Enzymol.* **167**:766–778.

Williams, J. G. K., and Szalay, A. A., 1983, Stable integration of foreign DNA into the chromosome of the cyanobacterium *Synechococcus* R2, *Gene* **24**:37–51.

Willison, J., Ahombo, G., Chabert, J., Magnin, J.-P., and Vignais, P., 1985, Genetic mapping of the *Rhodopseudomonas capsulata* chromosome shows non-clustering of genes involved in nitrogen fixation, *J. Gen. Microbiol.* **131**:3001–3015.

Willison, J. C., Magnin, J. P., and Vignais, P. M., 1987, Isolation and characterization of *Rhodobacter capsulatus* strains lacking endogenous plasmids, *Arch. Microbiol.* **147**:134–142.

Wolk, C. P., Vonshak, A., Kehoe, P., and Elhai, J., 1984, Construction of shuttle vectors capable of conjugative transfer from *Escherichia coli* to nitrogen-fixing filamentous cyanobacteria, *Proc. Natl. Acad. Sci. USA* **81**:1561–1565.

Wolk, C. P., Flores, E., Schmetterer, G., Herrero, A., and Elhai, J., 1985, Development of genetics of heterocyst-forming cyanobacteria, in: *Nitrogen Fixation Research* (H. J. Evans, P. J. Bottomley, and W. E. Newton, eds.), Martinus Nijhoff, Dordrecht, pp. 491–496.

Yen, H. C., Hu, N. T., and Marrs, B. L., 1979, Characterization of the gene transfer agent made by an overproducer mutant of *Rhodopseudomonas capsulata*, *J. Mol. Biol.* **131**:157–168.

Youvan, D. C., and Ismail, S., 1985, Light-harvesting II (B800–B850 complex) structural genes from *Rhodopseudomonas capsulata*, *Proc. Natl. Acad. Sci. USA* **82**:58–62.

Youvan, D. C., Elder, J. T., Sandlin, D. E., Zsebo, K., Alder, D. P., Panopoulos, N. J., Marrs, B. L., and Hearst, J. E., 1982, R-prime site-directed transposon Tn7 mutagenesis of the photosynthetic apparatus in *Rhodopseudomonas capsulata*, *J. Mol. Biol.* **162**:17–41.

Youvan, D. C., Ismail, S., and Bylina, E. J., 1985, Chromosomal deletion and plasmid complementation of the photosynthetic reaction centre and light-harvesting genes from *Rhodopseudomonas capsulata*, *Gene* **38**:19–30.

Yu, P.-L., Cullum, J., and Drews, G., 1981, Conjugational transfer systems of *R. capsulata* mediated by R plasmids, *Arch. Microbiol.* **128**:390–393.

Zinchenko, V. V., Babykin, M. M., and Shestakov, S. V., 1984, Mobilization of non-conjugative plasmids into *Rhodopseudomonas sphaeroides*, *J. Gen. Microbiol.* **130**:1587–1590.

Zsebo, K. M., and Hearst, J. E., 1984, Genetic-physical mapping of a photosynthetic gene cluster from *R. capsulata*, *Cell* **37**:937–947.

Zsebo, K. M., Wu, F., and Hearst, J. E., 1984, Tn5.7 construction and physical mapping of pRPS404 containing photosynthetic genes from *Rhodopseudomonas capsulata*, *Plasmid* **11**:182–184.

Genetic Manipulation of Photosynthetic Prokaryotes

5

C. NEIL HUNTER and NICHOLAS H. MANN

1. INTRODUCTION

The phototrophic bacteria have provided fascinating experimental models for the study of a wide range of biochemical topics. Obviously, their photosynthetic capabilities and ancillary phenomena such as the biogenesis of photosynthetic membranes and biosynthesis of tetrapyrroles and carotenoids have attracted the most attention. The metabolic diversity of these bacteria, which between the cyanobacteria and purple bacteria encompass photosynthesis, fermentation, dinitrogen fixation, aerobic and anaerobic respiration, hydrogen uptake, carbon dioxide fixation, and cellular differentiation, provides an array of interesting problems for physiologists and biochemists.

There is no area of interest of interest mentioned above which would not, or does not, benefit from the application of molecular genetic techniques. The isolation of the genes encoding important structural proteins such as reaction center or light-harvesting apoproteins can provide sequence information and a means to acquire further insights into structure–function relationships through site-directed mutagenesis—"protein engineering." Furthermore, our understanding of membrane biogenesis in these bacteria would obviously benefit from a knowledge of the sequence, arrangement, and regulation of genes encoding the biosynthesis of chlorophyll pigments, carotenoids, and phospholipids. In addition, the formation by many species of cyanobacteria of metabolically and morphologically

C. NEIL HUNTER ● Krebs Institute for Biomolecular Research, Department of Molecular Biology and Biotechnology, University of Sheffield, Sheffield S10 2TN, United Kingdom. NICHOLAS H. MANN ● Department of Biological Sciences, University of Warwick, Coventry CV4 7AL, United Kingdom. Dr. Hunter's contribution to this chapter was written in August 1989. A number of new developments in the field of genetic manipulation of photosynthetic prokaryotes have occurred since that time.

Photosynthetic Prokaryotes, edited by Nicholas H. Mann and Noel G. Carr. Plenum Press, New York, 1992.

differentiated cells responsible for dinitrogen fixation provides a fascinating system in which to study microbial differentiation.

In the past several years there has been a rapidly increasing interest in the molecular genetics of phototrophic bacteria, fuelled by the need of biochemists, biophysicists, and physiologists to acquire a new depth of understanding in their particular area of study. Moreover, as more and more tools become available, the increasing ease with which these bacteria can be manipulated genetically is attracting molecular biologists, who find that phototrophic bacteria may provide new insights into prokaryotic gene organization and expression.

It is true that many of the techniques currently in use have been borrowed or adapted from those used with other groups of bacteria. However, both the purple bacteria and the cyanobacteria have their unique characteristics which have influenced the ways in which these techniques have been implemented and have dictated which approaches have been most successful. Consequently, although this chapter attempts to deal with the genetic manipulation of these two groups of bacteria in a similar fashion, some differences in emphasis or organization are necessary to reflect each group's specialized features.

2. PROPAGATION OF RECOMBINANT DNA IN PURPLE BACTERIA

2.1. Broad-Host-Range Vectors

Some of the first attempts to construct vectors for purple bacteria exploited the fact that plasmids of incompatibility groups P-1 and Q in particular are capable of replication in a wide variety of Gram-negative bacteria (Franklin, 1985; Thomas and Smith, 1987). These groups are represented by plasmids RP1, RP4, and RK2 (P-1) and RSF1010 (Q); for the purposes of consideration as cloning vectors, plasmids of the former group are identical. Group P-1 plasmids are self-transmissible, whereas RSF1010 and derivatives are only mobilizable when transfer functions are provided *in trans*. In neither case is the need for a small plasmid with multiple antibiotic-resistance markers and a series of unique cloning sites satisfied; RP4 and RK2 are too large [50–60 kilobases (kb)] and RSF1010 derivatives possess relatively few unique restriction endonuclease sites with only one useful resistance marker, streptomycin. Thus, many attempts have been made to make smaller derivatives of RP4 and RK2 and to broaden the scope of RSF1010; much of this work was carried out by researchers primarily interested in *Pseudomonas* and *Rhizobium* species (Franklin, 1985). However, Miller and Kaplan (1978) reported efficient transfer of RP1 from *Escherichia coli* to *Rhodobacter sphaeroides* and subsequently it was demonstrated that RSF1010 could be mobilized into *Rb. sphaeroides* (Zinchenko *et al.*, 1984). An RSF1010 derivative, pU181, was introduced into *Rb. sphaeroides* by transformation (Fornari and Kaplan, 1982) (see also Section 2.4). More recently, it was demonstrated that the RSF1010 derivative pNH2, which contains a tetracycline-resistance gene in place of that for sulfonamide resistance, can be used to establish a bank of *Rb. sphaeroides* DNA fragments in *E. coli*, and that selected clones from such a bank are able to complement lesions in, for example, the biosynthesis of bacteriochlorophyll or the synthesis of light-harvesting or reaction center apoproteins (Coomber *et al.*, 1987; Ashby *et al.*, 1987; Hunter and Turner, 1988, Hunter and Coomber, 1988). Despite this, a new vector, pRK414, possesses the best range of capabilities as a vector (Keen *et al.*, 1988). It is derived from pRK404 (Ditta *et al.*, 1985), which is in turn derived from pRK290 (Ditta *et al.*, 1980). The parental plasmid is RK2. Since pRK414 is 10.5 kb in size, most of pRK2 has been "lost"; however, pRK414 has a polylinker cloning site, broad host range, and blue/white colony identification on IPTG/X-Gal agar, and is mobilizable. The use of pRK414 as a vector for purple bacteria has already been reported (Lee *et al.*, 1989). It is likely that RSF1010 and RK2 derivatives will be useful as vectors in a wide variety of purple bacteria, as shown by the recent studies of Gardiner *et al.* (1989) on *Rhodopseudomonas acidophila*, *Rhodopseudomonas palustris*, and *Rhodospirillum rubrum*.

Cosmid cloning has the potential to reduce a gene bank to a more manageable number of clones, by virtue of its large capacity for inserted DNA; fragments up to 40 kb can be cloned in this way. A general description of cosmid cloning is not appropriate here, but it should be noted that several cosmids have been used successfully to establish mobilizable banks of clones that can be used to complement mutations in genes of purple bacteria. Some of these cosmids, such as pLAFR1 and pLA2917, are related to RK2, mentioned earlier, and have been used to clone and mobilize genes from *Rb. sphaeroides* encoding ribulose bisphosphate carboxylase oxygenase (Weaver and Tabita, 1985) and components of the flagellar motor (Sockett *et al.*, 1990). It should also be mentioned that Pemberton and Harding (1986) were able to construct a cosmid bank in pHC79 (Hohn and Collins, 1980) which was mobilized triparentally in to *Rb. sphaeroides* mutants with lesions in carotenoid biosynthesis.

2.2. Nonreplicating Vectors

It would appear at first that such vectors would not be attractive for cloning and transferring DNA from *E. coli* to purple bacteria, since there is no possibility of maintenance of these plasmids outside the relatively narrow confines of the enteric bacteria. However, the elegant work of Simon *et al.* (1983) has greatly increased the usefulness of these vectors, which are based on the pBR322 family of plasmids. This was achieved in two ways. First, the recognition site for mobilization (*mob* site) was cloned from plasmid RP4 (see Section 5.1.2a) into pBR325, for example, to yield pSUP202. Second, RP4 along with its transfer functions was immobilized in the genome of *E. coli* to yield strain Sm10. More details on mobilization of plasmids into purple bacteria will be provided in Section 2.3. pSUP202 contains the single cloning sites *Pst*I, *Eco*R1, *Cla*I, *Hind*III, *Sal*I, and *Bam*HI and is 8.2 kb in size. It can be mobilized from strain Sm10 with extremely high efficiency, approaching 100% (Simon *et al.*, 1983). When genes encoding reaction center apoproteins of *Rb. sphaeroides* were cloned into pSUP202 they restored a reaction center-less mutant NF3 to photosynthetic growth with a frequency of 1.4×10^{-4} per recipient; this apparent lack of efficiency masks the high frequency of gene transfer and instead reflects the frequency with which the newly introduced *Rb. sphaeroides* DNA achieves its effect by recombination at a frequency of around 10^{-3} to 10^{-4} per recipient, since pSUP202 cannot maintain the cloned insert independently of the genome (Hunter and Turner, 1988). In subsequent work, a bank of *Rb. sphaeroides* DNA in *E. coli* was transferred to a series of reaction center-less or bacteriochlorophyll-less mutants of *Rb. sphaeroides*. Those clones that restored the mutants to photosynthetic growth were isolated and characterized by restriction endonuclease mapping and were found to overlap; in doing so, they form a cluster, 45 kb in length, which encompasses many of the genes essential for photosynthesis in this bacterium (Coomber and Hunter, 1989).

The main value of pSUP202 and its transposon Tn5 derivative, pSUP2021, however, lies in their use as vectors for localized and random transposon TN5 mutagenesis. These topics will be covered in Sections 3.1. and 3.2.

Finally, some studies by Taylor *et al.* (1983) on the mobilization of genes from *E. coli* to *Rhodobacter capsulatus* are included in this section. They were able to construct two pBR322 derivatives, pDPT42 and pDPT44, which could be mobilized triparentally so that when cloned inserts of DNA bearing genes for bacteriochlorophyll (Bchl) synthesis were inserted, Bchl⁻ strains of *Rb. capsulatus* were restored to photosynthetic growth. However, they noted some plasmid instability in their transconjugant strains together with a variety of recombination and complementation events. The fact that any stability at all was observed differs from the observation of pSUP202 instability in *Rb. sphaeroides*, although the plasmids constructed by Taylor *et al.* (1983) have not been tested in this bacterium.

2.3. Mobilizable Vectors

This section is included to bring together some topics already mentioned above. The vectors referred to are all mobilizable from *E. coli* strains

Sm10 and S17-1 in which the transfer functions of RP4 are immobilized in the genome (Simon *et al.*, 1983). Likewise, RP4 can supply this function as a plasmid, in a triparental system. However, in this case cointegrate structures can form as a result of the transposition of Tn*3*, which contributes the ampicillin-resistance gene to RP4; although these cointegrates may resolve into two plasmids after transfer to the recipient has taken place, RP4 will persist in the transconjugant strain because of its broad host range. This would be a severe problem for experiments involving transposon Tn*5* mutagenesis (see Section 3.1), since Tn*5* confers resistance to kanamycin on the recipient; this marker is also present on RP4. In fact, there are derivatives of RK2 that lack this marker (Meyer *et al.*, 1979), but another way around this problem would be to use a plasmid that mobilizes a vector triparentally, but is unable to replicate in the recipient. Such a plasmid, pRK2013, was constructed by Figurski and Helinski (1979); as a further refinement, the Kan^r gene was inactivated by insertion of transposon Tn*7*, which in turn confers resistance to streptomycin, spectinomycin, and trimethoprim (Leong *et al.*, 1982).

It should be mentioned that although *E. coli* strains SM10 and S17-1 seem to be perfect choices for cloning mobilizing genes, their use should probably be limited to the latter function, since they are relatively difficult to transform, and may not readily maintain stable cloned inserts in plasmids. S17-1 contains RP4::Mu immobilized in the genome, with transposon Tn*7* inserted to inactivate the Kan^r gene of RP4; this combination of elements appears to be unfavorable for the consistent introduction and maintenance of cloned DNA.

2.4. Shuttle Vectors

It has been known for some time that some purple bacteria possess one or more plasmids (Gibson and Niederman, 1970; Saunders *et al.*, 1976; Nano and Kaplan, 1984; Kuhl *et al.*, 1984, Willison *et al.*, 1987). This should open the way for the construction of shuttle cloning vectors, but this area has remained mostly unexplored. However, there is a report of plasmids from marine *Rhodopseudomonas* species being used to construct shuttle vectors (Matsunaga *et al.*, 1986).

2.5. Transduction

There are no generally useful methods of transferring DNA into purple bacteria by transducing phages. However, mention must be made of a gene transfer agent (GTA) apparently specific for *Rb. capsulatus*, which packages fragments of DNA and transfers them into cells of this organism (Yen *et al.*, 1979). This has been termed "capsduction" (Scolnik and Marrs, 1987). Fragments of the donor genome, no greater than 4.6 kb in length, are introduced into a recipient and then integrate into the chromosome by homologous recombination. It was the existence of this unique method of genetic exchange that started the exploration of the molecular genetics of purple bacteria, and which enabled Marrs and co-workers to produce detailed genetic maps of photopigment genes using GTA-mediated exchange (Yen and Marrs, 1976; Scolnik *et al.*, 1980; Taylor *et al.*, 1983). A further application of the GTA particle in interposon mutagenesis will be discussed in Section 3.3.

3. NONCLASSICAL MUTAGENESIS IN PURPLE BACTERIA

3.1. Random Transposon Mutagenesis

This is an extremely powerful tool for the isolation of genes, especially those which give rise to a readily detectable phenotype when disrupted by the insertion of a transposon. This applies to prokaryotic and eukaryotic systems. The most useful tool for purple bacteria was developed by Simon *et al.* (1983), in which the transposon Tn5 was introduced into their mobilizable "suicide" vector pSUP202. One particular insertion was termed pSUP2021. Upon transfer to hosts which do not permit the replication of this plasmid, for example, the genera *Rhizobium*, *Agrobacterium*, or *Rhodobacter*, Tn5 transposes at random into the bacterial genome since it cannot be maintained extrachromosomally. Transposon Tn5 encodes resistance to kanamycin and streptomycin in nonenteric genera; thus, isolation of Tn5 together with flanking DNA is relatively straightforward, and should result in cloning of the disrupted gene. There are several excellent reviews on this topic (Berg and Berg, 1983, Berg *et al.*, 1983; de Bruijn and Lupski, 1984; Noti *et al.*, 1987; De Bruijn, 1987). This technique has been used successfully in *Rhodobacter capsulatus* (Kaufmann *et al.*, 1984) and *Rhodobacter sphaeroides* (Hunter, 1988); in each case photosynthetically defective mutants were isolated and flanking regions characterized. The range of applications is being extended to include genes for motility (Sockett *et al.*, 1990), nitrogen fixation (Klipp *et al.*, 1988). C-4 dicarboxylate transport (Kelly *et al.*, 1990), and the fructose phosphotransferase system (Daniels *et al.*, 1988).

3.2. Localized Insertion Mutagenesis Using Transposons

This is a powerful tool for obtaining genetic and physical maps of genes clustered on a cloned stretch of the bacterial chromosome. The insert, cloned in vector pSUP202 (Simon *et al.*, 1983), is introduced into a strain of *E. coli* bearing Tn5 in the chromosome. "Random" transposition of Tn5 onto the pSUP202 recombinant occurs with a relatively low frequency of 10^{-3} to 10^{-4}, but is easily detected and isolated since the Tn5-bearing plasmids confer resistance to kanamycin on any sensitive strain of *E. coli*. Upon transfer to *Rb. sphaeroides*, for example (and assuming there is no insertion disrupting the mobilization function within pSUP202), double homologous recombination between the cloned DNA bearing Tn5 and the *Rb. sphaeroides* chromosome will result in disruption of the chromosome at a point corresponding to the position of Tn5 in the cloned insert. Thus, an array of transposition events along the cloned insert can each be defined physically by restriction mapping and genetically by homologous recombination in the bacterium. This apparent ease of analysis is hampered by two considerations. First, transposition by Tn5 is not completely random; the tetracycline-resistance gene of pSUP202 appears to be a "hot spot" for Tn5 insertion. Discussion of this point appears in Berg *et al.* (1983) and Noti *et al.* (1987). Second, insertion into the cloned DNA is not truly random either. For example, Hunter (1988) examined 85 insertions into a 7.2-kb region of cloned *Rb. sphaeroides* DNA which included the *puf* operon for reaction center and light-harvesting subunits. No insertions were found in a 2-kb region of this DNA at the 5' end of the operon. This reflects, in part, the fact that insertions into the end regions of a cloned insert are less likely to be detected phenotypically, since a double homologous recombination event may be required to produce a mutation, and this double event requires a reasonable length of DNA flanking the transposon, for example, 2–4 kb. The point is addressed in Noti *et al.* (1987). It is also the case that Tn5 will show a preference for certain sequences and a "dislike" for others—probably those most interesting to the investigator. Another point is that if only a single homologous recombination event occurs, a cointegrate structure, which included the pSUP202 vector, forms. This can certainly produce an insertion mutation, but in order to produce a "clean" transposon map these single recombination events, which can be easily screened for using a hybridization assay, should be eliminated from the analysis. Hunter (1988) investigated this point quantitatively and demonstrated that double homologous recombination was around 10% of the total (single plus double). This technique has been used to map genes for pigment synthesis and for reaction center and light-harvesting apoproteins in *Rb. sphaeroides* (Coomber *et al.*, 1990).

Other transposons have also been used in a similar manner, and for similar purposes. These include Tn7 (Youvan *et al.*, 1982), which was used to produce a series of mutations in genes affecting photosynthetic function in *Rb. capsulatus*. No map was produced from this study, and Tn7 has not been widely used since, possible because it displays a markedly nonrandom insertion. In order to counteract this, Zsebo *et al.* (1984) constructed a plasmid Tn5.7 which used the inverted repeat ends of Tn5 and the streptomycin-resistance genes of Tn7. However, Tn5 confers streptomycin resistance upon a nonenteric host in any case. In a subsequent publication, they reported the isolation and analysis of 45 insertions into the photo-synthetic gene cluster of *Rb. capsulatus,* which was born on the R′ plasmid pRPS404 (Zsebo and Hearst, 1984). This extensive work confirmed the location of many genes for the biosynthesis of carotenoids, bac-teriochlorophyll, and proteins of the photosynthetic apparatus, and identi-fied several more.

There are situations that require the insertion of an element at a specific point. Localized Tn5 mutagenesis cannot accomplish this. How-ever, there are several widely used techniques for site-directed insertion mutagenesis, which are dealt with below.

3.3. Site-Directed Insertion and Deletion/Insertion Mutagenesis

These techniques are extremely valuable, both from an analytical point of view and, in the latter case, for constructing a genetic background which is a prerequisite for undertaking a program of protein engineering.

There is a variety of insertion elements ("interposons") available for mutagenesis. In all cases, the procedure requires that the element be intro-duced by ligation into a specific site in the DNA under analysis; this DNA is then mobilized into the wild-type purple bacterium, where double homolo-gous recombination directs it to the appropriate point in the genome. In this respect, there is no difference between this technique and localized transposon Tn5 mutagenesis (Section 3.2.). The vector that transfers the interposon-mutated DNA can be either pSUP202 (Simon *et al.,* 1983) or additionally, in the case of *Rb. capsulatus,* the GTA particle. Interposons generally carry an antibiotic-resistance marker and can be constructed from the wide variety of transposons available, starting with removal of transposition function. A selection of some interposons appears in Youvan *et al.* (1985); there is a concentration on those markers that function well in purple bacteria, such as streptomycin and kanamycin. Prentki and Krisch (1984) published the construction of an interposon, Ω, which encodes streptomycin and spectinomycin markers and is flanked by useful re-striction sites and transcription/translation terminators. As an elegant re-finement, it is possible to excise the antibiotic markers for subsequent use, leaving behind the still powerfully disruptive terminators. Interposon mu-tagenesis has been used for analysis of genes, mainly in *Rb. capsulatus* (Scolnik and Haselkorn, 1984; Guiliano *et al.,* 1988).

It is also possible to use deletions as recipients for interposons, so that the gene under analysis is replaced, rather than merely disrupted. This is properly termed deletion/insertion, rather than just deletion mutagenesis, and has been used by Davis *et al.* (1988) to examine the effects of replacing *pufQ* with a kanamycin-resistance gene to investigate the control and func-tion of the *puf* operon encoding reaction center and light-harvesting LH1 polypeptides of *Rb. sphaeroides.* Other examples include the effect of such mutations on *puhA* encoding the reaction center H subunit (Sockett *et al.*, 1989) and on the *puc* operon encoding the LH2 polypeptides of *Rb. sphaeroides* (Lee *et al.*, 1989).

Finally, the replacement of a given gene with an antibiotic-resistant "cassette" provides an ideal genetic background for the study of site-directed mutations on the protein encoded by that gene. This is because only the mutated coding sequence is present and expressed in the deletion/insertion host strain, giving the investigator some confidence that the new phenotype arises from the engineered change. Youvan *et al.* (1985) constructed deletion/insertion strains of *Rb. capsulatus* for the purpose of protein engineering in reaction center and light-harvesting complexes. Subsequently, other laboratories have adopted this approach in *Rb. sphaeroides* (Farchaus and Oesterhelt, 1989; Burgess *et al.*, 1989a) (see Chapter 4).

3.4. Site-Directed Point Mutations

It is worthwhile to allot this topic to a separate heading, since it is likely to be one of the most exciting and productive areas of research in purple bacteria. The techniques used to construct the mutations are to some extent routine, and the genetic background has been mentioned in Section 3.3. However, Youvan *et al.* (1985) found that further modification of the *puf* operon was necessary in order to facilitate the removal, mutation, and reconstruction of a particular gene. Thus, they embarked on a series of mutagenesis steps to eliminate and create restriction sites flanking the genes of interest. Subsequently, Youvan and co-workers have created an array of mutations designed to probe the structure and function of the bacterial reaction center (Bylina and Youvan, 1988; Kirmaier *et al.*, 1988). This research has been given its impetus largely because of the availability of a high-resolution structure for the reaction center, albeit for *Rhodopseudomonas viridis* and *Rb. sphaeroides* rather than *Rb. capsulatus* (Deisenhofer *et al.*, 1985; Allen *et al.*, 1987).

It should be emphasized that with respect to the reaction center, the purple bacteria have furnished a structure with valuable implications for membrane protein biochemistry in general. It is not surprising, therefore, that there is a great deal of interest in the construction of deletion/insertion strains and site directed changes in *Rps. viridis* and *Rb. sphaeroides*.

A variety of vector/deletion systems are possible for the reaction center (*puf*) operon. There are choices to be made as to whether it is best to replace the single gene under study with an antibiotic-resistance cassette or to replace the entire operon. This in turn has implications for the restriction sites that need to be engineered. Also, the vector that delivers mutagenized genes from *E. coli* to the deletion/insertion strain of the purple bacterium has to be considered. It may appear attractive at first to use a broad-host-range system that is maintained independently of the chromosome, rather than a "suicide" system that necessarily recombines with the genome, since this should increase the production of altered complexes through a copy number effect. It remains to be seen whether or not this poses regulatory problems for the cell which may arise from disruption of unforeseen operon structures such as the *bch–puf* "superoperon" (Young *et al.*, 1989). These may require *cis* complementation, rather than it being *in trans* (Burgess *et al.*, 1989b).

4. ANALYSIS OF GENE EXPRESSION IN PURPLE BACTERIA

4.1. Promoter Probe Vectors

There are situations where it would be desirable to examine the effect of putative promoters not in the normal environment, but fused instead to a promoterless reporter gene encoding an easily assayable protein. There are many such vectors for *E. coli* and *Pseudomonas* and *Rhizobium* species, for example. Many of these rely on the use of *lacZY* constructions so that the color assay for β-galactosidase can be exploited. Modifications for use in purple bacteria require some means of introducing the vector into the bacterium, and possibly maintaining it independently of the genome. Nano *et al.* (1985) reported the use of a broad-host-range system which conferred oxygen-dependent β-galactosidase activity when introduced into *Rb. sphaeroides*. These vectors are hampered by the fact that the assays cannot be conducted *in vivo*, since there is no transport mechanism for lactose and analogues in these bacteria. However, assays can be performed on cell extracts. Bauer *et al.* (1988) have used fusions of the oxygen-regulated *puf* promoter region and *lacZY* to analyze the effects of deletions on promoter activity. Several other attractive systems beckon, including the use of promoter–*neo* fusions and promoter–*lux* fusions.

4.2. Expression Vectors

These vectors can drive the expression of genes above and beyond their normal levels. This is an important area, since one of the appeals of recombinant DNA technology is that it offers the prospect of greatly enhanced levels of protein production through a combination of promoter strength and plasmid copy number. Many attempts are being made to enhance the expression of purple bacterial genes in *E. coli* and no attempts will be made here to review the vast array of options for overexpression in that bacterium. It should be mentioned, however, that expression of *Rb. sphaeroides* genes in *E. coli* has not been notably successful so far. The other possibility is to overexpress a gene in the parent purple bacterium. A family of vectors was constructed with the strong oxygen-regulated *puf* promoter in a mobilizable, broad-host-range system based upon pRK404 (Johnson *et al.*, 1986). Another pair of vectors, pNF2 and pNF3,utilize the *nifHDK* promoter of *Rb. capsulatus* cloned into a pBR322-based plasmid. This appears to function well in *Rb. capsulatus* and allows the manipulation of protein levels by variation of the type of nitrogen source in the medium. This is a promising area of research and carries with it the possibility that genes for the biosynthesis of important structures such as chlorophylls and carotenoids could be produced in amounts large enough to initiate purification studies.

4.3. Analysis of Transcription and Translation

These analyses are an essential part of a thorough investigation of gene structure and function. Accordingly, Chory and Kaplan (1982) developed an *in vitro* transcription-translation system based on an "S-30" cell-free extract of *Rb. sphaeroides*. This has been used to investigate the synthesis of reaction center and light-harvesting polypeptides directed by *puf* and *puc* genes, respectively (Hoger *et al.*, 1986; Kiley *et al.*, 1987; Kiley and Kaplan, 1987).

Another development was a report of the purification of RNA polymerase by Forrest and Beatty, (1987). This area of research in purple bacteria is likely to grow, since the participation of environmentally regulated factors in photosynthetic gene expression has been proposed, although not yet demonstrated.

5. PROPAGATION OF RECOMBINANT DNA IN CYANOBACTERIA

5.1. Introduction of Recombinant DNA into Cyanobacteria

Any discussion of genetic manipulation in the cyanobacteria is complicated by the fact that for a long time there has been no formally agreed upon and authorized classification. Consequently, in the literature a variety of epithets have frequently been applied to the same strain. To avoid confusion, the Pasteur Culture Collection classification system (Rippka *et al.*, 1979) is used, where possible, throughout, regardless of the name used in the original literature, and where appropriate the alternative name is appended in parenthesis. A table of alternative strain designations has been compiled by Houmard and Tandeau de Marsac (1988).

Three potential routes exist for the introduction of recombinant DNA molecules into bacteria—transformation, conjugation, and transduction—of which only transformation and conjugation are applicable so far to the cyanobacteria. Plasmid transformation has been reported for a number of strains belonging to the genera of unicellular cyanobacteria, *Synechococcus* and *Synechocystis*. In the majority of strains the development of competence is a natural phenomenon, though in the case of one strain of *Synechocystis*, competence for transformation of chromosomal marker was induced by cold shock and calcium chloride treatment (Devilly and Houghton, 1977). One important feature of the transformation of cyanobacteria with plasmids encoding drug-resistance selectable markers is the remarkably long phenotypic lag before selection can be applied. The development of transformation systems and selection procedures have been reviewed by Porter (1986).

In the case of the filamentous species of cyanobacteria conjugation has proved to be the most effective route for the introduction of recombinant DNA. A conjugation system for the introduction of plasmid vectors into *Anabaena* species has been developed (Wolk *et al.*, 1984) and this approach has been successfully applied to a range of other filamentous cyanobacterium. Recently, it has been shown that the filamentous cyanobacterium *Anabaena* sp. strain M131 could be transformed with shuttle vector pRL6 by electroporation (Thiel and Poo, 1989) and it seems likely that this approach may be widely applicable within the cyanobacteria.

One of the problems facing genetic manipulation in the cyanobacteria is the widespread distribution of restriction endonucleases (Houmard and Tandeau de Marsac, 1988). This may limit the efficiency with which recombinant molecules propagated in *E. coli* or other hosts can be introduced into cyanobacteria. Another consequence is that DNA isolated from cyanobacteria is frequently resistant to cutting with a variety of restriction enzymes (van den Hondel *et al.*, 1983; Lambert and Carr, 1984; Herrero *et al.*, 1984). This resistance to restriction could result from either an absence of restriction sites or the presence of modified bases. It has been demonstrated for species from three genera, *Anabaena*, *Plectonema*, and *Synechococcus*, that their DNA contains a high proportion of N^6-methyladenine and 5-methylcytosine and that *dam*-like and *dcm*-like methylases are responsible as well as the site-specific methylase counterparts of the restriction systems (Padhy *et al.*, 1988).

5.2. Shuttle Vectors

The majority of cyanobacterial strains so far examined contain one or more plasmids ranging in size from 1.3 to greater than 1000 kb (Houmard and Tandeau de Marsac, 1988). However, none of the endogenous cyanobacterial plasmids so far characterized confer convenient selectable markers. *E. coli* plasmid vectors such as pBR322 and broad-host-range plasmids like RSF1010 are not capable of autonomous replication in cyanobacteria (van den Hondel *et al.*, 1980; Kuhlemeier and van Arkel, 1987), although a recent report (Kreps *et al.*, 1990), describes the conjugative transfer and autonomous replication of a promiscuous IncQ plasmid in the *Synechocystis* sp. PCC6803. Consequently, the two routes taken to developing plasmid vectors capable of being introduced into cyanobacteria by transformation have been the construction of integrative vectors (see Section 2.4) and shuttle vectors. The number of cyanobacterial species into which DNA can be introduced by transformation is still somewhat limited. *Synechococcus* sp. PCC 7942 (*Anacystic nidulans* R2) and *Synechococcus* sp. PCC 7002 (*Agmenellum quadruplicatum* PR-6) are transformable with plasmid DNA at comparatively high efficiencies. Transformation of a range of other species with chromosomal DNA has been reported (Shestakov and Reaston, 1987). Plasmids capable of replication in both *E. coli* and a cyanobacterial host have been developed for a number of strains of both *Synechococcus* and *Synechocystis;* however, substantive vector development has largely been confined to *Synechococcus* sp. PCC 7942 and PCC 7002 and *Synechocystis* sp. PCC6803. It is worth noting that *Synechococcus* sp. PCC 7942 and *Synechocystis* sp. PCC 6803 are capable of photoheterotrophy on glycerol and glucose, respectively. A comprehensive survey of plasmid vectors for use in the cyanobacteria has been compiled by Houmard and Tandeau de Marsac (1988).

The principles involved in the construction of these vectors and features associated with their use can be illustrated by considering some of the vectors developed from the small endogenous cryptic plasmid pUH24 of *Synechococcus* sp. PCC 7942 (*Anacystis nidulans* R2). pUC1, which is a derivative of pUH24 produced by insertion of Tn*901* and deletion of one of the Tn*901* inverted repeats (van den Hondel *et al.*, 1980), was ligated into the *E. coli* vector pACYC184 to generate pUC104 (Kuhlemeier *et al.*, 1981). Although pUC104 could confer both ampicillin and chloramphenicol resistance on its cyanobacterial host, ampicillin resistance is a poor selective marker. pUC104 was also used to produce a cosmid shuttle vector by ligating in the Lambda *cos* site (Tandeau de Marsac *et al.*, 1982). A surprising feature of the transformation efficiency of these shuttle vectors is that it is unaffected by plasmid size (Kuhlemeier *et al.*, 1984a,b) and that, in the absence of endogenous pUH24, it is more efficient with monomeric plasmids (Chauvat *et al.*, 1983). In addition to shuttle vectors derived from pAYC184, a variety of pBR-based vectors have been developed (Sherman and van de Putte, 1982; Golden and Sherman, 1983; Gendel *et al.*, 1983), including vectors with a polylinker site to increase their versatility (Gendel *et al.*, 1983a,b). A series of vectors have also been developed from the large endogenous plasmid pUH25, of *Synechococcus* sp. PCC 7942 (Laudenbach *et al.*, 1985), which, since pUH24 and pUH25 are compatible, would permit the simultaneous maintenance of two fragments of cloned DNA on separate replicons within the same host cell.

A variety of problems are associated with the use of shuttle vectors. First, there is the wide distribution of type II restriction endonucleases in cyanobacteria. *Synechococcus* sp. PCC 7942, though at one time thought to lack a restriction system, does in fact produce *Ani*I (Gallagher and Burke, 1985), which recognizes sequences containing the *E. coli dam* methylase site (Weisbeek *et al.*, 1985). Consequently, the problem of restriction is largely overcome by carrying out manipulations in *dam*$^+$ strains of *E. coli*. Second, there are the related problems of recombination between incoming recombinant vector and both the resident plasmid pUH24 and the chromosome. To overcome the problem of interplasmidic recombination, Kuhlemeier *et al.* (1983) produced a strain (R2-SPc) cured of pUH24. However, the problem still exists, when cloning homologous DNA sequences, of recombination with the chromosome, and the development of a recombination-deficient host strain is still awaited. In this context, the complete nucleotide sequence of the *rec*A gene of *Synechococcus* sp. PCC7002 has recently been obtained and *rec*A mutants isolated. (Murphy *et al.*, 1990).

5.3. Integrative Vectors

When homologous DNA sequences cloned into either *E. coli* plasmids such as pBR322 or into shuttle vectors capable of autonomous replication within the cyanobacterial host are transformed into cyanobacteria, a variety of recombinational events can occur. Where a single crossover occurs between the homologous sequences on the incoming circular plasmid and the chromosome, then integration of the complete plasmid into the chromosome will occur. This latter process has been extensively used for the stable propagation of cloned genes in cyanobacteria when they cannot be stably maintained in an autonomously replicating vector and a range of integrative vectors have been developed specifically for this purpose. In addition, this process of recombinational integration has been utilized for mutagenesis (see Section 6.2) and for promoter analysis using reporter genes (see Section 7.1).

Williams and Szalay (1983) constructed a set of plasmids composed of three distinct elements; a fragment of *Synechococcus* sp. PCC 7942 chromosomal DNA, a fragment from pAYC184 encoding chloramphenicol resistance interrupting the chromosomal fragment, and plasmid pBR322, which flanked the chromosomal DNA fragment. When pKW1061 was used to transform *Synechococcus* sp. PCC 7942 three distinct types of transformant were obtained. Type I transformants, the most abundant, contained the Cm^r gene inserted into the host chromosome. Type II transformants contained the Ap^r gene inserted into the host chromosome, and in type III transformants both the Cm^r and Ap^r genes had inserted. They proposed that type I transformants arose from nonreciprocal recombination (gene conversion) and type II transformants from a similar process of nonreciprocal recombination in concert with a single crossover. Type III transformants were thought to arise by a single reciprocal recombination event.

Further characterization of this integrative system (Kolowsky *et al.* 1984; Kolowsky and Szalay 1986) showed that the efficiency with which type I transformants were produced and their subsequent stability was determined by their position within the segment of *Synechococcus* DNA in the transforming vector. Furthermore, heterologous DNA of up to 20 kb was completely stable in type I transformants. Scanlan (1988) has established that for *Synechococcus* sp. PCC 7942 a minimum size of 0.5 kb for the homologous DNA is required to obtain reasonable recombination efficiency.

A variety of integrative vectors have been constructed (Elanskaya *et al.*, 1985; Shestakov *et al.*, 1985a,b) and in principle it appears possible to introduce a variety of heterologous DNA sequences stably into different sites on the same host chromosome by using an appropriate combination of such vectors.

A novel approach to the integration of cloned genes into the host chromosome based on "integration platforms" has been developed (van der Plas *et al.*, 1990). The integration platform consists of an incomplete *bla* gene and the pBR322 origin of replication separated by a gene encoding either streptomycin or kanamycin resistance inserted into the chromosome of *Synechococcus* sp. PCC 7942. When such a strain is transformed with any pBR-derived plasmid, recombination between the incoming plasmid and the integration platform leads to the restoration of a complete *bla* gene loss of the Km^r or Sm^r gene and its replacement by the insert of the donor plasmid. The presence of the pBR322 origin of replication allows for "plasmid rescue" of the integrated sequences.

5.4. Mobilizable Vectors

The first evidence that plasmid DNA could be conjugally transferred into cyanobacteria was provided by Delaney and Reichelt (1982), who observed a low-frequency transfer of the IncP plasmid R68.45 from *E. coli* to the unicellular cyanobacterium *Synechococcus* sp. PCC 6301. However, R68.45 was not maintained as an autonomous plasmid in its cyanobacterial host, but rather had become integrated into the chromosome. Wolk *et al.* (1984) made use of the fact that the IncP plasmid RP4 is able to mobilize the nonconjugative plasmid ColE1 and related plasmids provided they contain a *bom* site and also encode, via the *mob* gene, a DNA-nicking activity specific for this site. A series of shuttle vectors were constructed based on the pDU1 plasmid of *Nostoc* sp. PCC 7524 and the *E. coli* vector pBR322. It should be pointed out that pBR322 is derived from pMB1, a plasmid closely related to ColE1 and containing a *bom* site. The shuttle vectors were engineered to eliminate *Ava*I and *Ava*II sites. Conjugal transfer of these plasmids to several *Anabaena* strains including PCC 7120 and PCC 7118 was effected by a triparental mating. Two *E. coli* strains were used, one containing the shuttle vector and also a "helper" plasmid which would provide the essential *mob* function *in trans,* the other containing RP4. These *E. coli* strains were incubated together with the cyanobacterial recipient prior to plating out on selective media; highly efficient transfer of the shuttle vector was observed. Subsequently this approach was shown to also be applicable to a range of facultatively heterotrophic, nitrogen-fixing *Nostoc* strains (Flores and Wolk, 1985) and a cosmid vector has been constructed (Wolk *et al.,* 1988). The problem of restriction of incoming DNA during transfer to *Anabaena* species expressing either or both the *Ava*I and *Ava*II restriction systems has been overcome by Elhai and Wolk (1988), who developed a technique for the methylation of the cloned DNA in *E. coli* using the *Ava*I and *Ava*II methylases which largely overcomes this problem. A detailed methodology for the application of this conjugative transfer technique has been described (Elhai and Wolk, 1988). In addition, a biparental mating system based on an *E. coli* strain (S17-1) which has a derivative of RP4 integrated into the chromosome has been described (McFarlane *et al.,* 1987).

6. NONCLASSICAL MUTAGENESIS IN CYANOBACTERIA

6.1. Random Transposon Mutagenesis

Random transposon mutagenesis has not been widely utilized with the cyanobacteria. The TEM β-lactamase transposon Tn*901,* located on the *E. coli* plasmid pRI46, was used to introduce a selectable marker into the pUH24 plasmid of *Synechococcus* PCC 7942 to generate plasmid pCH1 (van den Hondel *et al.,* 1980). Subsequently, Tandeau de Marsac *et al.* (1982) utilized transposition of Tn*901* carried by plasmid pCH1 to mutagenize *Synechococcus* PCC 7942 and isolated a mutant defective in methionine biosynthesis. The same technique was employed by Kuhlemeier *et al.* (1984a) to isolate a nitrate reductase mutant of *Synechococcus* PCC 7942 and by Madueno *et al.* (1988) to generate a variety of mutants of *Synechococcus* PCC 7942 affected in a range of nitrate assimilation genes. Friedberg and Folkman (1991) have reported the use of bacteriophage Mu derivatives to mutagenize *Synechocystis* sp. PCC 7942.

Random transposon mutagenesis of two shuttle vectors for *Synechococcus* sp. PCC 7942 was carried out while the vectors were being propagated in an *E. coli* host carrying Tn5 in the chromosome (Gendel, 1987). On subsequent transfer to *Synechococcus* sp. PCC 7942 it was found that there was a high rate of transposon excision from inserts within a 4-kb region of the hybrid plasmid that was associated with replication within the cyanobacterial host. More recently, Borthakur and Haselkorn (1989) mutagenized *Anabaena* sp. PCC 7120 with Tn5 carried on plasmid pBR322*bla*::Tn5 and isolated a mutant unable to grow without combined nitrogen. In this case it was observed that methylation of the plasmid in *E. coli* with the *Ava* methylases prior to conjugative transfer increased the efficiency of Tn5 transfer; Tn5 has four *Ava*I restriction sites. Analysis of neomycin-resistant isolates by Southern blotting revealed that no pBR322 sequences could be detected, while a Tn5-specific probe hybridized to a different DNA fragment in each mutant tested, indicating that neomycin-resistant clones arose from random transposition and not from integration of pBR322*bla*::Tn5 into the chromosome.

6.2. Recombinational Mutagenesis

The possibility of recombination between homologous DNA sequences introduced by transformation and the chromosome, which was exploited (see Section 5.3) to insert cloned genes into the chromosome, can also be utilized to mutagenize the chromosome. A technique of random recombinational mutagenesis (ectopic mutagenesis) was developed for *Synechococcus* sp. PCC 7002 (Buzby *et al.*, 1985). The process involved the ligation of random fragments of total *Synechococcus* DNA produced by partial digestion with *Sau*3A to the Tn*1* Ap^r gene fragment derived from plasmid pDS1106 by digestion with *Bam*H1 and *Pvu*II. The products of the ligation, linear molecules incapable of autonomous replication, were transformed into *Synechococcus* sp. PCC 7002 and selection was made for ampicillin resistance. Among the recombinants were a variety of mutants as judged by their altered colony morphology or pigmentation. A similar process was developed for *Synechocystis* sp. PCC 6803 (Labarre *et al.*, 1989) in which a chloramphenicol- or a kanamycin-resistance marker was randomly ligated to chromosomal *Hae*II or *Ava*II fragments. Analysis of the drug-resistant transformants obtained suggested that the resistance markers had been incorporated in the genome by a process of gene conversion between two *Hae*II or *Ava*II sites with a precise deletion of the chromosomal DNA between these two sites.

In mutagenesis procedures of this type the recombinational events following transformation will be determined by whether the transforming DNA is linear or circular (see Section 5.3), but even in the case of circular DNA the gene replacement event is likely to be more common than a single crossover leading to integration of the whole plasmid and duplication of the target gene.

6.3. Localized and Site-Specific Mutagenesis

Once specific cyanobacterial genes, or groups of genes, have been cloned they may be subjected to localized mutagenesis and the mutated gene may then be used to replace its chromosomal counterpart by transformation back into the cyanobacterial host and subsequent recombination. In principle the gene may be mutagenized either by the insertion of a drug-resistance cassette into a specific site within the gene or random transposon insertions may be obtained. Golden *et al.* (1986) cloned the *psbA* I, II, and III genes of *Synechococcus* sp. PCC 7942 and in the case of the *psbA*II gene a region between two *Bst*EII was removed and replaced by a cassette encoding spectinomycin resistance. The plasmid was then transformed back into *Synechococcus* sp. PCC 7942 and selection made for spectinomycin-resistant transformants. Southern blot analysis confirmed the disruption of the *psbA*II at its native chromosomal site. In contrast, the *psbA*I gene was mutagenized by random Tn*5* insertion prior to its return to the cyanobacterial host. In principle this approach is equally applicable to genes which have undergone site-directed mutagenesis.

7. ANALYSIS OF GENE EXPRESSION IN CYANOBACTERIA

7.1. Promoter Probe Vectors

Promoter vectors and reporter genes have been widely used in a variety of microorganisms for the purposes of analyzing the regulation of transcription of particular genes and also the isolation of genes with a particular pattern of transcriptional regulation. To date three reporter genes have been successfully employed with cyanobacteria, namely *lacZ* (encoding β-galactosidase), *cat* (encoding chloramphenicol acetyltransferase), and *lux* (encoding luciferase), and have been deployed in a variety of fashions to analyze the regulation of gene expression in cyanobacteria. Two approaches may be taken to the construction of recombinants containing a reporter gene downstream from a cyanobacterial regulatory sequence. Either the reporter gene is inserted into the host cell chromosome or it is inserted into an autonomously replicating vector. In the latter case gene dosage effects are likely to make expression of the reporter gene more easily detectable. Schmetterer *et al.* (1986) demonstrated that expression of the *lux* genes from both *Vibrio harveyi and Vibrio fischeri* could be achieved in *Anabaena* sp. strain M-131 and that expression could be enhances by driving it from the strong promoter for ribulose 1,5-bisphosphate carboxylase/oxygenase from *Anabaena* sp. PCC 7120. Expression of β-galactosidase in *Synechococcus* sp. PCC 7002 has been described (Buzby *et al.*, 1985) and applied to the assessment of the effect of light intensity and nitrogen availability on expression of a *cpc* (phycocyanin) gene promoter (Gasparich *et al.*, 1987). Scanlan *et al.* (1990) utilized the *lacZ* gene to make transcriptional fusions with cyanobacterial promoters using both shuttle vectors and integration vectors and were able to identify CO_2-regulated promoters. A plasmid vector containing a multiple cloning site followed by a promoterless *cat* gene, protected by transcription terminators and mobilizable by conjugation in *Anabaena* sp. PCC 7120, has been constructed by Lang and Haselkorn (1991).

The choice of reporter gene is also important for the ease with which its expression can be detected. Qualitative expression of the *cat* gene may be easily detected by selecting for chloramphenicol resistance; however, quantitative studies are more difficult and require a spectrophotometric assay on cell-free extracts. In the case of the *lux* genes colonies emitting light can be visually identified and quantitative studies can be carried out by scintillation counting. The use of *lacZ* as a reporter gene in cyanobacteria in combination with the chromogenic substrate X-gal is complicated by the fact that the pigmentation of the colonies obscures that of the indicator. However, the use of the substrate methylumbelliferyl galactoside (Youngman, 1987), which is hydrolyzed to yield a highly fluorescent methylumbelliferone, overcomes this problem and may be used with cyanobacterial colonies or quantitatively with cell-free extracts (Scanlan, *et al.*, 1990). A detailed description of the use of these reporter genes is provided by Friedberg (1988).

7.2. Expression Vectors

No plasmid vectors specifically designed to maximize expression of a cloned gene within a cyanobacterial host and based on strong cyanobacterial promoters have so far been constructed. However, Friedberg and Seijffers (1986) have constructed shuttle plasmids in which a chloramphenicol acetyltransferase *(cat)* gene was linked to a phage Lambda operator-promoter under the control of the temperature-sensitive cI1857 repressor and demonstrated that expression of the *cat* gene could be controlled by temperature shifts.

Recently Price and Badger (1989) achieved expression of a human carbonic anhydrase gene (HCAII) in *Synechococcus* sp. PCC 7942. Plasmid pHCAII contains a *tac* promoter and a consensus *E. coli* ribosome-binding site upstream of the HCAII gene. The plasmid also contains the *lacIQ* gene in order to achieve repression of the *tac* promoter in the absence of the inducer IPTG. A *BamH*1 fragment from the *E.coli/Synechococcus* shuttle vector pSK6B was inserted into the *Bgl*II site of pCHAII to enable its replication and selection in *Synechococcus*. The *BamH*1 fragment of pSK6B carried a chloramphenicol-resistance gene transcribed from the *psbA* chloroplast promoter and also the origin of replication from the small 7.9-kb plasmid of *Synechococcus* sp. PCC 7942. The degree of expression of the human carbonic anhydrase gene in *Synechococcus* was calculated to be 0.3% of total protein. Chungjatupornchai (1990) has reported the expression of the mosquitocidal protein genes of *Bacillus thuringiensis* subsp. *israelensis* and the herbicide-resistance gene *bar* in *Synechocystis* PCC 6803.

7.3. Analysis of Transcription and Translation

Only limited progress has so far been made in the analysis of the molecular nature of the key signals in transcription and translation in cyanobacteria. The limited number of cyanobacterial promoters so far characterized (Tandeau de Marsac and Houmard, 1987) does not permit the deduction of consensus sequences and so far only limited homology with the *E. coli* −10 region has been detected in the case of some promoters. With regard to translation, sequences resembling Shine–Dalgarno sites have been observed approximately 10 bases upstream from the translational start codon in the case of some genes, while other exhibit no typical ribosome-binding site (see Tandeau de Marsac and Houmard, 1987). Wolk *et al.* (1991) used a derivative of Tn*5* to generate transcriptional fusions of promoterless bacterial luciferase genes in the genome of *Anabaena* sp. PCC 7120 to identify environmentally responsive genes.

8. PROSPECTS FOR THE FUTURE

It was not so long ago that the prospects for a thorough and wide-ranging genetic analysis of the photosynthetic prokaryotes seemed rather limited. This is no longer the case and there are few obstacles to hinder rapids progress in this area, which should contribute enormously to our understanding, particularly of photosynthesis, membrane biogenesis, and microbial differentiation. Of particular interest could be the analysis of the regulatory mechanisms by which photosynthetic organisms adjust their transcriptional and translational activity to respond to alterations in the quantity and quality of illumination. Among the hoped for developments in the near future would be efficient transformation and/or conjugation systems for all the commonly studied species, and electroporation may well provide the route. In addition, high-efficiency expression systems are desirable.

Recent developments, particularly in the area of genome mapping and the construction of overlapping cosmid libraries for some species of photosynthetic prokaryotes, ensure that these organisms remain at the forefront of microbiological and biochemical research, a position justified by their intrinsic interest.

REFERENCES

Allen, J. P., Feher, G., Yeates, T. O., Komiya, H., and Rees, D. C., 1987, Structure of the reaction centre from *Rhodobacter sphaeroides* R-26: The protein subunits, *Proc. Natl. Acad. Sci. USA* **84**:6162–6166.
Ashby, M. K., Coomber, S. A., and Hunter, C. N., 1987, Cloning, nucleotide sequence and

transfer genes for the B800–850 light harvesting complex of *Rhodobacter sphaeroides*, *FEBS Lett.* **213**:245–248.

Bauer, C. E., Young, D. A., and Marrs, B. L., 1988, Analysis of the *Rhodobacter capsulatus puf* operon, *J. Biol. Chem.* **263**:4820–4827.

Berg, D. E., and Berg, C. M., 1983, The prokaryotic transposable element Tn5, *Biotechnology* **1**:417–435.

Berg, D. E., Schmandt, M. A., and Lowe, J. B., 1983, Specificity of transposon Tn5 insertion, *Genetics* **105**:813–828.

Borthakur, D., and Haselkorn, R., 1989, Tn5 mutagenesis of *Anabaena* sp. strain PCC 7120: Isolation of a new mutant unable to grow without combined nitrogen, *J. Bacteriol.* **171**:5759–5761.

Burgess, J. G., Ashby, M. K., and Hunter, C. N., 1989a, Characterization and complementation of a mutant of *Rhodobacter sphaeroides* with a chromosomal deletion in the light harvesting (LH2) genes, *J. Gen. Microbiol.* **135**:1809–1816.

Burgess, J. G., Olsen, J. D., Gibson, L., and Hunter, C. N., 1989b, Relationship between ORFS R and Q and bacteriochlorophyll biosynthesis, in: *Abstracts from Symposium on Molecular Biology of Membrane-Bound Complexes in Phototrophic Bacteria*, Frieburg, Germany, p. 64.

Buzby, J. S., Porter, R. D., and Stevens, S. E., Jr., 1985, Expression of the *Escherichia coli lacZ* gene on a plasmid vector in a cyanobacterium, *Science* **230**:805–807.

Bylina, E. J., and Youvan, D. C., 1988, Directed properties of the primary donor in the photosynthetic reaction center, *Proc. Natl. Acad. Sci. USA* **85**:7226–7230.

Chauvat, F., Astier, C., Vedel, F., and Joset-Espardellier, F.,1983, Transformation in the cyanobacterium *Synechococcus* R2: Improvement of efficiency; role of the pUH24 plasmid, *Mol. Gen. Genet.* **191**:39–45.

Chory, J., and Kaplan, S., 1982, The *in vitro* transcription-translation of DNA and RNA templates by extracts of *Rhodopseudomonas sphaeroides*, *Meth. Enzymol.* **23A**:696–705.

Chungjatupornchai, W., 1990, Expression of the mosquitocidal-protein genes of *Bacillus thuringiensis* subsp. *israelensis* and the herbicide-resistance gene *bar* in *Synechocystis* PCC 6803, *Curr. Microbiol.* **21**:283–288.

Coomber, S. A., and Hunter, C. N., 1989, Construction of a physical map of the 45kb photosynthetic gene cluster of *Rhodobacter sphaeroides*, *Arch. Microbiol.* **151**:454–458.

Coomber, S. A., Ashby, M. K., and Hunter, C. N., 1987, Cloning and oxygen regulated expression of genes for the bacteriochlorophyll biosynthetic pathway in *Rhodopseudomonas sphaeroides*, in: *Progress in Photosynthesis Research*, Volume 4 (J. Biggins, ed.) Martinus Nijhoff, Dordrecht, pp. 737–740.

Coomber, S. A., Chaudri, M., Connor, A., Britton, G., and Hunter, C. N., 1990, Localised transposon Tn5 mutagenesis of the photosynthetic gene cluster of *Rhodobacter sphaeroides*, *Molec. Microbiol.* **4**:977–989.

Daniels, G. A., Drews, G., and Saier, M. H., Jr., 1988, Properties of a Tn5 insertion mutant defective in the structural gene (*fru*A) of the fructose-specific phosphotransferase system of *Rohodobacter capsulatus* and cloning of the *fru* regulon, *J. Bacteriol.* **170**:1698–1703.

Davis, J., Donohue, T. J., and Kaplan, S., 1988, Construction, characterization and complementation of a Puf⁻ mutant of *Rhodobacter sphaeroides*, *J. Bacteriol.* **170**:320–329.

De Bruijn, F. J., 1987, Transposon Tn5 mutagenesis to map genes, *Meth. Enzymol.* **154**:175–196.

De Bruijn, F. J., and Lupski, J. R., 1984, The use of transposon Tn5 mutagenesis in the rapid generation of correlated physical and genetic maps of DNA segments cloned into multicopy plasmids, *Gene* **27**:131–149.

Deisenhofer, J., Epp, O., Miki, K.,Huber, R., and Michel, H., 1985, Structure of the protein subunits in the photosynthetic reaction centre of *Rhodopseudomonas viridis* at 3Å resolution, *Nature* **318**:618–624.

Delaney, S. F., and Reichelt, B. Y., 1982, Integration of the R plasmid, R68.45, into the genome of *Synechococcus* PCC6301, *Abstr. Int. Symp. Photosynthetic Prokaryotes* **4**:D5.

Devilly, C. I., and Houghton, J. A., 1977, A study of genetic transformation in *Gloeocapsa alpicola, J. Gen. Microbiol.* **98:**277–280.

Ditta, G., Stanfield, S., Corbin, D., and Helinski, D., 1980, Broad host range cloning system for Gram-negative bacteria: construction of a gene bank of *Rhizobium meliloti, Proc. Natl. Acad. Sci. USA* **77:**7347–7351.

Ditta, G., Schmidhauser, T., Yakobsen, E., Lu, P., Liang, K. W., Finlay, D. R., Guiney, D., and Helinski, D. R., 1985, Plasmids related to the broad host range vector, pRK290, useful for gene cloning and for monitoring gene expression, *Plasmid* **13:**149–153.

Elanskaya, I. V., Morzunova, I. B., and Shestakov, S. V., 1985, Recombinant plasmids capable of integration with the chromosome of cyanobacterium *Anacystis nidulans* R2, *Mol. Genet. Mikrobiol. Virusol.* **11:**20–24.

Elhai, J., and Wolk, C. P., 1988, Conjugal transfer of DNA to cyanobacteria, in *Meth. Enzymol.* **167:**747–754.

Farchaus, J. W., and Oesterhelt, D., 1989, A *Rhodobacter sphaeroides puf* L, M and X deletion mutant and its complementation in *trans* with a 5.3 kb *puf* operon shuttle fragment, *EMBO J.* **8:**47–54.

Figurski, D., and Helinski, D. R., 1979, Replication of an origin-containing derivative of plasmid RK2 dependent on a plasmid function provided *in trans, Proc. Natl. Acad. Sci. USA* **76:**1648–1652.

Flores, E., and Wolk, C. P., 1985, Identification of facultatively heterotrophic, N_2-fixing cyanobacteria able to receive plasmid vectors from *Escherichia coli* by conjugation, *J. Bacteriol.* **162:**1339–1341.

Fornari, C. S., and Kaplan, S., 1982, Genetic transformation of *Rhodopseudomonas sphaeroides* by plasmid DNA, *J. Bacteriol.* **152:**89–97.

Forrest, M. E., and Beatty, J. T., 1987, Purification of *Rhodobacter capsulatus* RNA polymerase and its use for *in vitro* transcription, *FEBS Lett.* **212:**28–34.

Franklin, F. C. H., 1985, Broad host range cloning vectors for Gram negative bacteria, in: *DNA Cloning; A practical Approach,* Volume 1 (D. M. Glover, ed.), IRL Press, Oxford, pp. 165–188.

Friedberg, D., 1988, Use of reporter genes in cyanobacteria, *Meth. Enzymol.* **167:**736–747.

Friedberg, D., and Seijffers, J., 1986, Controlled gene expression utilizing Lambda phage regulatory signals in a cyanobacterium host, *Mol. Gen. Genet.* **203:**505–510.

Friedberg, D., and Folkman, N., 1991, Mu phage as a genetic tool in cyanobacteria, in: *Abstracts of the VII International Symposium on Photosynthetic Prokaryotes,* Amherst, Massachusetts, p. 117.

Gallagher, M. L., and Burke, W. F., 1985, Sequence-specific endonuclease from the transformable cyanobacterium *Anacystis nidulans* R2, FEMS Microbiol. Lett. **26:**317–321.

Gardiner, A. T., Gibson, L., Hunter, C. N., and Cogdell, R. J., 1989, Mobilization of cloning vectors in *Rhodospirillaceae,* in: *Abstracts from the Symposium on Molecular Biology of Membrane-Bound Complexes in Phototrophic Bacteria,* Frieburg, Germany, p. 9.

Gasparich, G., Buzby, J. S., Bryant, D. A., Porter, R. D., and Stevens, S. E., Jr., 1987, The effects of light intensity and nitrogen starvation on the phycocyanin promoter in the cyanobacterium *Synechococcus* PCC7002, in: *Progress in Photosynthetic Research,* Volume IV (J. Biggins, ed.), Nijhoff, Dordrecht, pp. 761–765.

Gendel, S. M., 1987, Instability of Tn5 inserts in cyanobacterial cloning vectors, *J. Bacteriol.* **169:**4426–4430.

Gendel, S., Strauss, N., Pulleyblank, D., and Williams, J., 1983a, A novel shuttle cloning vector for the cyanobacterium *Anacystis nidulans, FEMS Microbiol. Lett.* **19:**291–294.

Gendel, S., Strauss, N., Pulleyblank, D., and Williams, J., 1983b, Shuttle cloning vectors for the cyanobacterium *Anacystis nidulans, J. Bacteriol.* **156:**148–154.

Gibson, K. D., and Niederman, R. A., 1970, Characterisation of two circular satellite species of deoxyribonucleic acid in *Rhodopseudomonas sphaeroides, Arch. Biochem. Biophys.* **143:**471–484.

Golden, S. S., and Sherman, L. A., 1983, A hybrid plasmid is a stable cloning vector for the cyanobacterium *Anacystis nidulans* R2, *J. Bacteriol.* **155:**966–972.

Golden, S. S., Brusslan, J., and Haselkorn, R., 1986, Expression of a family of *psbA* genes encoding a photosystem II polypeptide in the cyanobacterium *Anacystis nidulans* R2, *EMBO.* **5:**2789–2798.

Guiliano, G., Pollock, D., Stapp, H., and Scolnik, P. A., 1988, A genetic-physical map of the *Rhodobacter capsulatus* carotenoid biosynthesis gene cluster, *Mol. Gen. Genet.* **213:**78–83.

Herrero, A., Elhai, J., Hohn, B., and Wolk, C. P., 1984, Infrequent cleavage of cloned *Anabaena variabilis* DNA by restriction endonucleases from *A. variabilis, J. Bacteriol.* **160:**781–784.

Hoger, J., Chory, J. C., and Kaplan, S., 1986, *In vitro* biosynthesis and membrane association of photosynthetic reaction center subunits from *Rhodopseudomonas sphaeroides, J. Bacteriol.* **165:**942–950.

Hohn, B., and Collins, J., 1980, A small cosmid for efficient cloning of large DNA fragments, *Gene* **11:**291–298.

Houmard, J., and Tandeau de Marsac, N., 1988, Cyanobacterial genetic tools: Current status, *Meth. Enzymol.* **167:**808–847.

Hunter, C. N., 1988, Transposon Tn5 mutagenesis of genes encoding reaction centre and light-harvesting LH1 polypeptides of *Rhodobacter sphaeroides, J. Gen. Microbiol.* **134:**1481–1489.

Hunter, C. N., and Coomber, S. A., 1988, Cloning and oxygen-regulated expression of the bacteriochlorophyll genes *bch* E, B, A, and C of *Rhodobacter sphaeroides, J. Gen. Microbiol.* **134:**1471–1480.

Hunter, C. N., and Turner, G., 1988, Transfer of genes coding for apoproteins of reaction centre and light-harvesting LH1 complexes to *Rhodobacter sphaeroides, J. Gen. Microbiol.* **134:**1471–1480.

Johnson, J. A., Wong, W. K. R., and Beatty, J. T., 1986, Expression of cellulase genes in *Rhodobacter capsulatus* by use of plasmid expression vectors, *J. Bacteriol.* **167:**604–610.

Kaufmann, N., Hudig, H., and Drews, G., 1984, Transposon Tn5 mutagenesis of genes for the photosynthetic apparatus in *Rhodopseudomonas capsulata, Mol. Gen. Genet.* **198:**153–158.

Keen, N. T., Tamaki, S., Kobayashi, D., and Trollinger, D., 1988, Improved broad-host-range plasmids for DNA cloning in Gram-negative bacteria, *Gene* **70:**191–197.

Kelly, D. J., Hamblin, M. J., and Shaw, J. G., 1990, Physiology and genetics of C4-dicarboxylate transport in *Rhodobacter capsulatus*, in: *Molecular Biology of Membrane-Bound Complexes in Phototrophic bacteria* (G. Drews, ed.), Plenum Press, New York, pp. 453–462.

Kiley, P. J., and Kaplan, S., 1987, Cloning, DNA sequence, and expression of the *Rhodobacter sphaeroides* light-harvesting B800–850 alpha and B800–850 beta genes, *J. Bacteriol.* **169:**3268–3275.

Kiley, P. J., Donohue, T. J., Havelka, W. A., and Kaplan, S., 1987, DNA sequence and *in vitro* expression of the B875 light-harvesting polypeptides of *Rhodobacter sphaeroides, J. Bacteriol.* **169:**742–750.

Kirmaier, C., Holten, D., Bylina, E. J., and Youvan, D. C., 1988, Electron transfer in a genetically modified bacterial reaction center containing a heterodimer, *Proc. Natl. Acad. Sci. USA* **85:**7562–7566.

Klipp, W., Masepohl, B., and Puhler, A., 1988, Identification and mapping of nitrogen fixation genes of *Rhodobacter capsulatus:* Duplication of a *nifA–nifB* region, *J. Bacteriol.* **170:**693–699.

Kolowsky, K. S., and Szalay, A. A., 1986, Double-stranded gap repair in the photosynthetic prokaryote *Synechococcus* R2, *Proc. Natl. Acad. Sci. USA* **83:**5578–5582.

Kolowsky, K. S., Williams, J. G. K., and Szalay, A. A., 1984, Length of foreign DNA in chimeric plasmids determines the efficiency of its integration into the chromosome of the cyanobacterium *Synechococcus* R2, *Gene* **27**:289–299.

Kreps, S., Ferino, F., Mosrin, C., Gerits, J., Mergeay, M., and Thuriaux, P., 1990, Conjugative transfer and autonomous replication of a promiscuous IncQ plasmid in the cyanobacterium *Synechocystis* PCC 6803, *Mol. Gen. Genet.* **221**:129–133.

Kuhl, S. A., Wimer, L. T., and Yoch, D. C., 1984, Plasmidless, photosynthetically incompetent mutants of *Rhodospirillum rubrum*, *J. Bacteriol.* **159**:913–918.

Kuhlemeier, C. J., and van Arkel, G. A., 1987, Host-vector systems for gene cloning in cyanobacteria, *Meth. Enzymol.* **153**:199–215.

Kuhlemeier, C. J., Borrias, W. E., van den Hondel, C. A. M. J. J., and van Arkel, G. A., 1981, Construction and characterization of two hybrid plasmids capable of transformation to *Anacystis nidulans* R2 and *Escherichia coli* K12, *Mol. Gen. Genet.* **184**:249–254.

Kuhlemeier, C. J., Thomas, A. A. M., van der Ende, A., van Leen, R. W., Borrias, W. E., van den Hondel, C. A. M. J. J., and van Arkel, G. A., 1983, A host-vector system for gene cloning in the cyanobacterium *Anacystis nidulans* R2, *Plasmid* **10**:156–163.

Kuhlemeier, C. J., Logtenberg, T., Stoorvogel, W., van Heugten, H. H. A., Borrias, W. E., and van Arkel, G. A., 1984a, Cloning of nitrate reductase genes from the cyanobacterium *Anacystis nidulans*, *J. Bacteriol.* **159**:36–41.

Kuhlemeier, C. J., Teeuwsen, V. J. P., Janssem, M. J. T., and van Arkel, G. A., 1984b, Cloning of a third nitrate reductase gene from the cyanobacterium *Anacystis nidulans* R2 using a shuttle gene library, *Gene* **31**:109–116.

Labarre, J., Chauvat, F., and Thuriaux, P., 1989, Insertional mutagenesis by random cloning resistance genes into the genome of the cyanobacterium *Synechocystis* PCC 6803, *J. Bacteriol.* **171**:3449–3457.

Lambert, G. R., and Carr, N. G., 1984, Resistance of DNA from filamentous and unicellular cyanobacteria to restriction endonuclease cleavage, *Biochem. Biophys. Acta* **781**:45–55.

Lang, J. D., and Haselkorn, R., 1991, A vector for analysis of promoters in the cyanobacterium *Anabaena* sp. PCC 7120, *J. Bacteriol.* **173**:2729–2731.

Laudenbach, D. E., Straus, N. A., and Williams, J. P., 1985, Evidence for two distinct origins of replication in the large endogenous plasmid of *Anacystis nidulans* R2, *Mol. Gen. Genet.* **199**:300–305.

Lee, J. K., Kiley, P. J., and Kaplan, S., 1989, Posttranscriptional control of *puc* operon expression of B800–850 light-harvesting complex formation in *Rhodobacter sphaeroides*, *J. Bacteriol.* **171**:3391–3405.

Leong, S. A., Ditta, G. S., and Helinski, D. R., 1982, Heme biosynthesis in *Rhizobium*. Identification of a cloned gene coding for delta-aminolevulinic acid synthetase from *Rhizobium meliloti*, *J. Biol. Chem.* **257**:8274–8730.

Madueno, F., Borrias, W. E., van Arkel, G. A., and Guerrero, M. G., 1988, Isolation and characterization of *Anacystis nidulans* R2 mutants affected in nitrate assimilation: Establishment of two new mutant types, *Mol. Gen. Genet.* **213**:223–228.

Matsunaga, T., Matsunaga, N., Tsubaki, K. and Tanaka, T., 1986, Development of a gene cloning system for the hydrogen-producing marine photosynthetic bacterium *Rhodopseudomonas* sp., *J. Bacteriol.* **168**:460–463.

McFarlane, G. J. B., Machray, G. C., and Stewart, W. D. P., 1987, A simplified method for conjugal gene transfer into the filamentous cyanobacterium *Anabaena* sp. ATCC 27893, *J. Microbiol. Meth.* **6**:301–305.

Meyer, R. J., Figurski, D., and Helinski, D. R., 1979, Properties of the plasmid RK2 as a cloning vehicle, in: *DNA Insertion Elements, Plasmids and Episomes* (A. I. Bukhari, A. Shapiro, and F. L. Adhya, eds.), Cold Spring Harbor Laboratory, Cold Spring Harbor, New York, pp. 559–566.

Miller, L., and Kaplan, S., 1978, Plasmid transfer and expression in *Rhodopseudomonas sphaeroides*, *Arch. Biochem. Biophys.* **187**:229–234.

Murphy, R. C., Gasparich, G. E., Bryant, D. A., and Porter, R. D., 1990, Nucleotide sequence and further characterization of the *Synechococcus* sp. strain PCC 7002 *rec*A gene: Complementation of a cyanobacterial *rec*A mutation by the *Escherichia coli rec*A gene, *J. Bacteriol.* **172:**967–976.

Nano, F. E., and Kaplan, S., 1984, Plasmid rearrangements in the photosynthetic bacterium *Rhodopseudomonas sphaeroides, J. Bacteriol.* **158:**1094–1103.

Nano, F. E., Shepherd, W. D., Watkins, M. M., Kuhl, S. A., and Kaplan, S., 1985, Broad-host-range plasmid vector for the *in vitro* construction of transcription/translational *lac* fusions, *Gene* **34:**219–226.

Noti, J. D., Jadadish, M. N., and Szalay, A. A., 1987, Site directed Tn5 and transplacement mutagenesis: Methods to identify symbiotic nitrogen fixation genes in slow-growing *Rhizobium, Meth. Enzymol.* **154:**197–217.

Padhy, R. N., Hottat, F. G., Coene, M. M., and Hoet, P. P., 1988, Restriction analysis and quantitative estimation of methylated bases of filamentous and unicellular cyanobacterial DNAs, *J. Bacteriol.* **170:**1934–1939.

Pemberton, J. M., and Harding, C. M., 1986, Cloning of carotenoid biosynthesis genes from *Rhodopseudomonas sphaeroides, Curr. Microbiol.* **14:**25–29.

Porter, R. D., 1986, Transformation in cyanobacteria, *CRC Crit. Rev. Microbiol.* **13:**111–132.

Prentki, P., and Krisch, H. M., 1984, *In vitro* insertional mutagenesis with a selectable DNA fragment, *Gene* **29:**303–313.

Price, G. D., and Badger, M. R., 1989, Expression of human carbonic anhydrase in the cyanobacterium *Synechococcus* PCC7942 creates a high CO_2-requiring phenotype, *Plant Physiol.* **91:**505–513.

Rippka, R., Deruelles, J., Waterbury, J. B., Herdman, M., and Stanier, R. Y., 1979, Generic assignments, strain histories and properties of pure cultures of cyanobacteria, *J. Gen. Microbiol.* **111:**1–61.

Saunders, V. A., Saunders, J. R., and Bennett, P. M., 1976, Extrachromosomal deoxyribonucleic acid in wild type and photosynthetically incompetent strains of *Rhodopseudomonas sphaeroides, J. Bacteriol.* **125:**1180–1187.

Scanlan, D. J., 1988, Biochemical and molecular genetic approaches to studying protein export by cyanobacteria, Ph.D. thesis, University of Warwick, Warwick, United Kingdom.

Scanlan, D. J., Bloye, S. A., Mann, N. H., Hodgson, D. A., and Carr, N. G., 1990, Construction of *lacZ* promoter probe vectors for use in *Synechococcus:* Application to the identification of CO_2 regulated promoters, *Gene* **90:**43–49.

Schmetterer, G., Wolk, C. P., and Elhai, J., 1986, Expression of luciferases from *Vibrio harveyi* and *Vibrio fischeri* in filamentous cyanobacteria, *J. Bacteriol.* **176:**411–414.

Scolnik, P., and Haselkorn, R., 1984, Activation of extra copies of genes coding for nitrogenase in *Rhodopseudomonas capsulata, Nature* **307:**289–292.

Scolnik, P. A., and Marrs, B. L., 1987, Genetic research with photosynthetic bacteria, *Annu. Rev. Microbiol.* **41:**703–726.

Scolnik, P. A., Walker, M. A., and Marrs, B. L., 1980, Biosynthesis of carotenoids derived from neurosporene in *Rhodopseudomonas capsulata, J. Biol. Chem.* **255:**2427–2432.

Sherman, L. A., and van de Putte, P., 1982, Construction of a hybrid plasmid capable of replication in the bacterium *Escherichia coli* and the cyanobacterium *Anacystis nidulans, J. Bacteriol.* **150:**410–413.

Shestakov, S. V., and Reaston, J., 1987, Gene-transfer and host vector systems of cyanobacteria, *Oxford Surv. Plant Molec. Biol. Cell Biol.* **4:**137–166.

Shestakov, S. V., Elanskaya, I. V., and Bibikova, M. V., 1985a, Integrative vectors for the cyanobacterium *Synechocystis* sp. 6803, *Dokl. Akad. Nauk SSSR* **282:**176–179.

Shestakov, S. V., Elanskaya, I., and Bibikova, M., 1985b, Vectors for gene cloning in *Synechocystis* sp. 6803, *Abstr. Int. Symp. Photosynthetic Prokaryotes* **5:**109.

Simon, R., Preifer, U., and Puhler, A., 1983, A broad host range mobilization system for *in vivo* genetic engineering: Transposon mutagenesis in Gram negative bacteria, *Biotechnology* **1**:784–787.

Sockett, R. E., Donohue, T. J., Varga, A. R., and Kaplan, S., 1989, Control of photosynthetic membrane assembly in *Rhodobacter sphaeroides* mediated by *puh*A and flanking sequences, *J. Bacteriol.* **171**:436–446.

Sockett, R. E., Foster, J. C. A., and Armitage, J. P., 1990, Molecular biology of the *Rhodobacter sphaeroides* flagellum, in: *Molecular Biology of Membrane-Bound Complexes in Phototrophic Bacteria* (G. Drews, ed.), Plenum Press, New York, pp. 473–478.

Tandeau de Marsac, N., and Houmard, J., 1987, Advances in cyanobacterial molecular genetics, in: *The Cyanobacteria* (P. Fay and C. Van Baalen, eds.), Elsevier, Amsterdam, pp. 251–302.

Tandeau de Marsac, N., Borrias, W. E., Kuhlemeier, C. J., Castets, A. M., van Arkel, G. A., and van den Hondel, C. A. M. J. J., 1982, A new approach for molecular cloning in cyanobacteria: Cloning of an *Anacystis nidulans met* gene using a Tn901-induced mutant, *Gene* **20**:111–119.

Taylor, D. P., Cohen, S. N., Clark, W. G., and Marrs, B. L., 1983, Alignment of genetic and restriction maps of the photosynthesis region of the *Rhodobacter capsulata* chromosome by a conjugation-mediated marker rescue technique, *J. Bacteriol.* **154**:580–590.

Thiel, T., and Poo, H., 1989, Transformation of a filamentous cyanobacterium by electroporation, *J. Bacteriol.* **171**:5743–5746.

Thomas, C. M., and Smith, C. A., 1987, Incompatibility group P plasmids: Genetics, evolution and use in genetic manipulation, *Annu. Rev. Microbiol.* **41**:77–102.

Van den Hondel, C. A. M. J. J., Verbeek, S., van der Ende, A., Weisbeek, P. J. Borrias, W. E., and van Arkel, G. A., 1980, Introduction of transposon Tn901 into a plasmid of *Anacystis nidulans:* Preparation for cloning in cyanobacteria, *Proc. Natl. Acad. Sci. USA* **77**:1570–1574.

Van den Hondel, C. A. M. J. J., van Leen, R. W., van Arkel, G. A., Duyvesteyn, M., and de Waard, A., 1983, Sequence-specific nucleases from the cyanobacterium *Fremyella diplosiphon*, and a peculiar resistance of its chromosomal DNA towards cleavage by other restriction enzymes, *FEMS Microbiol. Lett.* **16**:7–12.

van der Plas, J., Hegeman, H., de Vrieze, G., Tuyl, M., Borrias, M., and Weisbeek, P., 1990, Genomic integration system based on pBR322 sequences for the cyanobacterium *Synechococcus* sp. PCC 7942: Transfer of genes encoding plastocyanin and ferredoxin, *Gene* **95**:39–48.

Weaver, K. E., and Tabita, F. R., 1985, Complementation of a *Rhodopseudomonas sphaeroides* ribulose bisphosphate carboxylase-oxygenase regulatory mutant from a genomic library, *J. Bacteriol.* **164**:147–154.

Weisbeek, P. J., Teerstra, W., van Dijk, M., Bloemheuvel, G., de Boer, D., van der Plas, J., Borrias, W. E., and van Arkel, G. A., 1985, Modification and sequence analysis of the small plasmid pUH24 of *Anacystis nidulans* R2, *Abstr. Int. Symp. Photosynthetic Prokaryotes* **5**:328.

Williams, J. G. K., and Szalay, A. A., 1983, Stable integration of foreign DNA into the chromosome of the cyanobacterium *Synechococcus* R2, *Gene* **24**:37–51.

Willison, J. C., Magnin, J. P., and Vignais, P. M., 1987, Isolation and characterization of *Rhodobacter capsulatus* strains lacking endogenous plasmids, *Arch. Microbiol* **147**:134–142.

Wolk, C. P., Vonshak, A., Kehoe, P., and Elhai, J., 1984, Construction of shuttle vectors capable of conjugative transfer from *Escherichia coli* to nitrogen-fixing filamentous cyanobacteria, *Proc. Natl. Acad. Sci. USA* **81**:1561–1565.

Wolk, C. P., Cai, Y., Cardemil, L., Flores, E., Hohn, B., Murry, M., Schmetterer, G., Schrautemeier, B., and Wilson, R., 1988, Isolation and complementation of mutants of *Anabaena* sp. strain PCC7120 unable to grow aerobically on dinitrogen, *J. Bacteriol.* **170**:1239–1244.

Wolk, C. P., Cai, Y., and Panoff, J-M., 1991, Use of a transposon with luciferase as a reporter to identify environmentally responsive genes in a cyanobacterium, *Proc. Natl. Acad. Sci. USA* **88:**5355–5359.

Yen, H. C., and Marrs, B. L., 1976, Map of genes for carotenoid and bacteriochlorophyll biosynthesis in *Rhodopseudomonas capsulata, J. Bacteriol.* **126:**619–629.

Yen, H. C., Hu, N. T., and Marrs, B. L., 1979, Characterization of the gene transfer agent made by an overproducer mutant of *Rhodopseudomonas capsulata, J. Mol. Biol.* **131:**157–168.

Young, D. A., Bauer, C. A., Williams, J. C., and Marrs, B. L., 1989, Genetic evidence for superoperonal organisation of genes for photosynthetic pigments and pigment-binding proteins in *Rhodobacter capsulata, Mol. Gen. Genet.* **218:**1–12.

Youngman, P., 1987, Plasmid vectors for recovering and exploiting Tn917 transpositions in *Bacillus* and other Gram-positive bacteria, in *Plasmids—A Practical Approach* (K. G. Hardy, ed.), IRL Press, Oxford, pp. 93–98.

Youvan, D. C., Elder, J. T., Sandlin, D. E., Zsebo, K., Alder, D. P. Panopoulos, N. J., Marrs, B. L., and Hearst, J. E., 1982, R-prime directed transposon Tn7 mutagenesis of the photosynthetic apparatus in *Rhodopseudomonas capsulata, J. Mol. Biol.* **162:**16–41.

Youvan, D. C., Ismail, S., and Bylina, E. J., 1985, Chromosomal deletion and plasmid complementation of the photosynthetic reaction center and light harvesting genes from *Rhodopseudomonas capsulata, Gene* **38:**19–30.

Zinchenko, V. V., Babykin, M. M., and Shestakov, S. V., 1984, Mobilization of non-conjugative plasmids into *Rhodopseudomonas sphaeroides, J. Gen. Microbiol.* **130:**1587–1590.

Zsebo, K. M., and Hearst, J. E., 1984, Genetic-physical mapping of a photosynthetic gene cluster from *Rhodopseudomonas capsulata, Cell* **37:**937–947.

Zsebo, K. M., Wu, F., and Hearst, J. E., 1984, Physical mapping and Tn5.7 mutagenesis of pRPS404 containing photosynthesis genes from *Rhodopseudomonas capsulata, Plasmid* **11:**182–184.

Mass Culture of Cyanobacteria   6

AMOS RICHMOND

1. AN OUTLINE OF USES FOR CYANOBACTERIAL MASS

1.1. Human Food and Therapy

The history of *Spirulina* as a staple for the human diet is fascinating. As recounted by Furst (1978), Fray Toribio de Benavente reached the Valley of Mexico in 1524, 3 years after the fall of the Aztecs. He described a harvest of *tecuitlatl:* There breeds upon the water of the lake of Mexico a kind of very fine mud and at certain time of year when it is thickest the Indians collect it with a very fine-meshed net until their acales are filled with it; on shore they make on the earth or the sand some very smooth beds, two or three brazas (3.4–5.1 m) wide and a little less in length, and they cast it down to dry, sufficient to make a cake two dedos (3.6 cm) thick. In a few days it dries to the thickness of a worn ducat and they slice this cake like wide bricks; the Indians eat much of it and enjoy it well, this product is treated by all the merchants of the land, as cheese is among us; those who share the Indians' condiments find it very savory, having slightly salty flavor. (Furst, 1978).

Identification of tecuitlatl, however, came not from Mexico, but from Africa. Jean Leonard, a civilian botanist on a Belgian military expedition crossing the Sahara from the Atlantic to the Red Sea in 1940, became interested in the "dihe" cakes eaten by the Kanembu of Lake Chad and the blue-green alga from which the cakes were made (Furst, 1978). During the 1950s a world-wide interest in novel sources of protein to feed the growing human population led researchers to investigate the possibilities of large-scale algaculture. In 1963, the French Petroleum Institute became interested in reports of the dried alga cake "dihe," and Clement *et al.* (1967) laid the foundations for mass cultivation of *Spirulina.*

AMOS RICHMOND ● Microalgal Biotechnology Laboratory, The Jacob Blaustein Institute for Desert Research, Ben-Gurion University at Sede-Boker, Israel 84993.

Photosynthetic Prokaryotes, edited by Nicholas H. Mann and Noel G. Carr. Plenum Press, New York, 1992.

Spirulina powder has the highest protein content (60–70%) of any natural food, far more than fish flesh (15–20%), soybean (35%), dried milk (35%), peanuts (25%), fresh eggs (12%), or grains (8–14%), (Hendrickson, 1989). Feeding tests rank protein by the net protein utilization value (NPU) determined by the amino acid quality, its digestibility as indicated by the proportion absorbed by the intestinal tract, as well as its biological value, which related to the proportion of the protein retained by the body. Dried eggs have the highest protein value (94), followed by milk products (82), fish (80), and meat (67). *Spirulina*, with an NPU of 62, is similar to many grains and ranks higher than nuts. A special value of *Spirulina* as plant food relates to the absence of cellulose in its cell walls. Cell protein is thus readily digested and assimilated by the human body. This feature is important for people suffering from intestinal malabsorption or older people on strict diets who have difficulty in digesting complex protein (Hendrickson, 1989).

In addition to protein, the composition of *Spirulina* powder shows 20% carbohydrates, 5% fats, 7% minerals, and 3–6% moisture (Hendrickson, 1989). *Spirulina* is thus a low-fat, low-calory, cholesterol-free source of protein, unlike fat-loaded meat and dairy protein. The vitamin content of *Spirulina* reflects another important benefit as a human food, e.g., 10 g of powder contains 460% of the U.S. recommended daily allowance (RDA) of β-carotene. In addition, this amount contains some 21% of the U.S. RDA for thiamine and riboflavin, as well as 533% of that for vitamin B_{12}, *Spirulina* being the richest natural source for B_{12}. *Spirulina* may be of particular value for people who need an iron supplement in their diet. Also, an important nutritional feature of *Spirulina* concerns its essential fatty acid composition, and indeed, after a false start at the late 1970s as a "natural appetite suppressant," *Spirulina* is receiving today growing attention for its unique nutritional characteristics: It is sold today mainly in "health stores" in the form of powder, granules, or flakes and as tablets and capsules (Hendrickson, 1989).

There is some, albeit insufficient, clinical evidence for the therapeutic quality of *Spirulina*, which was summarized by Allan Jassby (personal communication, cited in Richmond, 1986). It is recognized as having an effect on lowering cholesterol levels in humans as well as alleviating premenstrual syndrome (PSM). Hendrickson (1989) reported recently on several physiological effects with clinical potential for *Spirulina;* there is evidence for lowering of cholesterol in humans and in rats, reducing oral cancer in hamsters, building a healthy lactobacillus population in the gut of rats, reducing kidney damage by drugs and heavy metals in rats, providing anti-liver cancer tumor activity as well as stimulation of the immune system in mice, correcting iron anemia in both rats and humans, and fostering recovery from malnutrition as well as usefulness in treating nutritional deficiencies in humans. Also, there are reports concerning the treatment of external wounds in humans, and one report gives some evidence for a fat-lowering effect of *Spirulina* pills, observed in a rather limited experiment (Becker *et al.*, 1986). Of particular interest is the finding that *Spirulina* spp. are relatively rich in the rare fatty acid γ-linolenic acid (GLA), reported to have wide therapeutic properties. It is approximately 170-fold more effective in lowering the plasma cholesterol level than linoleic acid. (Cohen *et al.*, 1987), the major constituent of most polyunsaturated oils. In addition, tests on children have shown that GLA is of benefit in treating atopic eczema (Wright and Burton, 1982), while in women it appears to reduce the severity of premenstrual syndrome (Horrobin, 1983). GLA has also been claimed to have a positive effect in heart diseases, Parkinson's disease, and multiple sclerosis. Direct provision of GLA could thus have a potentially important role in human nutrition (Cohen *et al.*, 1987). Recently, an extensive study showed that it was possible to prevent experimental oral cancer in rats by extract of *Spirulina* and *Dunaliella* (Schwartz *et al.*, 1988). All the control animals treated with 7,12-dimethylbenz(a)anthracene (DMBA) suffered from gross tumors of the right buccal pouch, whereas comparable animals fed with the *Spirulina* mixture exhibited a complete absence of tumors.

1.2. Animal Feed

A great number of nutritional studies have been designed to test the nutritional quality of *Spirulina platensis* as animal feed. Dehydrated *S. platensis* was evaluated as a protein replacement source in swine starting diet, and satisfactory animal preformance was observed with dehydrated *Spirulina* making up to 9% (the maximum level utilized) of the total diet without any apparent toxicity (Hugh *et al.*, 1985). Becker *et al.* (1982) summarized many studies concerning the nutritional value of sun-dried *Spirulina*. It had a protein efficiency ratio (PER) of between 1.70 and 2.20 at a level of 10% of the total protein in the diet. Unlike other microalgae (*Scenedesmus, Chlorella,* and *Coelastrum*), the biological value of *Spirulina* was not much affected by the method of drying, and sun-dried *Spirulina* had approximately the same or greater biological value than drum-dried or air-dried *Spirulina. Spirulina* had a significant PER as a supplement to the main feed, compared to the PER obtained by feeding grains such as corn or wheat alone. Thus, corn had a PER of 1.23 and net protein utilization (NPU) of 30.5, whereas three portions of *Spirulina* and one portion of corn had a PER of 1.80 and an NPU of 37.2. Similar results were obtained with *Spirulina* mixed with wheat and with rice. Using *Micractinium,* grown in waste-water ponds, as replacement of 50% of the protein in poultry diet resulted in a PER of 2.00, practically the same as the control (2.04) (Mokadey *et al.,* 1979). Positive results were also obtained with ruminants fed with *Spirulina* which replaced all the soybean meal (21% of the diet) without significant effect on weight gain or feed conversion efficiency (Calderon *et al.*, 1976). In other studies, *Spirulina* served well in the diet for silkworms (Hou and Chen, 1981), and very good results were obtained in feeding *Tilapia* fish with *Spirulina*. While the conventional feed formula which included fish meal and soybean meal could not be totally replaced with *Spirulina* and ground corn (1:4 w/w) at the early stage of fish development, the *Spirulina* and corn mix could well replace all the fish meal in the feed once the fish reached the weight of 100 g (Granoth and Porath, 1984).

1.3. Natural Products

A large array of natural products of economic potential may be produced from cyanobacteria. Phycobiliproteins may be used as pigments or colorants in the food industry as well as for cosmetics; Padgett and Krogmann (1987) describe a simple procedure for the purification of large amounts of phycocyanin and allphycocyanin from the cyanobacterium *Microcystis aeruginosa*. The Dai Nippon Inks and Chemical Co. of Japan markets phycocyanin powder under the trade name of "Linablue." U.S. Federal Drug Administration regulations restricting artificial colorings provide an added incentive to develop natural pigments from microorganisms. A number of reserve products taking the form of granular inclusions in cells such as cyanophycin and glycogen granules have been researched in cyanobacteria; the cell contents of these were greatly affected by various physiological and nutritional factors (see Allen, 1984).

Moore *et al.* (1987) reported on alkaloids possessing antibacterial and antimycotic action in *Hapalosiphon fontinalis* isolated from soil. They described the structure of hapalindole A, a novel chlorine- and isonitrile-containing indole alkaloid which is responsible for most of the antibacterial and antimycotic activity associated with *H. fontinalis*.

Other unique chemicals in cyanobacteria which exhibit antibiotic activity have also been described, supporting the conjecture that many potentially useful products of novel properties in cyanobacteria may yet await discovery. For further details see Chapter 8.

1.4. Production of Ammonia

Ramos *et al* (1987), as well as Guerrero *et al.* (1982), reported extensively on the photosynthetic production of ammonia by cyanobacteria. An effective photoproduction of ammonia under aerobic conditions from either nitrate or atmospheric nitrogen has been achieved by altering chemically the key ammonia assimilation pathway. Treatment with L-methionine-D-L-sulfoximine (MSX), a specific irreversible inhibitor of glutamine synthetase, causes the cells to export to the medium most (about 90%) of the ammonia resulting from light-driven reduction of nitrate of N_2. Production of ammonia by MSX-treated cell suspensions in 50-liter containers was rather stable, and proceeded at satisfactory rates with a good overall efficiency. The authors pointed out the feasibility of using this photobiological energy transduction and storage system for industrial generation of ammonia (Ramos *et al.*, 1987). For further details see Chapter 8.

1.5. N_2 Biofertilizers

Cyanobacteria play a major role among the microorganisms which reduce dinitrogen to ammonia and a range of cyanobacteria species known to be N_2-fixing exist in rice field ecosystems. Virtually all the dominant cyanobacteria in rice fields are N_2-fixing, and therein lies the explanation of how rice, which provides the staple diet for about one-half of the world's population, has been grown continuously in paddy soils for many centuries without addition of fertilizer (Watanabe *et al.*, 1987; Roger and Watanabe, 1984).

According to Venkataraman (1986), the concept of utilizing cyanobacteria as fertilizer for fixing nitrogen in paddy fields was developed by De (1939) in India, who recognized that field fertility was maintained by the proliferation of cyanobacteria in the soil, an observation which was later confirmed by Singh (1942) and Watanabe *et al.* (1951).

Research on the practical utilization of cyanobacteria in rice has been mostly described toward inoculation with new strains. In most soils the density of endogenous N_2-fixing cyanobacteria was usually higher than that attained by applying the recommended soil inoculum and Roger *et al.* (1987) suggested that attention should also be given to agricultural practices that enhance the growth of endogenous strains already adapted to local environmental conditions.

The importance of enhancing the development of endogenous N_2-fixing cyanobacteria as well as attempting inoculation of selected strains is amplified by the observation that when urea or ammoniacal N_2-fertilizers are applied to the flood water of rice crops, fertilizer use-efficiency is often low due to a high pH (Tirol *et al.*, 1982). According to Roger and Watanabe (1985), an algal bloom may have a value of about 15–25 kg nitrogen Ha^{-1}, imparting to cyanobacterial mass an agronomic significance which is visible even to the naked eye. The potential is clearly very much higher; Fontes *et al.* (1987) concluded by extrapolation that the N_2 fixation rate in outdoor cultures of *Anabaena variabilis* could be greater than 3 tons nitrogen Ha^{-1} $year^{-1}$.

The use of nitrogen-fixing cyanobacteria as nitrogen fertilizers is fraught with major problems; foremost is the inability to produce good inocula at an economical price. In addition, there is insufficient understanding of the environmental conditions prevailing in rice ecosystems, which at times may severely inhibit the development of inoculated cyanobacteria. Also, the efficiency of utilization of the fixed nitrogen by rice plants is often low (Watanabe, 1984), and efforts are therefore being exerted to isolate suitable strains of cyanobacteria that would be prolific producers of fixed nitrogen and would excrete it continuously, making it available to the growing rice plantlets (Fontes *et al.*, 1987; Boussiba, 1988). This topic is dealt with in greater detail in Chapter 8.

1.6. Production of Hydrogen

Numerous species of cyanobacteria have demonstrated a capability for light-induced hydrogen evolution (Kerfin and Boger, 1982; Kumazawa and Mitsui, 1981), one of the most successful of which is the filamentous nonheterocystous marine cyanobacterium *Oscillatoria* sp. (Miami BG 7 strain), which has been shown to produce hydrogen at relatively high rates for sustained periods of time (Phlips and Mitsui, 1981). Depletion of combined nitrogen from the growth medium was a prerequisite for the initiation of hydrogen production, maximum hydrogen-producing capability coinciding with the end of the linear phase of growth. The rates of H_2 production obtained to date are, however, too low for commercial exploitation, the maximum under laboratory conditions being 5–6 ul H_2 (mg dry weight)$^{-1}$ hr^{-1}, which is probably equivalent to a maximum of 25 ml H_2 m^{-2} day^{-1}, or 250 liters H_2 Ha^{-1} day^{-1} (~70–90 m^3 H_2 Ha^{-1} year^{-1}), far from covering the cost of production. Another approach which is currently being investigated concerns H_2 production by immobilized cyanobacteria, e.g., *Anabaena azolle* (Broures and Hall, 1986) or *Oscillatoria* spp. (Phlips and Mitsui, 1984). For further details see Chapter 8.

An altogether different possibility to utilize cyanobacteria as an energy crop is to convert the biomass to methane by anaerobic digestion. Samson and LeDuy (1986) converted *Spirulina maxima* biomass to methane using continuously stirred tank digesters with an energy conversion efficiency of 59%. At present, the prospects for using cyanobacteria as an energy crop seem very remote.

1.7. Clearing of Municipal Wastes

Numerous works relate to the production of *Spirulina* on effluents from biological treatment plants with the dual purpose of clearing water and obtaining cyanobacterial mass. Cyanobacteria readily remove nitrogen and phosphorus from effluents (Kosaric *et al.*, 1974; de la Noue *et al.*, 1984), but in addition they also may accumulate very large quantities of heavy metals. The economic potential of this system is debatable. The treatment of waste water is dealt with in detail in Chapter 8.

1.8. Cyanobacterial Flocculants

Fixed to their substrate, benthic cyanobacteria cannot escape unfavorable environmental conditions; their survival and profileration thus depend on a complex set of adaptations to the varying conditions prevailing at their sites of attachment. These include production of extracellular products capable of flocculating and sedimenting suspended clay particles, thereby allowing light to reach the benthic interface in turbid waters. Bar-Or and Shilo (1988) described the production, purification, and properties of benthic cyanobacterial flocculants from *Phormidium* sp. and *Anabaenopsis circularis*. They suggested that the diversity found in cyanobacterial flocculants indicated the possibility that cyanobacteria may produce flocculants of varying properties which could have promising potential for utilization in a broad range of industrial applications, such as waste water treatment, clarification of water in reservoirs, and sedimentation of colloids in the chemical and food industries.

2. BASIC PRINCIPLES IN MASS CULTIVATION

2.1. The Outdoor Environment

The effects of light on outdoor cultures of cyanobacteria, as well as other photoautotrophs, are not well understood. This is because just about all studies concerning the characterization of the photosynthetic response to light (e.g., Foy and Gibson, 1982; Zevenboom and Mur, 1984; Van Liere and Mur, 1979; Lee *et al.* 1985) have been conducted under carefully controlled laboratory conditions, in which temperature and irradiance were constant, and the intensity of the irradiance used use significantly lower than that of peak solar irradiance prevailing outdoors. Indeed, outdoor conditions are very difficult to simulate: the photon flux density (PFD) impinging on cultures grown outdoors changes within a few hours from zero at dawn to over 2000 μE m^{-2} sec^{-1} at noon, i.e., a rate several times above saturation. Within a few additional hours thereafter, the PFD declines to zero. This very large and fast change in irradiance during the day shifts the culture from being light-limited in the morning to being growth-inhibited at summer noon and thereafter light-limited once again later in the afternoon. At the same time, the temperature in open raceways is also changing and in the case of many cyanobacteria such as *Spirulina* sp. which have temperature optima of 37°C, the temperature of the culture in the morning may be growth-limiting throughout the year even in the tropics. Indeed, a continuously changing interaction of light and temperature takes place outdoors throughout the day, complicating the analysis of cell response to environmental factors in general and to light in particular.

Some measure of temperature control to secure a more favorable temperature for growth and productivity is quite feasible. First, the species selected for mass cultivation outdoors must be well adapted to the temperature range existing at the production site throughout the year. Various sources of waste heat may be used to heat the cultures via simple heat exchangers, and in addition, open raceways could be covered by transparent polyethelene sheets in winter, causing a rise of the maximal daily temperature of some 6–8° C. Finally, the culture could, in principle, be grown in closed systems in which a good measure of temperature regulation during the day may be readily achieved (see following section). Thus, the great challenge confronting researchers on outdoor mass cultivation of cyanobacteria concerns solar irradiance: how to use it most efficiently and how to minimize its ill effects during the periods in which it achieves very high rates (Richmond, 1987). Therein lies the importance of controlling the population density and ensuring optimal stirring in outdoor cultures.

2.2. The Effect of Light

The influence of light intensity on the growth rate of cyanobacteria is commonly described in terms of a Monod-like relationship in which, typically for many cyanobacteria, light inhibition becomes evident at relatively low light intensities (Mur, 1983). The effect of light, however, is unique among the various inputs in a photoautotrophic system, for light alone reaches the culture only at its surface. A superficial reference to the light response curve may lead to the perception that when temperature and nutrients are not growth-limiting, the most important parameter affecting growth is the total incident light energy at the culture surface. Furthermore, one might assume that any energy level above saturating light intensity, growth in the culture is light-sufficient. Such interpretations of the light curve are incorrect: as evident from work on the outdoor cultivation of *Spirulina* (Richmond and Vonshak, 1978; Vonshak *et al.*, 1982; Richmond and Grobbelaar, 1986), an optimal population density (OPD) is clearly manifested, as may be observed in a set of continuous cultures maintained at several population densities under identical conditions. At the OPD, the net output of biomass per unit area is greatest, representing, in effect, the highest net photosynthetic efficiency. At OPD, however, the specific growth rate μ is approximately one-half the maximal, because growth is strongly light-limited.

The maximal net output rate of photoautotrophic mass outdoors does not coincide with the highest specific growth rate; the latter occurs at a relatively low cell density, when mutual shading and thus light limitation are minimal. As the population density is increased, the specific growth rate declines, since the culture becomes increasingly light-limited, even though the solar energy impinging on the surface of the culture may be several times above saturation (Fig. 1). Furthermore, at very high cell densities, e.g., thrice or quadruple the optimal, net growth in a highly illuminated culture surface outdoors would cease altogether due to photoinhibition (Samuelsson *et al.*, 1985; Vonshak, 1988). Clearly, what matters in

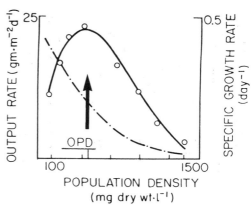

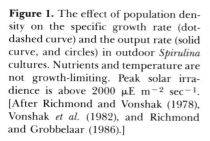

Figure 1. The effect of population density on the specific growth rate (dot-dashed curve) and the output rate (solid curve, and circles) in outdoor *Spirulina* cultures. Nutrients and temperature are not growth-limiting. Peak solar irradience is above 2000 μE m^{-2} sec^{-1}. [After Richmond and Vonshak (1978), Vonshak *et al.* (1982), and Richmond and Grobbelaar (1986).]

practice in outdoor production of biomass with respect to light is not only the total incident light energy at culture surface, but mostly a parameter which may be termed "light per cell." Measurement of this parameter is quite involved. At any sustainable population density, outdoor cultures are light-limited since, due to mutual shading, the cells are illuminated intermittently. "Light per cell" must thus be described in terms which relate to the duration of each light exposure of the average cell arriving in the photic zone at culture surface, as well as the relative length of the exposure in relation to the length of the exposure in the dark layers which prevail below the photic zone. At OPD, the volume of the photic zone may comprise 10–20% of the culture volume. The parameter "light per cell," then, is meaningful only in terms of the entire light regime (LR) to which the average cell in the culture is exposed. In species or strains sensitive to photoinhibition, a pertinent feature in the LR is the frequency of exposure to high PFD at the very top layer of the photic zone, where photoinhibition would be most severe for many cyanobacteria, as well as the average duration of each such harmful exposure. Considering the complexity of the LR parameter, there is obviously difficulty in measuring this elusive quantity, yet the full effect of solar irradiance on outdoor mass cultures cannot be comprehended unless perceived in terms of the light regime for the average cell in the culture.

Controversy over the effect of stirring and over the importance of turbulent streaming in photoautotrophic mass cultures stems, it seems, from disregarding the LR. Yet when light is limiting growth, as is common in outdoor cultures, stirring represents the most practical means by which to attempt to distribute solar energy evenly to all cells in the culture, i.e., induce a favorable light regime which is the dominant prerequisite for efficient utilization of solar energy. An essential point to consider in this respect is that at the optimal population density in, e.g., *Spirulina*, only approximately 20% of the cells receive, at any given instant, irradiant energy at a rate above the compensation point. Thus, the importance of turbulence in the culture as a means by which light is in effect distributed to the entire cell population should be self-evident. When stirring is insufficient, as seems to be the usual case in large-scale commercial open raceways, the efficiency of solar energy utilization must decline as the pattern of flow in the culture becomes increasingly laminar. Also, as follows from all the above, the positive effect of stirring on the output of biomass should increase as the population density increases (Fig. 2). Richmond and Grobbelaar (1986) showed that relatively slow stirring, e.g., considerably less than 30 cm sec^{-1} in an open raceway, plays havoc with the output rate when the population density is maintained far above the optimal—a situation which readily occurs in large reactors due to practical pressures to maintain high population densities and thereby secure high harvesting efficiency, i.e., retaining a high proportion of the filtered biomass on the screen. The finding that an increase in the rate of stirring shifts the OPD to a higher value illuminates the mode of action of stirring (Richmond and Vonshak, 1978). As the output rate is a function of both the specific growth rate and cell density, such a shift results in a considerable increase in the output rate. Appropriately, the more intense the stirring, the smaller is the relative effect of the population density on the output rate of biomass (Richmond and Grobbelaar, 1986).

Since cyanobacteria become light-saturated at a fraction of peak solar-light intensity, much potentially useful solar energy is, in effect, wasted. Thus, the light saturation constant I_s, which represents the intrinsic capacity of cells to utilize light energy, should be given special attention in mass cultures of photoautotrophs, as already pointed out by Goldman (1979). A high saturation constant will bestow a double advantage on photoautotrophs exposed to high PFD: it increases the efficiency by which high PFD is utilized and concomitantly diminishes the occurrence and magni-

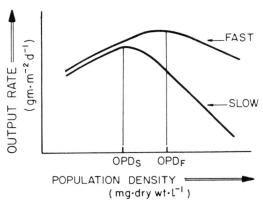

Figure 2. The interrelationships of the population density, the rate of stirring, and the output rate. [After Richmond and Vonshak (1978), Vonshak *et al.* (1982), and Richmond and Grobbelaar (1986).]

tude of photoinhibition, which would substantially reduce the output rate of sensitive strains in outdoor cultures (Vonshak, 1987). I_s is greatly affected by the temperature and there is evidence that in thermophiles, light and temperature adaptations are typically manifested by a lack of inhibition by bright light as well as by saturation of photosynthesis and growth at a relatively high PFD (Castenholz, 1969). This has practical relevance to outdoor cultures grown in closed reactors in which the temperature is observed.

One of the important factors affecting the OPD is the height of the water column in the reactor. Decreasing the water column in a *Spirulina* culture from 15 to 7.5 cm effected a doubling of the OPD. It had no effect, however, on the maximal areal output obtained in cultures maintained at OPD (Richmond and Grobbelaar, 1986). While the depth of the water column in the reactor does not exert a marked effect on the overall areal output at the proper OPD, decreasing the areal volume as much as would be technically feasible has definite advantages in mass cultivation. It should reduce the cost of production in general and that of harvesting in particular.

2.3. The Effect of Temperature

The basic effect of temperature on the biomass yield is reflected in the fact that the highest growth rate can be obtained only at the optimal temperature for growth. While this effect is obvious and well documented under laboratory conditions, the magnitude of the effect of temperature on the annual production of biomass outdoors seems not to be sufficiently appreciated. In many production sites outside the tropical regions, the differences in the maximum and minimum pond temperatures through the diurnal cycle may be close to 20°C. Thus, even when the maximal day temperature is optimal, the morning temperature could be 15 or 10°C below the optimal for *Spirulina*, preventing the full exploitation of 4–6 hr of morning radiation in the summer. Unpublished data from the author's laboratory indicate that raising the temperature in a *Spirulina* pond by some 10°C in summer mornings results in a 20% increase in the output rate. During the winter *Spirulina* cannot be grown in open ponds except in the tropics, but even there yields are significantly lower in winter. When the ponds in a subtropical region are covered in the winter with some transparent sheet or glass, the cultures survive, but production is greatly reduced, because the temperature is too far from the optimal throughout all or most of the day and because the cover reduces the PFD significantly. Thus, the diurnal variations in temperature which occur in several production sites of *Spirulina* result in temperature-limited growth during a significant fraction of the day. Recently Lee *et al.* (1985) showed that the light yield (g biomass kJ^{-1}) increased some 3- to 5-fold with increase in the temperature in cultures of *Chlorella,* an effect which is clearly indicated in other microorganisms. Collins and Boylen (1982) exposed *Anabaena variabilis* to shifts in environmental factors, showing that a rise in temperature greatly increased the saturating light intensity (E m^{-2} sec^{-1}) constant. Also, temperature was found to exert a decisive effect on the maintenance energy coefficient (kJ g biomass^{-1} hr^{-1}). Thus, a rise in temperature up to the optimal exerts a two fold effect on productivity: increasing the light yield and decreasing the energy spent on maintenance (Pirt, 1986).

Tomaselli *et al.* (1987) grew *Spirulina platensis* M-2 in light-limited continuous cultures at elevated temperatures to examine the influence of high growth temperatures, such as those reached in closed photobioreactors, on its chemical composition. At 40° a significant decrease in the protein content (22%) and a marked increase in lipids (43%) and in carbohydrates (30%) were observed. Also, the fatty acid composition was modified to a higher degree of saturation.

The temperature of the culture at night is another important factor which affects the net output rate of biomass. In itself, the effect of temperature on dark respiration is of course very well documented, but experience shows that the magnitude of the night loss in biomass as affected by temperatures, as well as by other factors, varies greatly. This process is not sufficiently understood in outdoor cultures. One practical aim in cyanobacterial biomass production thus emerges: maintain a temperature as close to the optimal for growth as possible during the entire light period, then let temperature decline quickly with the onset of darkness.

3. REACTORS FOR MASS CULTIVATION OF PHOTOAUTOTROPHS

3.1. Open Raceways

The shallow mixed pond for microalgae production was introduced in the early 1960s by Oswald (Dodd, 1986). The principle is based on the use of a pond in a meandering configuration with channels utilizing paddle wheel mixers of various designs with low shearing forces. To eliminate hydraulic losses and solid deposition, a long, single-loop configuration was introduced using suitable baffles to direct the streaming.

Almost all commercial reactors for production of *Spirulina* are based on shallow raceways in which the cultures are mixed in a turbulent flow sustained by a paddle wheel (Fig. 3). One important exception is at Sosa Taxacoco (Mexico), where facilities for producing sodium bicarbonate in an alkaline lake were adapted to harvest naturally occurring *Spirulina*. At most commercial sites, in, for, example, the United States, Israel, India, Thailand, and Taiwan, two general types of open raceway ponds are used:

Figure 3. A paddle wheel operating in a 5000-m² pond at the Earthrise Farms in Southern California.

in one, the lining is made of concrete; in the other, a shallow earthen tunnel in lined with 1- to 2-mm-thick sheets of polyvinyl chloride (PVC) or some other durable food-grade plastic material. Since the capital investment in the reactor forms an important component of the overall cost of production, the cost and durability of the lining (which is a major component in the cost of the reactor) have a clear impact on the economic feasibility of mass production of cyanobacteria (Richmond, 1986; Vonshak and Richmond, 1988).

The size of commercial ponds varies from 1,000 to 5,000 m². Stirring is usually accomplished with a paddle wheel, with large-diameter wheels (approximately 2 m) which work at a low rate of revolutions per minute (10 rpm), these being preferable to other designs, e.g., wheels of smaller diameter which rotate faster. Beating the culture vigorously is harmful since it results in excessive formation of foam. Stirring is employed in almost all mass cultures of cyanobacteria and its beneficial effect is due to a variety of factors: it continuously provides nutrients, including CO_2 at the cell surface; it helps remove excessive oxygen from the medium to the atmosphere; and most importantly, stirring, as already elucidated, improves the light regime, facilitating improved utilization of solar irradiance (Richmond, 1986; Vonshak and Richmond, 1988).

An important drawback of the open raceway is the limited possibility of temperature control. This is particularly serious for cyanobacteria, which usually have an optimal temperature in the region of 35°C, a temperature which is not commonly reached in the open raceways employed except during a short period at midday. Another important drawback of the open-channel raceway is that the water column in the pond should not be reduced below 15 cm, to prevent severe reduction of the flow and turbulence. This mandates an areal volume of at least 150 liters m^{-2}, requiring the maintenance of a rather low population density (approximately 500 mg dry weight of *Spirulina* mass per liter). A low population density in turn increases the cost of harvesting and the cost of pond maintenance.

3.2. Closed Tubular Systems

The possibility of growing photosynthetic microorganisms in closed systems rather than in the common open raceway has been pioneered by John Pirt and co-workers (1983), who proposed the theory and design for a tubular photobioreactor. Gudin and Chaumont (1983) were the first to develop a tubular system expanding to 1000 m² for the cultivation of *Porphyridium* spp. (Fig. 4). Florenzano and Materassi at the University of

Figure 4. A closed photobioreactor at the Centre d'Etudes Nucleaires de Cadarache, France. The solar receptor is made of polyethylene tubes of 64 mm diameter and 1500 m length. A double layer of polyethylene tube is used; temperature control is achieved by floating or submerging the upper layer (containing the algal culture) on or in the pool of water by adjustment of the amount of air in the lower layer. The pilot facility is composed of five identical units of 20 m²; the total culture volume is 6500 liters. [Photograph by kind permission of Dr. D. Chaumont.]

Florence pioneered the development of a closed photobioreactor for the production of cyanobacteria, particularly *Spirulina platensis* (Torzillo *et al.*, 1986). The first photobioreactor developed in Florence was made with transparent plexiglass tubes 13 cm in diameter, a size selected to achieve a surface-to-volume ratio similar to that of the open ponds (about 100 liters of culture suspension per m² of illuminated surface). Each photobioreactor was made up of several tubes laid side by side on a white polyethelene sheet joined by a connection to form a loop, each connection incorporating a narrow tube for oxygen release. At the exit of the tubular circle the culture suspension flowed into receiving tanks. A diaphragm pump raised the culture to a feeding tank containing a siphon that allowed an intermittent discharge of 340 liters into the photobioreactors at 4-min intervals, sustaining a flow rate of 0.6 m sec^{-1}. The maximal length of the circle in Florence was 500 m, corresponding to a volume of 8000 liters and a surface area of 80 m².

Figure 5. A tubular photobioreactor developed at the Algal Biotechnology Laboratory at the Blaustein Institute for Desert Research, Sede-Boker, Israel. It features UV=resistant polycarbonate tubing (of 32mm diameter), with circulation provided by an air lift. Cooling is provided by water spray which is controlled by a temperature sensor.

One difficulty with a closed system is that it heats up quickly, necessitating some cooling for at least a few hours during midday. One simple method is to cool the tube system by spraying water on the surface when the temperature of the culture reaches a critical value (Fig. 5). In order to reduce the amount of water required for cooling, the group in Florence (Torzillo *et al.*, 1986) developed a thermotolerant strain of *Spirulina* (*Spirulina platensis* M-2) which was able to grow at up to 42°C and could tolerate a daily exposure of up of 46°C for at least 3 hr. Comparing the output rate in open raceways to tubular photobioreactors revealed a consistent increase of about 10–30% in the output rate of the photobioreactor above that of the open raceway. This increase in output was attributed to the effect of temperature, which was maintained close to optimal during long hours of daylight. No significant differences were found in the protein content and amino acid composition between the biomass grown in tubular photobioreactors and that grown in open raceways. Some differences, however, were found in the fatty acid composition, and biomass grown in open ponds showed a higher degree of unsaturation, due mainly to a higher content of γ-linolenic acid. This probably stems from the higher temperatures prevailing in the photobioreactor.

4. MANAGEMENT OF OUTDOOR CULTURES

For the successful maintenance of cultures of cyanobacteria aimed to obtain maximum productivity, one must monitor the culture and constantly assess its relative performance and "well-being." Indeed, the importance of obtaining detailed information to facilitate reliable and quick evaluation of the physiological state of the culture cannot be overemphasized. Warning signals must be recognized early to prevent the development of conditions which, within 1 or 2 days, could culminate in the entire loss of the culture. In essence, the basic requisite for correct maintenance is a consistent evaluation of the degree to which the photosynthetic as well as other metabolic systems of the cyanobacteria function in reference to the maximal potential obtainable under the given environmental conditions (Richmond, 1986).

Thus, the aim for management is to ensure that no nutritional limitations develop in the culture, so that growth is limited only by climatic factors, i.e., light and temperature. Ideally, solar irradiance should be the sole limitation to growth of photoautotrophs in outdoor cultures.

4.1. Controlling Nutrient Levels and pH

Routine tests should be performed to ensure that no deficiency in mineral nutrients is developing. The concentration of nutrient elements in the growth medium is constantly depleted in continuous cultures as new biomass is removed continuously from the reactor. In addition, some trace elements may gradually precipitate and nitrogen and ammonia may be lost due to the high pH which is usually maintained for cyanobacteria; nitrogen loss may occur by denitrification which may occur in anaerobic pockets formed in certain localities in the pond when mixing is inadequate or pond design is defective. One practical method to supplement the nutritional status in the culture is to monitor the level of nitrogen, using it as a guideline for adding, in equivalent amounts, the entire formula of the growth medium except for carbon and phosphorus, which are added separately. This practice is based on the assumption that the relative depletion of nitrogen from the medium is roughly equivalent to that of the rest of the nutrient elements in the growth medium. No doubt such a protocol is prone to error; the concentrations of some minor elements, in particular, may with time build up or become depleted, irrespective of cyanobacterial growth and nitrogen utilization. In fact, a nutritional imbalance in the growth medium may be at the root of the observation that *Spirulina* cultures grown continuously outdoors become "old," requiring a complete replacement of the growth medium every few weeks.

The pH in the pond will rise continuously as a result of depletion of carbon through photosynthesis, necessitating constant addition of carbon to a highly productive pond. In a *Spirulina* culture growing vigorously in midsummer the pH would rise by 0.10–0.15 pH units daily. The pH may be maintained by a continuous inflow of CO_2 or by the addition of mineral acid to a medium of high alkalinity. Sodium bicarbonate may be used in addition to CO_2 to adjust the pH for cyanobacteria such as *Spirulina*, the pH optimum of which is in the range of 8.5–10.0, and which thrive at high (i.e., 0.2 M) concentrations of bicarbonate. Carbon nutrition represents one of the major components in the operating cost of commercial production.

4.2. Measurement of Growth

Growth may be estimated quickly by measuring changes in the overall turbidity of the culture. This, however, provides only a rough estimation of growth and should be followed routinely by measurements of the dry weight. The buildup in the concentration of particulate carbon is likewise a dependable parameter to estimate growth. The concentrations of cell chlorophyll and protein are often used to express growth in the culture, but these parameters must be used with caution in outdoor cultures, since they are strongly affected by environmental conditions. Chromatic adaptation or slight nitrogen deficiency may effect, within days, considerable changes in cell chlorophyll and protein content, respectively, which do not necessarily correspond to changes in the overall biomass concentration in the reactor.

The measurements of growth should be accompanied as a rule by a detailed microscopic evaluation. Environmental changes may cause morphological modifications in some species which should be carefully recorded and studied. The most important reason for microscopic examination, however, is to trace the development of foreign species as well as protozoa or fungi which may quickly proliferate and either take over the cultivated species or cause damage by competing with the desired species. Any buildup in the number of other organisms in the culture should be regarded as a warning signal that the cultured species may have come under some pressure. The temperature may have swayed too much from the optimal, bestowing a relative advantage on other species. Or it may signal that the concentration of a particular nutrient has declined below the optimal for the cultured species. Indeed, an important criterion of pond maintenance is based on the observation that any weakness in the growth of the cultured species provides an opportunity for foreign organisms to rapidly increase their population. A useful practice is to record any modification in the culture's population by routine photomicrography. With time, the experience accumulated thereby would permit dependable diagnosis and interpretation of various malfunctions that repeatedly develop in outdoor cultures.

4.3. Dissolved Oxygen

Unlike "high-rate" algal ponds which are used for waste-water clearance and in which the O_2 produced by the photosynthesizing biomass is taken up by heterotrophic microorganisms (Soeder, 1986), very high concentrations of dissolved oxygen (DO) may build up in dense (e.g., 5–15 mg chlorophyll liter^{-1}) cultures of photoautotrophs. Concentrations of 35–43 mg O_2 liter^{-1}, representing maximal supersaturation values of 400–500%, are common in *Spirulina* cultures grown in large (e.g., 1000 m^2) open raceways in which mixing is not sufficiently intense or in closed systems, in which the DO concentration may build up to the maximum within a 20-m flow in a closed tube. The full array of effects that very high concentrations of DO exert on an outdoor culture of cyanobacteria is not sufficiently known. There is, however, good evidence that excessive DO would promote, under suitable conditions, photooxidation followed by quick death of the culture (Abeliovich and Shilo, 1972). Also, there is good evidence that the DO affects CO_2 assimilation, as CO_2 and O_2 are in direct competition for photosynthetically generated reductants (Radmer and Kok, 1976). Also, DO effects the protein content in *Spirulina*. Forty-five percent of O_2 in the gas phase lowered the protein content from 48% dry weight in the control to 22% (Torzillo *et al.*, 1984). In an open raceway, vigorous stirring effects a significant decrease in the level of supersaturated O_2, bestowing an additional advantage on this process.

The DO is readily measured with an oxygen electrode and is a reliable and sensitive indicator of conditions that relate to growth and productivity. There are many indications from the author's work that an early detection of an inexplicable decrease in oxygen or a drop of the normal rate of the daily increase (under given conditions) of DO in outdoor mass cultures of *Spirulina* serves as a reliable warning signal that the culture is stressed and may quickly deteriorate if corrective measures are not taken (Richmond, 1986).

4.4. Maintaining the Optimal Population Density

The population density represents a major parameter in production of photoautotrophic cyanobacterial mass, exerting far-reaching effects on the general performance and productivity of the culture. The optimal population density may be defined as the cell concentration which yields the highest output rate under given environmental conditions, or which brings the highest return on the investment capital. Experimentation to elucidate the relationships between the specific growth and output rates and between the optimum population density indicated that the latter varied somewhat between seasons (Fig. 6). Plotting the output rate of the pond

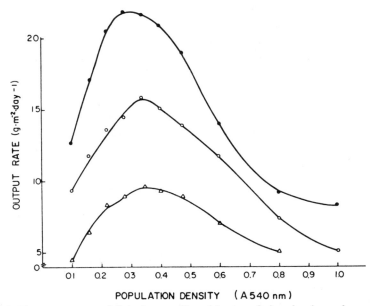

Figure 6. The output rate of *Spirulina* as affected by population density and season of the year. (○) summer (June–September), (□) autumn and spring (October–November, and April–May), (△) winter (December–February). The data represent averages obtained from several experiments carried out throughout the year. [From Vonshaak *et al.* (1982)].

against the population density during the course of the year revealed that the higher the temperature and the availability of irradiance per cell, the more pronounced became the dependence of the output rate on the population density (Fig. 2). Thus in summer, the maximum output rate reached in a *Spirulina* culture maintained at optimum density was about three times as high as that obtained in a similar culture maintained at the highest population density used in our experiment. In winter, when, due to temperature limitations, the output was a fraction of that obtained in summer, the maximum output from a culture maintained at optimal density was only double that obtained from a culture kept at a very high population density (Fig. 2). A decline in the output rate of *Spirulina* was always associated with very high population densities (i.e., approximately 1000 mg liter^{-1} at a pond water column of 12–15 cm). This decline cannot be explained by merely the decreased growth rate that corresponds to rise in density. But, as the PFD per cell decreases, the relative cell expenditure on maintenance energy increases. Likewise, decreasing the population density below the optimal range always resulted in a significant decrease in the output rate, apparently due to increased photoinhibition, which would naturally be more severe as mutual shading decreases. The experience with mass *Spirulina* culture in the author's laboratory indicates that a practical procedure would be to harvest the culture each time the population density reaches a concentration of about 500 mg dry weight liter^{-1} (in a 15-cm water column), removing some 25% of the biomass in the reactor on each harvest. When productivity of the biomass in the reactor on each harvest. When productivity is at its peak, a *Spirulina* culture must be harvested every 2 days to sustain the range of optimal cell density (Richmond, 1988).

4.5. Maintaining Cultures of Monospecies

Microorganisms different from the cultured species, i.e., contaminants, represent a major limitation to productivity in open outdoor cultures of *S. platensis*. Many years of experience with such cultures indicate that the only contaminants which seriously impede and endanger the culture are other species of cyanobacteria or certain microalgae. Of the less important contaminants, diatoms appear frequently both in winter and summer. Diatoms often became temporarily abundant when a new culture of *Spirulina* was initiated, but were never observed to become a serious pest, disappearing usually in a few days. In general, many observations reveal that at the lower population density (i.e., approximately 400 mg dry weight liter^{-1}), much more contamination was prevalent than at the higher population densities (i.e., 600 mg liter^{-2} and above). This was particularly true for the two most damaging contaminants in outdoor cultures of *Spirulina platensis*: *Spirulina minor* and *Chlorella* spp.

A small species of *Spirulina*, measuring 10 μm width and between 80 and 240 μm in helix length and which was tentatively identified as *Spirulina minor* (Richmond, 1986), constitutes a very harmful contaminant in Israel. It may rapidly overtake a culture of *Spirulina platensis* harvested by filtration, as a result of a continuous enrichment of the small filaments readily passing through the harvesting screen (325 mesh). When filtration as a method for removal of biomass from the culture was replaced by "bleeding" the culture (i.e., removing a certain volume from the culture and replacing it with a fresh medium), the ratio between *S. platensis* and *S. minor* gradually changed, and *S. platensis* regained dominance within 2–3 weeks. The temperature optimum for this strain bestows on it an advantage over *S. platensis* when the maximum daily temperature range is 22–27°C, and thus *S. minor* in open raceways has usually been most harmful in the spring.

Maintaining the population density at relatively high areal volumes and harvesting by bleeding were found to be the only useful preventative methods to arrest the development of *S. minor* and ensure the dominance of *S. platensis*. Clearly, the development of a commercial filter that in harvesting would separate the entire algal biomass from the medium would be very useful for mass production of *Spirulina*.

Chlorella spp. represent a most harmful contaminant in *Spirulina* cultures (Richmond *et. al.*, 1982). Although pH optimum for most *Chlorella* species is below 8.0, there exist alkalophilic types of *Chlorella* which thrive in a *Spirulina* medium. Contamination by *Chlorella* spp. is most severe when *Spirulina* growth is temperature-limited. *Chlorella* spp. under these circumstances have an obvious advantage over *S. platensis*, the optimal temperature for which is 10°C higher than the optimal for most species of *Chlorella*. A decline in temperature which coincides with an increase in the organic load in the medium imparts a clear advantage to the mixotrophic *Chlorella* spp., which may rapidly take over a culture of the strictly photoautotrophic *Spirulina*. This process is accelerated if intensive harvesting by screening takes place, *Chlorella* being approximately 3–5 μm in diameter; its population, like that of *Spirulina minor*, is steadily enriched in the course of harvesting by screening (Vonshak *et al.*, 1983). In cultures maintained at low population densities (i.e., high light per cell) *Chlorella* was particularly successful in becoming rapidly dominant, whereas in dense *Spirulina* cultures, *Chlorella* contamination has always been less severe. This is because cyanobacteria are as a rule very sensitive to high light intensity, but have extremely low maintenance energy (Mur, 1983). Cyanobacteria grown in dense cultures thus have an advantage over eukaryotic organisms such as *Chlorella* spp. which require a much higher maintenance energy and are thus placed at a disadvantage at low levels of light per cell.

Several strategies are available for the control of *Chlorella* contamination. First, high alkalinity as obtained by 0.2 M bicarbonate as well as a high pH of 10.3 or greater were shown to impede the growth of *Chlorella* (Richmond *et al.*, 1982). Repeated pulses of 1–2 mM NH_3, followed by a 30%

dilution of the culture, is also an effective treatment which is based on the differential sensitivity of *Spirulina* and *Chlorella* cells to NH_3, *Chlorella* spp. being significantly more sensitive to ammonia than are *Spirulina* spp. There are indications that the ammonia treatment is useful not only to control the growth of *Chlorella*, but also to check contamination by protozoa, as observed by Lincoln *et al.*, (1983). Indeed, outdoor cultures of *Spirulina* in 1000-m² ponds which were continuously fed with NH_4OH as the nitrogen source were found to be less susceptible to contamination.

In conclusion, contamination of *Spirulina* cultures grown outdoors may be controlled to a great extent. As a rule, contaminating organisms do not present a serious difficulty as long as vigorous *Spirulina* growth is maintained, and it is worth noting that no cyanophages attacking *Spirulina* have been observed. Nevertheless, considering the frequency with which it is necessary to replace the entire medium as well as the loss of desirable biomass due to competing contaminants, the latter are estimated to cause a loss of at least some 15% in overall productivity of *Spirulina* in open raceways.

4.6. Cyanobacteria Production Based on Local Resources

The production of *Spirulina* may be greatly simplified, avoiding the high-technology systems described so far. Cultivation may be carried out in ditches in which the lining is made of local materials; stirring may be provided by a simple device driven by wind energy or even by animals; harvesting may be readily performed using some suitable cloth for filtration; and the slurry obtained may be dehydrated in the sun. Becker and Venkataraman (1982) and Seshadri and Thomas (1979) pioneered this approach, experimenting with *Spirulina* growth media based on low-cost nutrients obtainable from rural wastes, such as urine, bone meal, and effluent from biogas digesters. The quality of the product obtained would be inferior to that attainable with more advanced technology, but the product could very well serve as animal or fish feed and even as human food. In one study, *Tilapia mossambica* was cultivated in artificial ponds with a relatively high stocking density and fed with solar-dried *Spirulina* which was cultivated and processed by rural means (Becker and Venkataraman, 1982). The *Spirulina* slurry was added to ground nut cakes and the yield of *Tiplapia* fed on *Spirulina* thus mixed was higher by 41% compared to that of fish grown on ground nut cake alone. There are several examples to show that *Spirulina* may serve as an excellent additive for animal feed, a relatively small amount of *Spirulina* significantly improving the biological value of the feed.

Fox (1988) has described systems that integrate sanitation, biogas generation, *Spirulina* production, composting, and fish culture. These have been designed for village conditions in the tropics, and several such units, according to Fox, have been tested in villages in developing countries with encouraging results (Fox, 1988). The idea is based on establishing communal latrines into which animal manure and plant wastes could be added and serve as raw material for biogas digestors which ferment waste, breaking it down to gas, liquid, and solid components. A simple gas separator resolves the biogas into methane and carbon dioxide. The liquid effluent is sterilized in a series of solar-heated pipes, becoming a safe mineral nutrient source for culturing *Spirulina* fed with the carbon dioxide released from the digestor. The sludge is removed from the digestor and composted before being spread on the soil. Fox provided an analysis showing that this integrated system can pay for itself in the first year (Fox, 1988).

The biggest constraint in the agriculture of developing countries is the nonavailability of chemical fertilizers at reasonable prices. Nitrogenous fertilizers in particular have been recognized as the limiting factor in good production. In India today, nitrogen-fixing cyanobacteria are grown in shallow earthen ponds producing nitrogen-fixing cyanobacterial mass for a cost of only one-third that of chemical fertilizers. Spreading this mass in the rice paddy is said to increase rice yield in India and other countries by some 15% (Venkataraman, 1986). The starter culture of the inoculum is normally a soil-based mixture of *Aulosira, Tylopothrix, Scytonema, Nostoc, Anabaena,* and *Plectonema.* The idea of using a mixture of algae is to offset the ecological or edaphic dangers to any one particular strain in a given area. Naturally, a chemical fertilizer could greatly further increase the yield but in most farming areas in Southeast Asia chemical fertilizers are simply not economically feasible (Venkataraman, 1986).

5. PROCESSING CYANOBACTERIAL MASS

5.1. Harvesting

The only cyanobacterium produced to date commercially is *Spirulina platensis.* The following relates to the experience accumulated with this species. Similar filtration devices are used for harvesting commercially produced *Spirulina.* Two types of screens are available: vibrating and stationary. The latter is usually 300–500 mesh with a filtration area of 2–4 m^2 per unit, capable of harvesting 10–18 m^3 of *Spirulina* culture per hour. Filtration provides a slurry of 8–10% dry weight, from which water may be further removed by various methods. Vibrating screens filter the same volume per unit time as the inclined stationary screens, but employ only one-third of the filtering area, and their harvesting efficiency is often very high (Vonshak and Richmond, 1988). Two main difficulties are recognized with the filtration devices used today for *Spirulina* harvesting. First, in the process of screening, the fragile *Spirulina* filaments may become physically damaged, thereby increasing the organic load of the medium which is returned to the pond after filtration. Also, the filtration process leads to continuous enrichment of the culture with unicellular microalgae or with *Spirulina minor,* which readily pass through the screen. The vibrating screen may cause greater mechanical damage to the filaments than the stationary screen, which therefore seems the most suitable for harvesting *Spirulina.*

The slurry obtained after filtration must be further dried, possibly by vacuum filtration using vacuum tables or vacuum belts. The vacuum treatment also effects removal of excessive salts from the slurry, which, if not washed properly, may amount to 20–30% of the dry weight. The washed stock is frequently homogenized before being dried (Richmond, 1986).

5.2. Drying

In commercial *Spirulina* production, drying is major economic consideration, in that it may constitute some 20% of the production cost. The various systems for drying differ both in the extent of capital investment and the energy requirements, and may have a marked effect on the nutritional value and the taste of the product. Experience with *Spirulina* production is that the harvested slurry (approximately 8% in dry weight) should be well rinsed in acid water at pH 4.0, to remove absorbed carbonates. It can then be stored at 0–2°C for several days, or frozen to -18°C for an indefinite time. The usual method for drying *Spirulina* is spray-drying, which yields a uniform powder from which pills may readily be formed. Drum-drying also yields a useful product, in the form of flakes which possess a special flavor and texture which some people find very appealing and for which commercial demand could be developed.

Direct drying of the *Spirulina* slurry in the sun is also feasible. Sun-drying, however, is not recommended for preparing an algal product intended for food, for two reasons: First, a rather unpleasant odor is frequently associated with sun-drying, due to the slowness of the dehydration process, which enables degradative processes to set in before drying is completed, and second, sun-drying does not include exposure to high heat (up to 120°C for a few seconds in drum-drying) and would thus generally lead to a higher bacterial count.

For the production of animal feed, sun-drying may be an acceptable method. In the author's laboratory, fish feed made of *Spirulina* that had been vacuum-dried to 20% dry matter was mixed with different proportions of corn meal and dried in the sun. Dehydration to about 10% (w/v) water was complete within 1 day. The resultant product took the form of irregularly shaped pieces of the *Spirulina*/corn meal mixture (1:4 w/w), which successfully served as the sole diet for vigorously growing *Tilapia* fish in tanks (Granoth and Porath, 1984).

6. FUTURE PROSPECTS

Producing mass cultures of cyanobacteria for industrial purposes represents a novel biotechnology which naturally prompts questions concerning the future for this endeavor. Two distinctly different modes of production seem to have emerged, each requiring a separate appraisal (Vonshak and Richmond, 1988). One relates to industrial production based on the most modern technology, i.e., fully controlled sophisticated production systems with selected strains engineered to produce desired products. The second mode of production which naturally evolves in developing countries is based on local resources and simple technology.

The difficulty with industrial production relates both to the small current demand for special products made from cyanobacteria as well as their cost of production. Since cyanobacteria such as *Spirulina* have proven to be safe and of high nutritional value both for humans and for animals, the market potential is not the limiting factor blocking intensive expansion of this industry. It is the very high cost of production compared with the cost of alternative products from conventional sources that impedes the expansion of industrial production of cyanobacteria. The high cost stems from two main factors: one is the very high investment capital which an industrial process requires and the other relates to our lack of understanding of how to utilize efficiently the high-intensity solar energy that prevails outdoors. The combination of high cost of investment capital together with the relatively low productivity per unit of reactor area, as well as the high cost of processing, combine to place the production cost of, for example, 1 kg of dried *Spirulina* powder in the range of 15–20$. Purely on a protein basis, this cost would imply approximately 30$ per kg of pure protein, which is very high compared to equivalent protein from meat or fish, and is totally out of range compared to the price of soybean meal protein. Thus, *Spirulina* for feed is confined to a very specialized market, e.g., colorful tropical fish which respond very well to *Spirulina* feed or as a minor additive in aquaculture for several animals at an early stage of development, as well as other special outlets for which the market is naturally small. There seem to be somewhat better prospects for *Spirulina* as a human food, particularly as a supplement to vegetarian diets. Indeed, most of the *Spirulina* produced today goes into pills and powder which are sold as "health food." Mainly due to the high price, this outlet is naturally limited and does not provide a basis for a large expansion of commercial *Spirulina* production. It seems, therefore, that as long as the high cost of production prevails, mass production of cyanobacteria would be a small-scale endeavor, limited to special natural products. In this market, the industrial prospects for cyanobacteria will be drastically improved when new strains are engineered in which the content of some high-cost natural products, e.g., γ-linoleic acid in *Spirulina* or flocculating agents in *Phormidium,* would be greatly increased. The future of industrialized cyanobacteria production thus depends on a significant improvement in three different aspects. One relates to engineered strains that on the one hand would contain greatly increased concentrations of desired products and on the other hand would respond much better to high PFDs, thus yielding much higher quantities of biomass per unit of capital invested in reactors. Therein lies an important advantage of cyanobacteria, which can be genetically manipulated with relative ease as compared to eukaryotes (Craig *et al.,* 1988). A *Spirulina* with a γ-linolenic acid content higher by an order of magnitude than that of the wild type would conceivably command a lucrative demand, similar to that enjoyed presently by *Dunaliella bardawil,* over 5% of the dry weight of which is β-carotene. Another aspect relates to the need to devise improved reactors that under proper climatic conditions would result in a much higher productivity, which is the single most important factor in reducing production cost. First and foremost, for any real progress to take place, our understanding of the unique physiology of dense mass cultures grown outdoors under fluctuating light and temperature must be vastly expanded. Viewing the short history of industrialized microalgaculture, the author feels that it is safe to expect that the future holds a promise for continuous development in all the three aspects on which the development of an industry with a sound economic basis depends.

The future for growing cyanobacteria for food, feed, and biofertilizer based on local resources and simple technology also seems promising as a means of improvement, albeit modest, of the quality of life in several rural areas around the world. It is well to remember that chronic malnutrition is most common in arid and semiarid regions, where human diets tend to be high in carbohydrates and sugars, but low in proteins as well as essential vitamins and minerals. Cyanobacteria such as *Spirulina* which grow well in many such areas may provide a suitable complement to the diet of both animals and humans. The most promising venture for local technology, however, concerns the utilization of nitrogen-fixing cyanobacteria as biofertilizer. Large-scale cyanobacterial inoculum-production at state seed farms and farmer training centers has already been initiated by the Agricultural Departments of the Indian states of Tamil Nadu and Uttar Pradesh. Several Asian countries (e.g., China, Burma, Vietnam, Thailand, and the Philippines) have also begun work to popularize this practice. The ultimate success of this technology will depend on cooperation among researchers, extension personnel, and the ultimate users—farmers. The present trend and interest seem to show that the application of cyanobacteria in rice cultivation is already well on the way to successful exploitation (Venkataraman, 1986).

Locally produced cyanobacterial biofertilizers could be readily used all over the world for rice paddy cultivation or for inoculating suitable soils to increase their productivity. Using cyanobacteria grown on local resources in this fashion offers economic opportunity without resorting to the negative elements of the so-called "green revolution" based on chemical fertilizers and pesticides, soil exhaustion, and, worst of all, dependence on costly imports and "know-how" from the developed world.

REFERENCES

Abeliovich, A., and Shilo, M., 1972, Photooxidative death in blue-green algae, *J. Bacteriol.* **111**:682.

Allen, M. M., 1984, Cyanobacterial cell inclusions, *Annu. Rev. Microbiol.* **33**:1–25.

Bar-Or, Y., and Shilo, M., 1988, Cyanobacterial flocculants, *Meth. Enzymol.* **167**:616–622.

Becker, E. W., and Venkataraman, L. V., 1982, Biotechnology and Exploitation of Algae: The Indian Approach, German Agency for Technical Cooperation, Eschborn, Germany.

Becker, E. W., Jakober, B., Luft, D., and Schmulling, R. M., 1986, Clinical and biochemical evaluations of the alga *Spirulina* with regards to its application in the treatment of obesity a double-blind cross-over, *Nutr. Rep. Int.* **33**:(4):565–574.

Boussiba, S., 1988, N₂-fixing cyanobacteria as nitrogen biofertilizer, a study with the isolate *Anabaena azollae, Symbiosis* **6**:129–138.

Broures, M., and Hall, D. O., 1986, Ammonia and hydrogen production by immobilized cyanobacteria, *J. Biochem.* **3**:307–321.

Calderon, J. F., Merino, H., and Barragan, M., 1976, Valor alimentico del algal espirulina *(Spirulina geiteri)* para ruminants, *Tec. Pecu. Mex.* **31**:42.

Castenholz, R. W., 1969, Thermophilic blue-green alga and the thermal environment, *Bacteriol. Rev.* **33**(4):476–504.

Clement, G., Rebeller, M., and Zarrouk, C., 1967, Wound treating medicaments containing algae, *Frivli Med.* **1967**:5279.

Cohen, Z., Vonshak, A., and Richmond, A., 1987, Fatty acid composition of *Spirulina* strains grown under various environmental conditions, *Phytochemistry,* **26**(8):2255–2258.

Collins, C. D., and Boylen, C. W., 1982, Ecological cosequences of long-term exposure of *Anabaena variabilis* (cyanophyceae) to shifts in environmental factors, *Appl. Environ. Microbiol.* **44**(1):141–148.

Craig, R., Reichelt, B. Y., and Reichelt, J. L., 1988, Genetic engineering of micro-algae, in: *Microalgal Biotechnology* (M. A. Borowitzka and L. J. Borowitzka, eds.), Cambridge University Press, Cambridge, pp. 415–455.

De, P. K., 1939, The role of bluegreen algae in nitrogen fixation in rice fields, *Proc. R. Soc. Lond. B* **127**:121.

De la Noue, J., Cloutier-Mantha, L., Walsh, P., and Picard, G., 1984, Influence of agitation and aeration modes on biomass production by *Oocystis* sp. grown on wastewaters, *Biomass* **4**:43–58.

Dodd, J. C., 1986, Elements of design and construction, in: *Handbook for Microalgal Mass Culture* (A. Richmond, ed), CRC Press, Boca Raton, Florida, pp. 265–283.

Fontes, A. G., Vargas, M. A., Moreno, J., Guerrero, M. G., and Losada, M., 1987, Factors affecting the production of biomass by a nitrogen-fixing blue-green alga in outdoor culture, *Biomass* **13**:33–43.

Fox, R. D., 1988, Nutrient preparation and low cost basin construction for village production of *Spirulina*, in: *Algal Biotechnology* (T. Stadler, J. Mollion, M. C. Verdus, Y. Karamanos, H. Morvan, and D. Christiaen, eds.), Elsevier, Amsterdam, pp. 355–364.

Foy, R. H., and Gibson, C. E., 1982, Photosynthetic characteristics of planktonic blue-green algae: Changes in photosynthetic capacity and pigmentation of *Oscillatoria redekei* van goor under high and low light, *Br. Phycol. J.* **17**:183–193.

Furst, P. T., 1978, *Spirulina, Hum. Nature* **60**:60–65.

Goldman, J. C., 1979, Outdoor algal mass cultures. II. Photosynthetic yield limitations, *Water Res.* **13**:119–160.

Granoth, G., and Porath, D., 1984, An attempt to optimize feed utilization by *Tilapia* in a flow-through aquaculture, in: *Proceedings of the International Symposium on Tilapia and Aquaculture* (E. Fishelzon, ed.), Nazereth, Israel, pp. 550–558.

Gudin, C., and Chaumont, D., 1983, Solar biotechnology study and development of tubular solar receptors for controlled production of photosynthetic cellular biomass for methane production and specific exocellular biomass, in: *Energy from Biomass*, Series E. Volume 5 (W. Palz and D. Pirrwitz, eds.), D. Reidel, Dordrecht, pp. 184–193.

Guerrero, M., Ramos, J. L., and Losada, M., 1982, Photosynthetic production of ammonia, *Experientia* **38**:53.

Hendrickson, R. (ed.), 1989, *Earth Food Spirulina*, Ronore Enterprises, Laguna Beach, California.

Horrobin, D. F., 1983, The role of essential fatty acids and prostaglandins in the premenstrual syndrome, *J. Reprod. Med.* **28**:465–468.

Hou, R. H., and Chen, R. S., 1981, The blue-green alga, *Spirulina platensis*, as a protein source for artificial rearing of *Bombyx mori, Appl. Entomol. Zool.* **16**:169–171.

Hugh, W. I., Dominy, and Duerr, E., 1985, Evaluation of dehydrated *Spirulina (Spirulina platensis)* as a protein replacement in swine started diets, *Research Extension Series*, **1985**:3–10.

Kerifn, W., and Boger, P., 1982, Light-induced hydrogen evolution by blue-green algae (cyanobacteria), *Physiol. Plant* **54**:93–98.

Kosaric, N., Nguyen, H. T., and Bergougnou, M. A., 1974, Growth of *Spirulina maxima* in effluents from secondary waste-water treatment plants, *Biotechnol. Bioeng.* **16**:881–896.

Kumazawa, S., and Mitsui, A., 1981, Characterization and optimization of hydrogen photoproduction by salt water blue-green alga, *Oscillatoria* sp. Miami BG7. I. Enhancement through limiting the supply of nitrogen nutrients, *Int. J. Hydrogen Energy* **6**(4):339–348.

Lee, Y. K., Tan, H. M., and Hew, C. S., 1985, The effect of growth temperature on the bioenergetics of photosynthetic algal cultures, *Biotechnnol. Bioeng.* **27**:555–561.

Lincoln, E. P., Hall, T. W., and Koopman, B., 1983, Zooplankton control in mass algal cultures, *Aquaculture* **32**:331.

Mokadey, S., Yannai, S., Einav, P., and Berk, Z., 1979, Algae grown on wastewater as source of protein for young chickens and rats, *Nutr. Rep. Int.* **20**:31–39.

Moore, R. E., Cheuk, Yang, X.-Q. G., Patterson, G. M. L., Bonjouklian, R., Smitka, T. A., Mynderse, J. S., Foster, R. S., Jones, N. D., Swartzendruber, J. K., and Deeter, J. B., 1987, Hapalindoles, antibacterial and antimycotic alkaloids from the cyanoophyte *Hapalosiphon fontinalis, J. Org. Chem.* **52**:1036.

Mur, L. R., 1983, Some aspects of the ecophysiology of cyanobacteria, *Ann. Microbiol.* **134B**:61–72.

Padgett, M. P., and Krogmann, D. W., 1987, Large scale preparation of pure phycobiliproteins, *Photosynth. Res.* **11**:225–235.

Phlips, E. J., and Mitsui, A., 1981, Characterization and optimization of hydrogen production by salt water blue-green alga, *Oscillatoria* sp. Miami BG7. II. Use of immobilization for enhancement of hydrogen production, *Int. J. Hydrogen Energy* **11**(2):83–89.

Phlips, E. J., and Mitsui, A., 1984, Development of hydrogen production activity in the marine blue-green alga *Oscillatoria* sp. Miami BG7 under natural sunlight conditions, in: *Advanced Photosynthesis Research*, Volume II (C. Sybesma, ed), Martinus Nijhoff/Dr. W. Junk, The Hague, pp. 801–804.

Pirt, S. J., 1986, Tansley review no. 4, The thermodynamic efficiency (quantum demand) and dynamics of photosynthetic growth, *New Phytol.* **102:**3–37.

Pirt, S. J., Lee, Y. K., Walach, M. R., Pirt, M. W., Balyuzi, H. H. M., and Bazin, M. J., 1983, A tubular bioreactor for photosynthetic production of biomass from carbon dioxide: Design and performance, *J. Chem. Technol. Biotechnol.* **33B:**35–58.

Radmer, R. J., and Kok, B., 1976, Photoreduction of O-2 primes, *Plant Physiol.* **58:**336.

Ramos, J. L., Guerrero, M. G., and Losada, M., 1987, Factors affecting the photoproduction of ammonia from dinitrogen and water by the cyanobacterium *Anabaena* sp. strain ATCC 33047, *Biotechnol. Bioeng.* **29:**566–571.

Richmond, A. (ed.), 1986, *CRC Handbook of Microalgal Mass Culture*, CRC Press, Boca Raton, Florida.

Richmond, A., 1987, The challenge confronting industrial microagriculture: High photosynthetic efficiency in large-scale reactors, *Hydrobiologia* **151/152:**117–121.

Richmond, A., 1988, *Spirulina*, in: *Microalgal Biotechnology* (A. Borowitzka and L. Borowitzka, eds.) Cambridge University Press, Cambridge, pp. 83–121.

Richmond, A., and Grobbelaar, J. U., 1986, Factors affecting the output rate of *Spirulina platensis* with reference to mass cultivation, *Biomass* **10:**253–264.

Richmond, A., and Vonshak, A., 1978, *Spirulina* culture in Israel, *Arch. Hydrobiol. Beih. Ergebn. Limnol.* **11:**274–280.

Richmond, A., Karg, S., and Boussiba, S., 1982, Effect of bicarbonate and carbon dioxide on the competition between *Chlorella vulgaris* and *Spirulina platensis*, *Plant Cell Physiol.* **23(8):**1411–1417.

Roger, P. A., and Watanabe, I., 1984, Algae and aquatic weeds as source of organic matter and plant nutrients for wetland rice, *Organic Matter and Rice*, International Rice Research Institute, Los-Banos, Philippines, pp. 147–168.

Roger, P. A., and Watanabe, I., 1985, Technologies for utilizing biological nitrogen fixation in wetland rice: Potentialities, current usage and limiting factors, *Fert. Res.* **9:**39–77.

Roger, P. A., Santiago-Ardales, S., P. M., and Watanabe, I., 1987, The abundance of heterocystous blue-green algae in rice soils and inocula used for application in rice fields, *Biol. Fert. Soils* **5:**98–105.

Samson, R., and LeDuy, A., 1986, Detailed study of anaerobic digestion of *Spirulina maxima* algal biomass, *Biotechnol. Bioeng.* **28:**1014–1023.

Samuelsson, G., Lonneborg, A. L., Rosenqvist, E., Gustafsson, P., and Oquist, G., 1985, Photoinhibition and reactivation of photosynthesis in the cyanobacteria *Anacystis nidulans*, *Plant Physiol.* **79:**992–995.

Schwartz, J., Shklar, S., Reid, S., and Trickler, D., 1988, Prevention of experimental cancer by extracts of *Spirulina-Dunaliella* algae, *Nut. Cancer* **11:**127–134.

Singh, R. N., 1942, The fixation of elementary nitrogen by some of the commonest bluegreen algae from the paddy field soils of the United Provinces and Bihar, *Indian J. Agric. Sci.* **2:**743.

Seshadri, C. V., and Thomas, S., 1979, Mass culture of *Spirulina* using low-cost nutrients, *Biotechnol. Lett.* **1:**287–291.

Soeder, C. J., 1986, An historical outline of applied algology, In: *Handbook of Microalgal Mass Culture* (A. Richmond, A., ed), CRC Press, Boca Raton, Florida, pp. 25–41.

Tirol, A. C., Roger, P. A., and Watanabe, I., 1982, Fate of nitrogen from a blue-green alga in a flooded rice soil, *Soil Sci. Plant Nutr.* **28:**559–569.

Tomasselli, L., Torzillo, G., Giovanneti, L., Pushparaj, B., Bocci, F., Tredici, M., Papuzzo, T., Baloni, W., and Materassi, R., 1987, Recent research on *Spirulina* in Italy, *Hydrobiologia* **151:**79–82.

Torzillo, G., Giovannetti, L., Bocci, F., and Materassi, R., 1984, Effect of oxygen concentration on the protein content of *Spirulina* biomass, *Biotechnol. Bioeng.* **26:**1134–1135.

Torzillo, G., Pushparaj, B., Bocci, F., Balloni, W., Materassi, R., and Florenzano, G., 1986, Production of *Spirulina* biomass in closed photobioreactors, *Biomass* **11:**61–74.

Van Liere, L., and Mur, L. R., 1979, Growth kinetics of *Oscillatoria agardhii* gomont in continuous culture, limited in its growth by the light energy supply, *J. Gen. Microbiol.* **115:**153–160.

Venkataraman, L. V., 1986, Blue-green algae as biofertilizer, in: *Handbook of Microalgal Mass Culture* (A. Richmond, ed.), CRC Press, Boca Raton, Florida, pp. 455–472.

Vonshak, A., 1987, Strain selection of *Spirulina* suitable for mass production, *Hydrobiologia* **151/152:**75–77.

Vonshak, A., 1988, Photoinhibition as a limiting factor in outdoor cultivation of *Spirulina platensis,* in: *Algal Biotechnology* (T. Stadler, J. Mollion, M. C. Verdus, Y. Karamanos, H. Morvan, and D. Christiaen, eds.), Elsevier, Amsterdam, pp. 365–370.

Vonshak, A., and Richmond, A., 1988, Mass production of the blue-green alga *Spirulina:* An overview, *Biomass* **15:**233–247.

Vonshak, A., Abeliovich, A., Boussiba, S., and Richmond, A., 1982, Production of *Spirulina* biomass: Effects of environmental factors and population density, *Biomass* **2:**175–186.

Vonshak, A., Boussiba, S., Abeliovich, A., and Richmond, A., 1983, Production of *Spirulina* biomass: Maintenance of microalgal culture outdoors, *Biotechnol. Bioeng.* **25:**341–351.

Watanabe, I., 1984, Use of symbiotic and free-living blue-green algae in rice culture, *Outlook Agric.* **13:**166–172.

Watanabe, A., Nishigaki, S., and Konishi, O., 1951, Effect of nitrogen fixing bluegreen algae on the growth of rice plants, *Nature* **168:**748.

Watanabe, I., De Datta, S. K., and Roger, P. A., 1987, Nitrogen cycling in wetland rice soils, in: *Advances in Nitrogen Cycling in Agriculture Ecosystems* (J. R. Wilson, ed.), C.A.B. International, United Kingdom, pp. 239–256.

Wright, S., and Burton, J., 1982, A controlled trial of the treatment of attopic eczema in adults with evening primrose oil (Efamol) *Lancet* **73:**372–375.

Zevenboom, W., and Mur, L. R., 1984, Growth and photosynthetic response of the cyanobacterium *Microcystis aeruginosa* in relation to photoperiodicity and irradiance, *Arch. Microbiol.* **139:**232–239.

Cyanobacterial Toxins 7

HANS UTKILEN

1. INTRODUCTION

The first scientific account of a cyanobacterial bloom causing the death of animals was published by Francis (1878), and subsequent incidents involving the intoxication of animals up until 1961 were excellently reviewed by Schwimmer and Schwimmer (1968). Ingram and Prestcott (1954) reviewed reports on the intoxication of animals and birds by cyanobacteria in Minnesota for the period 1882–1933; five species, *Anabaena flos aquae, Aphanizomenon flos-aquae, Coelosphaerium keutzigianum, Gloeotrichia echinulata,* and *Microcystis aeruginosa,* were involved in the reported cases. A similar report on Minnesota for the period 1948–1950 was published by Olson (1964). Today reports implicating cyanobacteria in the sickness and death of livestock, pets, birds, and wildlife come from all over the world (Gorham and Carmichael, 1988). Some reports indicate that human intoxication has occurred after consumption of water contaminated by cyanobacterial blooms (Schwimmer and Schwimmer, 1968; Dillenberg and Dehnel, 1960; Falconer *et al.*, 1983a).

Falconer *et al.* (1983a) reported an epidemiological study of a human population which obtained their drinking water from a reservoir exhibiting seasonal blooms of a hepatotoxic *Microcystis aeruginosa.* A comparison with a population obtaining their drinking water from another source unaffected by cyanobacterial blooms clearly indicated that that the pattern of admission of patients to hospital with liver complaints coincided with the seasonal cyanobacterial blooms. The lysis of dense cyanobacterial blooms associated with public water supplies has been implicated in gastrointestinal disorders in the United States, East Africa, and Australia (Falconer, 1989); the organisms involved are species of *Microcystis, Anabaena, Aphanizomenon,* and *Os-*

HANS UTKILEN ● Department of Environmental Medicine, National Institute of Public Health, Geitmyrsveien 75, 0462 Oslo 4, Norway.
Photosynthetic Prokaryotes, edited by Nicholas H. Mann and Noel G. Carr. Plenum Press, New York, 1992.

cillatoria. A variety of other reviews dealing with episodes of intoxication, toxigenic strains, and the properties and modes of action of cyanobacterial toxins have been published (Gorham, 1964a,b; Collins, 1978; Moore, 1981; Codd, 1984; Skulberg *et al.*, 1984; Carmichael *et al.*, 1985; Gorham and Carmichael, 1988; Codd *et al.*, 1989).

Veterinarians have been aware of the toxic effects of cyanobacteria for at least a century (Francis, 1878), while public health authorities and the water supply industry have only recently appreciated the possible health risk associated with cyanobacterial blooms.

In a recent issue of *The Toxicologist* (February 1990) several abstracts dealing with cyanobacterial toxins are published, covering uptake and localization of one particular cyanobacterial toxin, microcystin, and evaluation of chemoprotectants directed against it. This, taken together with the steady appearance of new reports about toxigenic cyanobacteria (Meriluoto *et al.*, 1989; Sivonen *et al.*, 1989a,b; Jinno *et al.*, 1989; Skulberg *et al.*, 1992), improved methods of toxin detection (Brooks and Codd, 1986; Meriluoto and Eriksson, 1988; Codd *et al.*, 1989), and uptake and toxic effects, indicate the rapidly awakening interest and concern regarding cyanobacterial toxins. This research activity should provide an understanding of the occurrence and properties of cyanobacterial toxins in freshwater environments and the necessary background information for the water industry and public health authorities to make informed decisions and the correct health risk assessments.

2. TOXIN-PRODUCING CYANOBACTERIA

There are approximately 1500 known species of cyanobacteria, though the number varies depending on the taxonomic authority (Gibson and Smith, 1982). However, the toxin-producing species constitute a relatively small proportion of the total (Table I). The most common toxigenic isolates in fresh- or brackish water include species of *Anabaena, Coelosphaerium, Gloeotrichia, Gomphospharia, Nodularia,* and *Oscillatoria.* In marine environments toxigenic strains of *Lyngbya, Schizothrix,* and *Oscillatoria* are most frequently encountered. The toxin-producing species usually posses gas vacuoles, which, under specific conditions, cause the cells to float at the surface, forming dense blooms. The toxin content of the water column will therefore often be highest at the surface and decrease with increasing depth during the course of a bloom. The density of a bloom will also be highest near the shore to which the wind is blowing and consequently toxin concentrations will decrease with increasing distance from that shore.

The number of bird and animal deaths caused by a cyanobacterial bloom is illustrated by the case of a bloom of *Anabaena flos-aquae* in Storm Lake, Iowa in 1952 (Rose, 1953) when 5000–7000 gulls, 560 ducks, 200

Table I. Species of Cyanobacteria Which Have Been Confirmed to Have Toxin-Producing Strains[a]

Species	Ref.
Coccogonophycideae	
Coelosphaerium kützingianum Näg.	Fitch *et al.* (1934)
Gomphosphaeria lacustris Chod.	Gorham and Carmichael (1988)
Gomphosphaeria nägeliana (Unger) Lemm.	Berg *et al.* (1986)
Microcystis aeruginosa Kütz.	Hughes *et al.* (1958)
Microcystis cf. *botrys* Teil.	Berg *et al.* (1986)
Microcystis viridis (A. Br.) Lemm.	Watanabe *et al.* (1986)
Microcystis wesenbergii Kom.	Gorham and Carmichael (1988)
Synechococcus Nägeli sp. (strain Miami BCII 6S)	Mitsui *et al.* (1983)
Synechococcus Nägeli sp. (strain ATCC 18800)	Amann (1977)
Synechocystis Sauvageau sp.	Lincoln and Carmichael (1981)
Hormogonophycideae	
Anabaena circinalis Rabenh.	May and McBarron (1973)
Anabaena flos-aquae (Lyngb.) Breb.	Porter (1887)
Anabaena hassallii (Kütz.) Wittr.	Andrijuk *et al.* (1975)
Anabaena lemmermanni P. Richt.	Fitch *et al.* (1934)
Anabaena spiroides var. contracta Kleb.	Beasly *et al.* (1983)
Anabaena variabilis Kütz.	Andrijuk *et al.* (1975)
Anabaenopsis milleri Woron.	Lanaras *et al.* (1989)
Aphanizomenon flos-aquae (L.) Ralfs	Jackim and Gentile (1968)
Cylindrospermum Kützing sp.	Sivonen *et al.* (1989b)
Cylindrospermopsis raciborskii (Wolos.) Seenaya & Subbu Raju	Hawkins *et al.* (1985)
Fischerella epiphytica Ghose	Ransom *et al.* (1978)
Gloeotrichia echinulata (J. E. Smith) P. Richter	Ingram and Prescott (1954)
Hormothamnion enteromorphoides Grun.	Gerwick *et al.* (1986)
Lyngbya majuscula Harvey	Grauer and Arnold (1961)
Nodularia spumigena Mertens	Francis (1878)
Nostoc linckia (Roth) Born. et Flah.	Ransom *et al.* (1978)
Nostoc paludosum Kütz.	Andrijuk *et al.* (1975)
Nostoc rivulare Kütz.	Davidson (1959)
Nostoc zetterstedtii Areschoug	Mills and Wyatt (1974)
Oscillatoria acutissima Kuff.	Barchi *et al.* (1984)
Oscillatoria agardhii/rubescens group	Østenvik *et al.* (1981)
Oscillatoria formosa Bory	Skulberg *et al.* (1992)
Oscillatoria nigroviridis Thwaites	Mynderse *et al.* (1977)
Oscillatoria Vaucher sp.	Sivonen *et al.* (1989b)
Pseudanabaena catenata Lauterb.	Gorham *et al.* (1982)
Schizothrix calcicola (Ag.) Gom.	Mynderse *et al.* (1977)
Scytonema pseudohofmanni Bharadw.	Moore *et al.* (1986)
Tolypothrix byssoidea (Hass.) Kirchn.	Barchi *et al.* (1983)
Trichodesmium erythraeum Ehrb.	Feldmann (1932)

[a]The references represent selected key papers and the earlier papers deal primarily with field investigations of cases of intoxication. This list was compiled and provided by O. Skulberg, Norwegian Institute for Water Research.

pheasants, 2 hawks, 50 squirrels, 18 muskrats, 15 dogs, 4 cats, and 2 pigs died: a toll reminiscent of that for an Edwardian shooting party. Mice fed with lakewater during the period of this bloom died within 15–20 min. Diseases associated with blooms have even been described for insects, there being one report that indicates such an intoxication of honeybees (May and McBarron, 1973).

Olson (1964) pointed out that of 92 cyanobacterial blooms investigated, 49 produced toxins. An investigation of a large number of freshwater blooms in Europe revealed toxin production in over 50% of cases (Codd *et al.*, 1989). There is, consequently, a good case to assume that when a cyanobacterial bloom occurs it is more likely than not to be toxin-producing.

Within an individual water body the toxin concentration per unit cyanobacterial biomass may vary widely from week to week at the same sampling station (Codd and Bell, 1985). The concentration of toxin varies not only for different strains of *Anabaena flos-aquae,* but also for different clones of the same isolate (Gorham *et al.*, 1964; Carmichael and Gorham, 1977). These workers found a mixed pattern of high and low toxicity per unit cyanobacterial biomass when sampling in a lake on a single occasion. An investigation over a 3-year period (1984–1986) (Benndorf and Henning, 1989) showed that each year *Microcyctis* sp. was nontoxic at the beginning of the growing season, but developed high toxicity during the first strong biomass increase in the summer. The increase in toxicity, measured in terms of LD50 for mice, was found to occur at different times each year.

These variations in the toxicity of cyanobacterial blooms contribute to the problem of hazard assessment associated with drinking water supplied to humans and animals and for human leisure use.

3. NATURE AND PROPERTIES OF THE TOXINS

Although a range of toxins are known to be produced by cyanobacteria, only a few have been fully chemically characterized (Fig. 1) and their mode of action established (Gorham and Carmichael, 1988). The possible pharmacological use of these toxins remains completely unexplored. The toxins so far isolated and identified include alkaloids, oligopeptides, lipopolysaccharides (LPS), debromoaplysia toxin, and lyngbyatoxin (Fig. 1). The alkaloids paralyze skeletal and respiratory muscles and can cause death by respiratory arrest within minutes of exposure. The peptide toxins act more slowly and their target is the liver, where they cause necrosis, resulting in death by hemorrhagic shock or liver failure, after a period ranging from a few hours to a few days (Gorham and Carmichael, 1988). The phenolic bislactones, aplysiatoxin, and debromoaplysiatoxin are skin irritants.

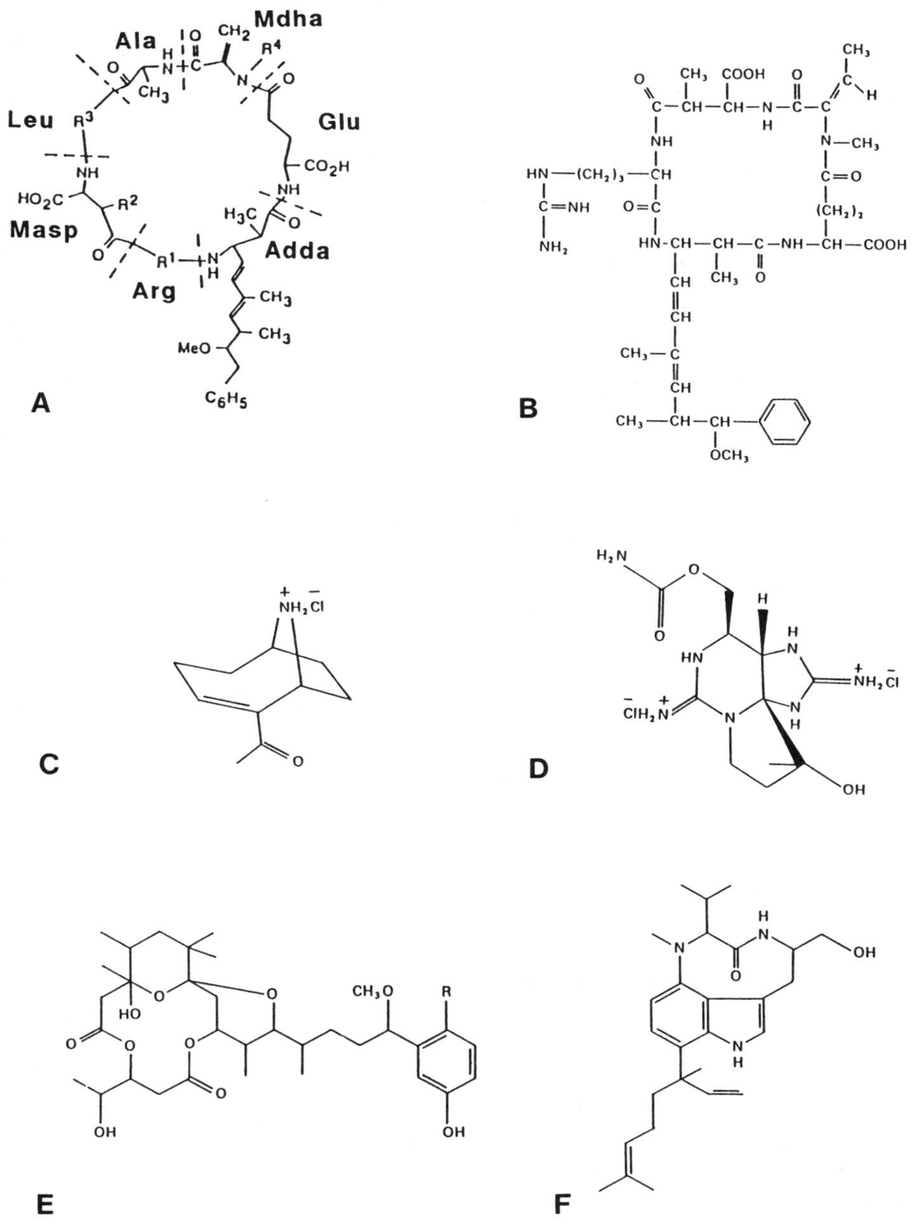

Figure 1. Structure of some cyanobacterial toxins. (A) Microcystin, (B) *Nodularia* toxin, (C) anatoxin-a, (D) aphanotoxin (saxitoxin), (E) lyngbiatoxin-a, (F) debromoaplysiatoxin. Arg, arginine; Masp, methyl iso-aspartate; Leu, leucine; Ala, alanine; Mdha, methyl-dehydroalanine; Glu, glutamate; Adda, 3-amino-9-methoxy-2,6,8-trimethyl-10-phenyldeca-4,6-dienoic acid.

Animals and birds have been intoxicated and killed by *Anabaena* blooms (Rose, 1953; Olson, 1964; Carmichael *et al.*, 1977; McBarron *et al.*, 1975). Six distinct toxins, called anatoxins, are produced by strains of *Anabaena* (Carmichael and Gorham, 1978). The anatoxins [Antx-a, Antx-a(s), Antx-b, Antx-b(s), Antx-c, and Antx-d] are distinguished on the basis of the strains which produce them and by the reactions exhibited by mice, rats, and chicks after interperitoneal injection (Carmichael and Gorham, 1978). The effects of Antx-a and Antx-b are more rapid than those of Antx-b and Antx-b(s), while Antx-c is the slowest acting of the anatoxins (Carmichael and Gorham, 1978). Antx-a (Fig. 1), from strain NRC-44-1, was the first toxin from a freshwater cyanobacterium to be characterized. It is a secondary amine with a molecular weight of 166 daltons and acts as a potent postsynaptic depolarizing neuromuscular blocking agent (Carmichael *et al.*, 1975; Spivak *et al.*, 1980; Aronstam and Witkop, 1981). The toxin has been chemically synthesized from cocaine by Campbell *et al.* (1977) and was shown to exhibit the same properties as the authentic toxin (Carmichael *et al.*, 1979). The lethal dose for Antx-a after interperitoneal administration to mice is 250 μg kg^{-1} (Devlin *et al.*, 1977), however, different species exhibit different sensitivities, the lethal oral dose for ducks and calves being about one-quarter that for mice and rats (Carmichael and Gorham, 1977). Carmichael *et al.* (1979) showed that different species responded differently to Antx-a and that the neurological effects were the same as those obtained with nicotine, muscarine, and decametonium. Antx-a(s) from *A. flos-aquae* strain 525-17 is different from Antx-a (Carmichael and Gorham, 1978) in that, although it causes some of the effects exhibited by Antx-a, it also causes viscous salivation and tear flow (sometimes bloody), as well as urinary incontinence and defecation in mice and rats prior to respiratory arrest (Gorham and Carmichael, 1988). The LD50 for Antx-a(s) after interperitoneal administration to mice is 40–50 μg kg^{-1} and the survival time is 30–60 min. The other four anatoxins are recognized by their different physiological effects (Carmichael and Gorham, 1978, 1981).

Blooms of *Aphanizomenon flos-aquae* have resulted in the intoxication of animals and fish (Prescott, 1948; Ingram and Prescott, 1954; Phinney and Peck, 1961; Gentile and Maloney, 1969). Toxins from *Aphanizomenon flos-aquae* blooms were first isolated by Sawyer *et al.* (1968) and were shown to have a more rapid toxic effect in mice than Antx-a. Injection of the most toxic fraction from *Aphanizomenon flos-aquae* in *Fundulus heteroclitus* resulted in rapid death of the fish and the lowest lethal dose was 29 mg kg^{-1} (Jackim and Gentile, 1968). Fish were shown to be more sensitive to *Aphanizomenon flos-aquae* toxin than mice (Gentile and Maloney, 1969). The toxin from *Aphanizomenon flos-aquae* exhibited the same toxic effects as that from the marine alga *Gonyaulax catenella* and another neurotoxin, tetrodotoxin (Sawyer *et al.*, 1968). In fact, *Aphanizomenon flos-aquae* was shown to produce two neurotoxic alkaloids, neosaxitoxin and saxitoxin

(Jackim and Gentile, 1968; Sawyer *et al.*, 1968; Sasner *et al.*, 1984; Mahmood and Carmichael, 1986). The LD50 for intraperitoneal administration to mice is 10 μg kg^{-1} for both toxins (Carmichael *et al.*, 1985). These toxins have a fast-acting neuromuscular action, inhibiting nerve conduction by blocking sodium channels, without affecting permeability to potassium, the transmembrane potential, and membrane resistance (Sasner *et al.*, 1981; Adelman *et al.*, 1982). Another cyanobacterium which has recently been found to produce a novel neurotoxin is a species of *Oscillatoria* isolated at the Norwegian Institute for Water Research (Skulberg *et al.*, 1991)

Toxic blooms of *Microcystis* spp. have probably resulted in more damage than any other cyanobacterium. *Microcystis* spp. have been involved in the intoxication of a wide range of animals throughout the world (Fitch *et al.*, 1934a,b; Ingram and Prescott, 1954; Schwimmer and Schwimmer, 1968; McDonald, 1960; Gorham, 1964a,b; McBarron and May, 1966; Soll and Williams, 1985). The deaths of fish are also associated with *Microcystis* spp. blooms (Shelubski, 1951; Yamagiski and Aoyama, 1972; Eriksson *et al.*, 1986).

Six chemically related, low-molecular-weight, toxic peptides have been shown to be produced by *Microcystis aeruginosa* and *Anabaena flos-aquae* (Gorham and Carmichael, 1988) and are generically refererred to as microcystins. Some strains of *Microcystis* produce a single toxin, others four, and some as many as six, but in all cases one or two toxins account for 90% of the total toxicity. A series of structural studies (Botes *et al.* 1982a,b, 1984, 1985; Krishnamurthy *et al.*, 1989) on five of the toxins indicated that they all belonged to a family of seven-residue cyclic peptides which had the general structure, cyclo-D-Ala-L-Xaa-*erythro*-β-methyl-D-isoaspartic acid-L-Yaa-Adda-D-isoglutamic acid-*N*-methyldehydroalanine, where Xaa and Yaa represent variable L-amino acids and Adda is 3-amino-9-methoxy-2,6,8-trimethyl-10-phenyldeca-4,6-dienoic acid. Microcystin-LR, in which the two variable amino acids are leucine and arginine, is the most commonly found species of microcystin.

The intraperitoneal LD50 for mice of the toxic peptides produced by *Microcystis* spp. and *Anabaena flos-aquae* is 50 μg kg^{-1} (Botes *et al.*, 1982a,b; Krishnamurthy *et al.*, 1986). The peptide toxin causes rapid and extensive centribular necrosis of the liver with loss of characteristic architecture of hepatic cords. Both hepatocytes and endothelial cells are destroyed with vesiculation of cell membranes (Østvenik *et al.*, 1981; Runnegar and Falconer, 1981; Foxall and Sasner, 1981). By using microcystin labeled with ^{125}I, Falconer *et al.* (1986) showed that the liver was the main target for accumulation and disposition of the toxin. Wheeler *et al.* (1942) and McDonald (1960) showed that intraperitoneal injection of fresh untreated *Microcystis* sp. did not kill mice, while injection of dried material caused death. This indicates that the cyanobacterial cells have to be damaged in order to release the toxin. The LD50 for microcystin is 30–50 times higher for oral compared to intraperitoneal administration (Konst *et al.*, 1965; Heaney, 1971).

Recently, there have been considerable advances in understanding the mechanism of action of the most commonly found variant of the microcystins. MacKintosh *et al.* (1990) demonstrated that microcystin-LR is an extremely specific and potent inhibitor of protein phosphatases 1 and 2A in eukaryotic cells from phyla as diverse as mammals, protozoa, and plants. The K_i values for microcystin-LR on these phosphatases is below 0.1 nM. It is of striking interest that microcystin-LR appears to act in the fashion of okadaic acid, a C_{44} polyketal fatty acid, by binding to the same site on protein phosphatases 1 and 2A. In this context it is important to note that okadaic acid is not only responsible for diarrhetic shellfish poisoning, but is also a very potent tumor promoter. It has not yet been established, but would appear possible, that microcystin-LR may have the same tumor-promoting activity. It is also interesting to speculate what the biological function of microcystin-LR is with the *Microcystis* cell since Mackintosh *et al.* (1990) were unable to detect any protein phosphatase activities resembling those of phosphatase 1 and 2A in nine different species of freshwater and marine cyanobacteria, although other protein phosphatase activities have been found in *Anabaena* sp. PCC7120 (N. H. Mann, personal communication).

Like *Anabaena flos-aquae* NRC-44, *Microcystis aeruginosa* produces a mixture of toxins, and toxin production is an unstable phenotype even within a single clone (Gorham, 1964a,b). Interestingly, it was shown that while domestic species of birds were sensitive to the toxin, wild species were unaffected (Gorham, 1964a,b). Indeed, ducks were shown to be more resistant than other animals and could tolerate doses of 16 g kg^{-1} (Konst *et al.*, 1965). This enormous increase in resistance makes it seem likely that birds living permanently on a lake subject to cyanobacterial blooms may evolve a resistance (or develop an immunity) to the toxin.

A diarrhea-inducing toxin has also been isolated from *Microcystis* spp. (Aziz, 1974; Malyarevskaya *et al.*, 1970), and fish deaths due to increased thiaminase activity, leading to vitamin B_1 shortage, have been reported (Malyarevskaya *et al.*, 1970). Cyanobacterial thiaminase accumulated by fish was suspected to be involved in Haff's disease, which was common among people in fish-eating communities along the Baltic coast in the 1920s and 1930s (Berlin, 1948). One report (Solomatina and Matchinskaya, 1972) implicates a cyanobacterial toxin in the inhibition of transamination reactions.

The toxins produced by *Anabaena flos-aquae*, *Aphanizomenon flos-aquae*, and *Microcystis aeruginosa* have been studied to a much greater extent than those produced by other cyanobacteria because they are the most common bloom-forming species which have been linked to animal deaths. However, there are other cyanobacteria which produce peptide and alkaloid toxins and, indeed, species which produce toxins of distinct structural natures. *Anabaena hassalii, Anabaena variabilis, Calothrix* spp., and *Nostoc linkia* have been shown to produce toxins (Andrijuk *et al.*, 1975). Amann (1972) demonstrated that *Synechococcus* sp. ATCC 18800 was toxic to fish and that the toxin was a pteridine, consisting of a chromatophore and a seven-amino acid oligopeptide that differed from the other toxic peptides characterized from cyanobacteria. The cyanobacteria *Fischerella epiphytica, Gloeotrichia echinulata*, and *Nostoc linkia* were found to be toxic for *Paramecium caudatum*.

Like other Gram-negative bacteria, the cell envelope of cyanobacteria contains lipopolysaccharide (LPS), the structure of which, in the few cyanobacteria so far studied, exhibits certain resemblances to the structural endotoxin of *E. coli* and *S. typhimurium*, but also several differences (Drews and Wechesser, 1982). Utkilen *et al.* (1991) examined LPS from species of *Oscillatoria, Anabaena*, and *Microcystis* and found in all cases that the LPS contained 3-hydroxyoctanoate (3-OH-18:O) as a major constituent (Fig. 2). This fatty acid is rare in the LPS from other Gram-negative bacteria (Wilkinson, 1988) and consequently, in environmental samples, this fatty acid probably can be used to distinguish cyanobacterial LPS from that of other Gram-negative bacteria.

The toxicity of cyanobacterial LPS seems to be much weaker than that from other groups of bacteria (Keleti *et al.*, 1981). However, in spite of this, the cyanobacterial LPS from blooms of *Anabaena* sp. and *Schizothrix calcicola* may have contributed to extensive outbreaks of human gastroenteritis, diarrhea, cramps, nausea, and dizziness in the United States (Lippy and Erb, 1976; Billings, 1981; Carmichael *et al.*, 1985).

Cohen and Reif (1953) described several episodes of erythematous papulovescicular dermatitis among swimmers after contact with *Anabaena* sp. Subsequently, there have been many accounts of dermatitis and/or irritation following contact with freshwater cyanobacteria (Skulberg *et al.*, 1984; Carmichael *et al.*, 1985; Codd and Bell, 1985) and some marine cyanobacteria have been reported to produce dermatotoxins (Moore, 1981). The lyngbyatoxin A (an indole alkaloid) and debromoaplysiatoxin (a phenolic alkaloid) are also potent tumor promoters, as are the related compounds from *Schizothrix calcicola* and *Oscillatoria nigroviridis* (Moore, 1984a,b; Fuijki *et al.*, 1984).

Many cyanobacterial toxins have not yet been isolated, but they are recognized by their pathological effects and survival times after ingestion. Gorham and Carmichael (1988) noted reports of a peptide toxin produced by a strain of *Microcystis aeruginosa* isolated in the Ukraine which produced neurotoxic effects on cardiac muscle and caused blood hemostasis. Watanabe *et al.* (1981) and Watanabe and Oishi (1982, 1985) reported a toxic peptide produced by *Microcystis aeruginosa* that has a molecular weight in the range 770–2950 daltons and which exhibits an LD100 of 2 mg kg^{-1} after interperitoneal administration to mice. Lincoln and Carmichael

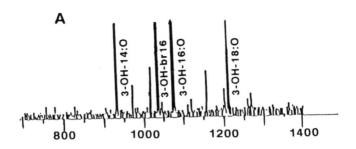

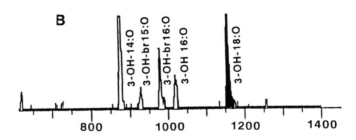

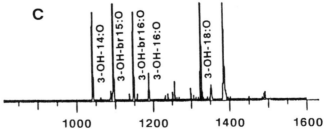

Figure 2. Mass spectrometry profiles of 3-OH-fatty acids from (A) *Anabaena flos-aquae,* (B) *Microcystis aeruginosa,* (C) *Oscillatoria bornetti.*

(1981) described a hepatotoxic species of *Synechocystis* which had a lethal dose of approximately 500 mg kg^{-1} and caused death some 5–7 hr after administration.

4. PURIFICATION AND QUANTIFICATION OF TOXINS

The conventional method for the purification of cyanobacterial peptide toxins involves freezing the cells, followed by solvent extraction and concentration of the toxin by absorption in C_{18} cartridges. The toxins are then purified by reversed-phase high-performance liquid chromatography (Siegelman *et al.*, 1984; Broooks and Codd, 1986; Dierstein *et al.*, 1988). The use of an internal-surface reversed-phase (ISRP) column for peptide toxin purification was introduced by Meriluoto and Eriksson (1988) as part of a small-scale procedure for toxin extraction utilizing sonication. Martin *et al.* (1990) developed a rapid isolation procedure for the peptide toxin from *Microcystis aeruginosa* by using octadecylsilyl solid-phase extraction. High-performance thin-layer chromatography has also been used for purification of the cyanobacterial peptide toxins (Poon *et al.*, 1987) and this approach has been applied to the sequential purification of peptide and alkaloid toxins from the same sample of *Anabaena flos-aquae* (Jamel Al-Layl *et al.*, 1988).

Until recently, both identification and quantification of cyanobacterial toxins were done by mouse bioassay, which has obvious shortcomings (Billings, 1981). Although the verification of the presence of a toxin still has to be done via bioassay, sensitive and specific assays are being developed, including chemical, cytotoxic, and immunological methods. HPLC-based methods have been developed to permit the rapid quantification of less than 1 μg of peptide or alkaloid toxins (Siegelman *et al.*, 1984; Brooks and Codd, 1986; Meriluoto and Eriksson, 1988). The presence of peptide toxins in natural or laboratory experiments can be screened for by using isolated rat hepatocytes (Aune and Berg, 1986; Berg and Aune, 1987). This *in vitro* test involves scoring toxin-induced morphological damage and enzyme leakage, and the testing has to be performed shortly after hepatocyte isolation.

Polyclonal and monoclonal antibodies have been produced against the peptide toxin from *Microcystis* sp. (Kfir *et al.*, 1986; Codd *et al.*, 1989) and immunoassays are being developed. An ELISA assay for the peptide toxin of *Microcystis aeruginosa* has been developed using polyclonal antisera raised in rabbits (Brooks and Codd, 1986). The antisera used for this ELISA assay were highly specific and did not cross-react with peptides from *Oscillatoria* or *Anabaena*, but gave a positive response with all toxic blooms of *Microcystis aeruginosa*.

5. FACTORS INFLUENCING TOXIN PRODUCTION

As mentioned earlier, the toxicity of cyanobacterial blooms may vary weekly (Codd and Bell, 1985), seasonally (Benndorf and Henning, 1989), and annually (Watanabe *et al.*, 1981). These variations can be explained by environmental factors that affect toxicity, alterations in the toxicity of individual strains, or alterations in the relative proportion of toxic and nontoxic strains in a population. Understanding the regulation of toxin synthesis may enable predictions to be made concerning toxic production in aquatic environments and should, therefore, constitute an important aspect of research on toxic cyanobacteria.

Toxin production in *Anabaena flos-aquae* NRC-44 is a function both of light intensity and growth temperature (Peary and Gorham, 1966). Light saturation of growth for this strain was found to occur at 3000 lux and the optimal growth temperature was 22°C (Gorham *et al.* 1964). Both toxin production and growth were enhanced regardless of the light intensity at 22.5°C compared to 15°C, though a light intensity of 5000 lux resulted in a 50% reduction in both growth and toxin production (the optimal light intensity for toxin production being 2000 lux). Toxin production in *Aphanizomenon* sp. is also a function of both temperature and light intensity (Gentile, 1971), with toxin production being reduced by about 50% at a light intensity of 5000 lux (Gentile and Maloney, 1969). Growth at 26°C enhanced toxin production compared to 20°C, and production was not influenced by the nitrogen source.

Again, in the case of *Microcystis aeruginosa* NRC-1, light and temperature play a role in determining toxin production, though the interrelationship is complex (Gorham, 1964b). At 25°C for a well-oxygenated culture toxin production was maximal at 25°C and independent of light intensity. Although the optimal growth temperature for this strain is in the range 25–32.5°C, toxin production decreased above 25°C, at high light intensity, becoming negligible at 32.5°C and in no case was maximum toxicity associated with maximum biomass. Indeed, in cultures grown at 25°C there was a marked decline in toxin concentration per cell between the fourth and fifth days of culture, without any evidence for a corresponding increase of toxin in the medium. Billings (1981) showed that for *Microcystis aeruginosa* UV-006 toxicity declined significantly at the onset of the stationary phase. In addition, toxin production was less at very high or very low light intensities and the optimal conditions for growth did not coincide with those for toxin production (van der Westhuizen and Eloff, 1985). Watanabe and Oishi (198), however, found the lowest degree of toxin production in the middle of exponential phase for *Microcystis aeruginosa* M228 and the toxicity of this strain was shown to increase about fourfold when the light intensity was increased from 7.53 to 30.1 E m^{-2} sec^{-1}, while decreases in the phosphorus or nitrogen in the medium had no significant effects

(Watanabe and Oishi, 1983). The reduction in toxicity of *Microcystis* sp. strains at sub- or supraoptimal growth temperatures has been confirmed by Codd *et al.* (1989) for a variety of isolates.

Continuous culture of a variety of toxic cyanobacteria, including species of *Anabaena, Aphanizomenon,* and *Microcystis,* has revealed that the most important factor affecting toxin production is light intensity (unpublished results from the author's laboratory). In addition, toxin production was enhanced at lower growth rates, and when nitrogen was limiting there was no change in the toxin/total protein ratio.

The amount of effort being expended on analysis of the regulation of toxin production is small compared to that invested searching for new toxins and establishing their mechanisms of action. Consequently, our ability to predict patterns of toxin production in aquatic environments is still extremely limited and restricts our ability to manage these systems properly as sources of drinking water or for recreational purposes.

6. EFFECTS ON HUMAN HEALTH

Possibly the first report of human intoxication by cyanobacteria is that of Farre (1844), who described the presence of an *Oscillatoria*-like organism in the feces of a patient suffering from severe abdominal cramps. Subsequently there has been a series of reports relating cyanobacteria to human disease (Tisdale, 1931; Heise, 1949; McElhenny *et al.,* 1962; Cohen and Reif, 1953; Dillenberg and Dehnel, 1960; Schwimmer and Schwimmer, 1964, 1968; Moiheka and Chu, 1981; Sargunar and Sargunar, 1979; Billings, 1981; Falconer *et al.,* 1983a; Hawkins *et al.,* 1985).

The cyanobacterium *Cylindrospermum raciborskii,* established as being hepatotoxic for mice, was found to be present in large quantities in a water supply prior to a severe outbreak of hepatoenteritis which affected 148 human consumers (Bourke *et al.,* 1983). Falconer *et al.* (1983a) demonstrated that, during the lysis of a *Microcystis* bloom caused by the addition of copper sulfate, the ensuing contamination of the water supply for the city of Armndale (Australia) resulted in liver injury to the regional population (Fig. 3). These findings show that cyanobacterial blooms in drinking water reservoirs can result in the intoxication of consumers. However, proper treatment of the water should obviate the potential problem, for as Falconer stated: "For municipal water supplies, removal of cyanobacterial toxins from potable water is a relatively straight-forward engineering proposition. Activated carbon has been found to be very effective in removing cyanobacterial peptide and alkaloid toxins" (Falconer *et al.,* 1983b).

There are routes, other than via the water supply, by which cyanobacterial toxins may lead to human disease. There are several reports of allergic reactions and dermatitis caused by cyanobacteria (Heise, 1949, 1951;

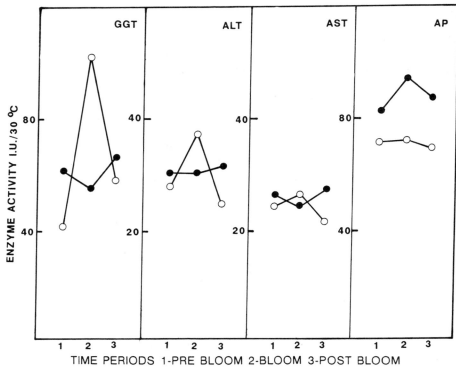

Figure 3. Serum enzyme activities of people living within the Armidale (Australia) water supply area (○) and outside it (△), before, during, and after a bloom of toxic *Microcystis* sp. in the water supply reservoir. GGT, Glutamyl transpeptidase; ALT, alanine aminotransferase; AST, aspartate aminotransferase; AP, alkaline phosphatase. [Redrawn from Falconer *et al.* (1983a).]

McElhenny *et al.*, 1962; Mittal *et al.*, 1979). Cyanobacteria were implicated in Haff's disease (see Section 3) that was common in the 1920s and 1930s among populations along the Baltic coast that had a substantial component of fish in their diet. This illustrates the point that the accumulation of cyanobacterial toxins in aquatic organisms which form part of the food chain to humans may represent another route for human exposure to cyanobacterial toxins that has received little attention. The importance of low-level exposure to cyanobacterial toxins may not be recognized; however, the strong resemblance of microcystin-LR to the potent tumor promoter okadaic acid in terms of mode of action (MacKintosh *et al.*, 1990) may necessitate a reappraisal of the epidemiological significance of long-term exposure.

REFERENCES

Adelman, W. J., Jr., Fohlmeister, J. F., Sasner, J. J., Jr., and Ikawa, M., 1982, Sodium channels blocked by aphatoxin obtained from the blue-green alga, *Aphanizomenon flos-aquae*, *Toxicon* **28**:513–516.

Amann, M., 1977, Untersuchungen uber ein pteridin als bestandteil des toxischen prinzips aus *Synechococcus*, Ph.D. thesis, Eberhard-Karls-University, Tubingen, Stuttgart.

Andrijuk, E. I., Kopteva, Z. P., Smirnova, M. M., Skopina, V. V., and Tantsjurenko, E. V., 1975, On problem of toxin-formation of blue-green algae, *Microbiol. Zh.* **37**:67–72.

Aronstam, R. S., and Witkop, B., 1981, Anatoxin-a interactions with cholinergic synaptic molecules, *Proc. Natl. Acad. Sci. USA* **70**:419–420.

Aune, T., and Berg, K., 1986, Use of freshly prepared rat hepatocytes to study toxicity of blooms of the blue-green algae *Microcystis aeruginosa* and *Oscillatoria agardhii*, *J. Toxicol. Environ. Health* **19**:325–336.

Aziz, K. M., 1974, Diarrhea toxin obtained from a waterbloom-producing species, *Microcystis aeruginosa* Kutz, *Science* **183**:1206–1207.

Barchi, J. J., Norton, T. R., Furusawa, E., Patterson, G. M. L., and Moore, R. E., 1983, Identification of cytotoxin from *Tolypothrix byssoidea* as tubercidin, *Phytochemistry* **22**:2851–2852.

Barchi, J. J., Moore, R. E., and Patterson, G. M. L., 1984, Acutiphycin and 20,21-didehydroacutiphycin, new antineoplastic agents from the cyanophyte *Oscillatoriaq acutissima*, *J. Am. Chem. Soc.* **106**:8193–8197.

Beasly, V. R., Cooppock, R. W., Simon, J., Ely, R., Buck, W. B., Coley, R. A., Carlson, D. M., and Gorham, P. R., 1983, Apparent blue-green algae poisoning in swine subsequent to ingestion of a bloom dominant by *Anabaena spiroides*, *J. Am. Vet. Med. Assoc.* **1982**:423–424.

Benndorf, J., and Henning, M., 1989, *Daphnia* and toxic blooms of *Microcystis aeruginosa* in Bautzen Reservoir (GDR), *Int. Rev. Gesamten Hydrobiol.* **74**:233–248.

Berg, K., and Aune, T., 1987, Freshly prepared rat hepatocytes used in screening the toxicity of blue-green algae, *J. Toxicol. Environ. Health* **20**:187–197.

Berg, K., Skulberg, O. M., Skulberg, R., Underdal, B., and Willen, T., 1986, Observations of toxic blue-green algae (Cyanobacteria) in some Scandinavian lakes, *Acta Vet. Scand.* **27**:440–452.

Berlin, R., 1948, Haff disease in Sweden, *Acta Med. Scand.* **129**:560–572.

Billings, W. H., 1981, Water-associated human illness in blue-green algae blooms, in: *The Water Environment: Algal Toxins and Health* (W. W. Carmichael, ed.), Plenum Press, New York, pp. 243–256.

Botes, D. P., Kruger, H., and Viljoen, C. C., 1982a, Isolation and characterization of four toxins from the blue-green alga *Microcystis aeruginosa*, *Toxicon* **20**:945–944.

Botes, D. P., Viljoen, C. C., Kruger, H., Wessels, P. L., and Williams, D. H., 1982b, Structure of toxins of the blue-green alga *Microcystis aeruginosa*, *S. Afr. J. Sci.* **78**:378–379.

Botes, D. P., Tuiman, A. A., Wessels, P. L., Viljoen, C. C., Kruger, H., Williams, D. H., Santikarn, S., Smith, R. J., and Hammond, S. J., 1984, The structure of cyanoginosin-LA, a cyclic heptapeptide toxin from the cyanobacterium *Microcystis aeruginosaul*, *J. Chem. Soc. Perkin Trans.* **1**:2311–2318.

Botes, D. P., Wessels, P. L., Kruger, H., Runnegar, M. T. C., Santikarn, S., Smith, R. J., Barna, J. C. J., and Williams, D. H., 1985, Structural studies on cyanoginosins-LR, -YR, -YH and -YM toxin, from *Microcystis aeruginosa*, *J. Chem. Soc. Perkin Trans.* **1**:2747–2748.

Bourke, A. T. C., Hawes, R. B., Neilson, A., and Stallmann, N. D., 1983, An outbreak of hepato-enteritis (the Palm Island Mystery Disease) possibly caused by algal intoxication, *Toxicon Suppl.* **3**:45–48.

Brooks, W. P., and Codd, G. A., 1986, Extraction and purification of toxic peptides from natural water blooms and laboratory isolates of the cyanobacterium *Microcystis aeruginosa*, *Lett. Appl. Microbiol.* **2**:1–3.

Campbell, H. F., Edwards, O. E., and Kolt, R. J., 1977, Synthesis of nor-anatoxin-a and anatoxin-a, *Can. J. Chem.* **55**:1372–1379.

Carmichael, W. W., and Gorham, P. R., 1977, Factors influencing the toxicity and animal susceptibility of *Anabaena flos-aquae* (Cyanophyta) blooms, *J. Phycol.* **13**:97–101.

Carmichael, W. W., and Gorham, P. R., 1978, Anatoxins from clones of *Anabaena flos-aquae* isolated from lakes of western Canada, *Mitt. Int. Ver. Limnol.* **21**:285–295.

Carmichael, W. W., and Gorham, P. R., 1981, The mosaic nature of toxic blooms of cyanobacteria, in: *The Water Environment: Algal Toxins and Health* (W. W. Carmichael, ed.), Plenum Press, New York, pp. 161–172.

Carmichael, W. W., Bigg, D. F., and Gorham, P. R., 1975, Toxicology and pharmacological action of *Anabaena flos-aquae* toxin, *Science* **187**:542–544.

Carmichael, W. W., Gorham, P. R., and Biggs, D. F., 1977, Two laboratory case studies on oral toxicity to calves of the freshwater cyanophyte (blue-green alga) *Anabaena flos-aquae* (NRC-44-1, *Can. Vet. J.* **18**:71–75.

Carmichael, W. W., Biggs, D. F., and Peterson, M. H., 1979, Pharmacology of anatoxin-a, produced by the freshwater cyanophyte *Anabaena flos-aquae* NRC-44-1, *Toxicon* **17**:229–236.

Carmichael, W. W., Jones, C. L. A., Mahmoood, N. A., and Theiss, W. C., 1985, Algal toxins and water-based diseases, *CRC Crit. Rev. Environ. Control* **15**:275–313.

Codd, G. A., 1984, Toxins of freshwater cyanobacteria, *Microbiol. Sci.* **1**:48–52.

Codd, G. A., and Bell, S. G., 1985, Eutrophication in freshwaters, *Water Pollut. Control* **84**:225–232.

Codd, G. A., Bell, S. G., and Brooks, W. P., 1989, Cyanobacterial toxins in water, *Water Sci. Technol.* **21**:1–13.

Cohen, S. G., and Reif, C. B., 1953, Cutaneous sensitization to blue-green algae, *J. Allergy* **24**:452-457.

Collins, M., 1978, Algal toxins, *Microbiol. Rev.* **42**:725–746.

Davidson, F. F., 1959, Poisoning of wild and domestic animals by a toxic waterbloom of *Nostoc rivulare* Kutz, *J. Am. Water Works Assoc.* **51**:1277–1287.

Devlin, J. P., Edwards, O. E., Gorham, P. R., Hunter, N. R., Pike, R. K., and Stavric, B., 1977, Anatoxin-a, a toxic alkaloid from *Anabaena flos-aquae* NRC-44-1, *Can. J. Chem.* **55**:1367–1731.

Dierstein, R., Kaiser, I., and Weckesser, J., 1988, Rapid determination of *Microcystis* sp. toxins by reversed-phase liquid chromatography, *FEMS Microbiol. Lett.* **49**:143–147.

Dillenberg, H. O., and Dehnel, M. K., 1960, Toxic waterblooms in Saskatchewan, 1959, *Can. Med. Assoc. J.* **83**:1151–1154.

Drews, G., and Weckesser, J., 1982, Function, structure and composition of cell walls and external layers, in: *The Biology of Cyanobacteria* (N. G. Carr and B. W. Whitton, eds.), Blackwell, Oxford, pp. 333–357.

Eriksson, J. E., Byland, G., and Lindholm, T., 1986, The toxin of the cyanobacterium *Microcystis aeruginosa* Kutz and its effects on common carp (*Cyprinus carpio* L.), in: *Abstracts of the Symposium on Toxicology* (A. Oikari and O. Pelkonen, eds.), University Joensuu Report Series, Finland, pp. 62–63.

Falconer, I. R., 1989, Effects on human health of some toxic cyanobacteria (blue-green algae) in reservoirs, lakes, and rivers, *Toxicity Assessment* **4**:175–184.

Falconer, I. R., Bereford, A. M., and Runnegar, M. T. C., 1983a, Evidence of liver damage by toxin from a bloom of the blue-green alga, *Microcystis aeruginosa*, *Med. J. Aust.* **1**:511–514.

Falconer, I. R., Runnegar, M. T. C., and Huynh, V. L., 1983b, Effectiveness of activated carbon in removal of algal toxin from potable water supplies: A pilot investigation, in: *Technical Papers of the 10th Federal Convention of the Australian Water and Wastewater Association*, Sydney, pp. 261–268.

Falconer, I. R., Buckley, T., and Runnegar, M. T. C., 1986, Biological half-life, organic distribution and excretion of ^{125}I labelled toxic peptides from the blue-green alga *Microcystis aeruginosa*, *Aust. J. Biol. Sci.* **39:**17–21.

Farre, A., 1844, On the minute structure of certain substances expelled from the human intestine, having the ordinary appearance of shred of lymph, but consisting entirely of filaments of confervoid type probably belonging to the genus *Oscillatoria*, *Trans. Microscop. Soc. Lond.* **1:**383–385.

Feldmann, J., 1932, Sur la biologie des *Trichodesmium* Ehrenberg, *Rev. Algol.* **6:**357–358.

Fitch, C. P., Bishop, L. M., Boyd, W. L., Gortner, R. A., Rogers, C. T., and Tilden, J. E., 1934, "Waterbloom" as a cause of poisoning in domestic animals, *Cornell Ver.* **24:**30–39.

Foxall, T. L., and Sasner, J. J., Jr., 1981, Effects of hepatic toxin from the cyanophyte *Microcystis aeruginosa*, in: *The Water Environment: Algal Toxins and Health* (W. W. Carmichael, ed.), Plenum Press, New York, pp. 365–387.

Francis, G., 1878, Poisonous Australian lakes, *Nature* **18:**11–12.

Fujiki, H., Suganama, M., Tahira, T., Yoshioka, A., Nakayasu, M., Endo, Y., Shudo, K., Takayama, S., Moore, R. E., and Sugimura, T., 1984, New classes of tumor promoters: Teleocidin, aplysiatoxin and palytoxin, in: *Cellular Interactions by Environmental Tumor Promoters* (H. Fujiki, ed.), Japanese Scientific Society Press/VNU Science Press, Utrecht, pp. 37–45.

Gentile, J. H., 1971, Blue-green and green algal toxins, in: *Microbial Toxins VII: Algal and Fungal Toxins* (S. Kadis, A. Ceigler, and S. J. Ajl, eds.), Academic Press, New York, pp. 27–66.

Gentile, J. H., and Maloney, T. E., 1969, Toxicity and environmental requirements of a strain of *Aphanizomenon flos-aquae* (L.) Ralfs, *Can. J. Microbiol.* **52:**75–87.

Gerwick, W. H., Lopez, A., van Duyne, G. D., Clardy, J., Ortiz, W., and Baez, A., 1986, Hormothamnoine, a novel cytotoxic styrylchromone from the marine cyanophyte *Hormothamnion enteromorphoides* Grunow, *Tetrahedron Lett.* **27:**1979–1982.

Gibson, C. E., and Smith, R. V., 1982, Freshwater plankton, in: *The Biology of Cyanobacteria* (N. G. Carr and B. A. Whitton, eds.), Blackwell, Oxford, pp. 463–489.

Gorham, P. R., 1964a, Toxic algae as a public health hazard, *J. Am. Water Works Assoc.* **15:**1481–1488.

Gorham, P. R., 1964b, Toxic algae, in: *Algae and Man* (D. F. Jackson, ed.), Plenum Press, New York, pp. 307–336.

Gorham, P. R., and Carmichael, W. W., 1988, Hazards of freshwater blue-green algae (cyanobacteria), in: *Algae and Human Affairs* (C. A. Lembi and J. R. Warland, eds.), Cambridge University Press, Cambridge, pp. 403–431.

Gorham, P. R., MacLachlan, J., Hammer, V. T., and Kim, W. K., 1964, Isolation and culture of toxic strains of *Anabaena flos-aquae* (Lyngb.) de Breb, *Verh. Int. Ver. Limnol.* **15:**796–804.

Gorham, P. R., McNicholas, S., and Allen, E. A. D., 1982, Problems encountered in searching for new strains of toxic planktonic cyanobacteria, *S. Afr. J. Sci.* **28:**357–362.

Grauer, F. H., and Arnold, H. L., 1961, Seaweed dermatities, *Arch. Dermatol.* **84:**720–732.

Hawkins, P. R., Runnegar, M. T. C., Jackson, A. R. B., and Falconer, I. R., 1985, Severe hepatotoxicity caused by the tropical cyanobacterium *Cylindrospermopsis raciborskii* (Wolz) Seenaya and Subla Raju isolated from a domestic water supply reservoir, *Appl. Environ. Microbiol.* **50:**1292–1295.

Heaney, S. J., 1971, The toxicity of *Microcystis aeruginosa* Kuts from some English reservoirs, *Water Treat. Exam.* **20:**235–244.

Heise, H. A., 1949, Symptoms of hay fever caused by algae, *J. Allergy* **20:**383–385.

Heise, H. A., 1951, Symptoms of hay fever caused by algae, *Ann. Allergy* **9:**100–101.

Hughes, E. O., Gorham, P. R., and Zehnder, A., 1958, Toxicity of a unialgal culture of *Microcystis aeruginosa*, *Can. J. Microbiol.* **4:**225–236.

Ingram, W., and Prescott, G., 1954, Toxic freshwater algae, *Am. Midl. Nat.* **52:**75–87.

Jackim, E., and Gentile, J. H., 1968, Toxins of blue-green algae: Similar to saxitoxin, *Science* **162:**916–916.

Jamel Al-Layl, K., Poon, G. K., and Codd, G. A., 1988, Isolation and purification of peptide and alkaloid toxins from *Anabaena flos-aquae* using high performance thin layer chromatography, *J. Microb. Meth.* **7:**251–258.

Jinno, H., Ando, M., Matsui, S., and Takeda, M., 1989, Isolation of toxins from natural blue-green algae and unicellular cultured *Microcystis aeruginosa*, *Environ. Toxicol. Chem.* **8:**493–498.

Keleti, G., Sykora, J. L., Maiolie, L. A., Doerfler, D. L., and Campbell, I. M., 1981, Isolation and characterization of endotoxin from cyanobacteria (blue-green algae), in: *The Water Environment: Algal Toxins and Health* (W. W. Carmichael, ed.), Plenum Press, New York, pp. 447–464.

Kfir, R., Johannsen, E., and Botes, D. P., 1986, Monoclonal antibody specific for cyanoginosin-LA: Preparation and characterization, *Toxicon* **24:**543–552.

Konst, H., McKercher, P. D., Gorham, P. R., Robertson, A., and Howell, J., 1965, Symptoms and pathology produced by toxic *Microcystis aeruginosa* NRC-1 in laboratory and domestic animals, *Can. J. Comp. Med. Vet. Sci.* **29:**221–228.

Krishnamurthy, T., Carmichael, W. W., and Sarver, E. W., 1986, Investigations of freshwater cyanobacteria (blue-green algae) toxic peptides. I. Isolation, purification, and characterization of peptides from *Microcystis aeruginosa* and *Anabaena flos-aquae* NRC-44-1, *Biochem. Physiol. Pflanz.* **62:**473–487.

Krishnamurthy, T., Szafraniec, L., Hunt, D. F., Shabanowitz, J., Yates III, J. R., Hauer, C. R., Carmichael, W. W., Skulberg, O., Codd, G. A., and Missler, S., 1989, Structural characterization of toxic cyclic peptides from blue-green algae by tandem mass spectrometry, *Proc. Natl. Acad. Sci. USA* **86:**770–774.

Lanaras, T., Tsitsamis, S., Chlichlia, C., and Cook, C. M., 1989, Toxic cyanobacteria in Greek freshwater, *J. Appl. Phycol.* **1:**67–73.

Lincoln, E. P., and Carmichael, W. W., 1981, Preliminary tests of toxicity of *Synechocystis* sp. grown on wastewater medium, in: *The Water Environment: Algal Toxins and Health* (W. W. Carmichael, ed.), Plenum Press, New York, pp. 223–230.

Lippy, E. C., and Erb, J., 1976, Gastrointestinal illness at Sewickley, Pa., *J. Am. Water Works Assoc.* **68:**606–610.

MacKintosh, C., Beattie, K. A., Klumpp, S., Cohen, P., and Codd, G. A., 1990, Cyanobacterial microcystin-LR is a potent and specific inhibitor of protein phosphatases 1 and 2A from both mammals and higher plants, *FEBS Lett.* **264:**187–192.

Mahmood, N. A., and Carmichael, W. W., 1986, The pharmacology of anatoxin-a(s), a neurotoxin produced by the freshwater cyanobacterium *Anabaena flos-aquae* NRC 525-17, *Toxicon* **24:**425–434.

Malyarevskaya, A. Y., Birger, T. I., and Arsan, O. M., 1970, Alterations in the biochemical composition of the Perch (*Perca*) under the influence of lethal concentrations of *Microcystis aeruginosa* Kutz, *Gidrobiol. Zh.* **6:**36–41.

Martin, C., Kaarina, S., Matern, U., Dierstein, R., and Wechesser, J., 1990, Rapid purification of peptide toxins microcystin-LR and nodularin, *FEMS Microbiol. Lett.* **68:**1–6.

May, V., and McBarron, E. J., 1973, Occurrence of the blue-green alga *Anabaena circinalis* Rabenh. in New South Wales and toxicity to mice and honeybees, *J. Aust. Inst. Agric.Sci.* **39:**264–266.

McBarron, E. J., and May, V., 1966, Poisoning of sheep in New South Wales by the blue-green *Anacystis cyanea* Dr. and Dail. *Aust. Vet. J.* **42:**449–453

McBarron, E. J., Walker, R. I., Gardner, I., and Walker, K. H., 1975, Toxicity to livestock of the blue-green alga *Anabaena flos-aquae*, *Aust. Vet. J.* **51:**587–588.

McDonald, D. W., 1960, Algal poisoning in beef cattle, *Can. Vet. J.* **1:**108–110.

McElhenny, T. R., Bold, H. C., Brown, R. M., and McGovern, J. P., 1962, Algae: A cause of inhalant allergy in children, *Ann. Allergy* **20:**739.

Meriluoto, J. A. O., and Eriksson, J. E., 1988, Rapid analysis of peptide toxins in cyanobacteria, *J. Chromatogr.* **438**:93–99.

Meriluoto, J. A. O., Sandstrom, A., Eriksson, J. E., Remaud, G., Craig, A. G., and Chattopadhyaya, J., 1989, Structure and toxicity of a peptide hepatotoxin from the cyanobacterium *Oscillatoria agardhii*, *Toxicon* **27**:1021–1034.

Mills, D. H., and Wyatt, J. T., 1974, Ostracod reactions to non-toxic and toxic algae, *Oecologica* (Berl.) **17**:171–177.

Mitsui, A., Rosner, D., Goodman, A., Reyes-Vasquez, G., Kusumi, T., Kodama, T., and Nomoto, K., 1983, Hemolytic toxins in marine cyanobacterium *Synechococcus* sp., in: *Proceedings of the International Red Tide Symposium*, Takamatsu, Japan, pp. 121–122.

Mittal, A., Agarwal, M. K., and Schivpuri, D. N., 1979, Repiratory allergy to algae: Clinical aspects, *Ann. Allergy* **42**:253–256.

Moiheka, S. N., and Chu, G. W., 1981, Dermatitis-producing alga *Lyngbya majuscula* Goment in Hawaii II: Biological properties of the toxic factor, *J. Phycol.* **7**:8–13.

Moore, R., 1981, Toxins from marine blue-green algae, in: *The Water Environment: Algal Toxins and Health* (W. W. Carmichael, ed.), Plenum Press, New York, pp. 15–23.

Moore, R. E., 1984a, Structure–activity studies of aplysiatoxin-type tumor promoters, in: *Cellular Interactions by Environmental Tumor Promoters* (H. Fujiki, ed.), Japanese Scientific Society Press/VNU Science Press, Utrecht, pp. 49–57.

Moore, R. E., 1984b, Public health and marine toxins from blue-green algae, in: *Seafood Toxins* (E. P. Ragilis, ed.), Amer. Chem. Soc. Washington, D.C., pp. 369–375.

Moore, R. E., Patterson, G. M. L., Mynderse, J. S., Barchi, J. J., Norton, T. R., Furusawa, E., and Furusawa, S., 1986, Toxins from cyanophytes belonging to the Scytonemataceae, *Pure Appl. Chem.* **58**:263–271.

Mynderse, J. S., Moore, R. E., Kashiwagi, M., and Norton, T. R., 1977, Antileukemia activity in the Oscillatoriaceae: Isolation of debromoaplysia toxin from *Lyngbya*, *Science* **196**:538–540.

Olson, T. A., 1964, Waterfowl tomorrow, in: *Blue-Greens* (J. K. Lindurska, ed.), U. S. Department of Interior Fish and Wildlife Service, Washington, D.C., pp. 349–356.

Østenvik, O., Skulberg, O. M., and Søli, N. E., 1981, Toxicity studies with blue-green algae from Norwegian inland waters, in: *The Water Environment: Algal Toxins and Health* (W. W. Carmichael, ed.), Plenum Press, New York, pp. 315–324.

Peary, J., and Gorham, P., 1966, Influence of light and temperature on growth and toxin production by *Anabaena flos-quae*, *J. Phycol.* **2**(Suppl):3–11.

Phinney, H., and Peck, C., 1961, Klamath Lake, an instance of natural enrichment. I: Algae and metropolitan wastes, in: *Transactions of 1960 Seminar, Robert A. Taft Sanitary Engineering Center, Cincinnati*, U.S. Department of Health, Education and Welfare, Washington, D. C., pp. 22–27.

Poon, G. K., Priestley, I. M., Hunt, S. M., Fawell, J. K., and Codd, G. A., 1987, Purification procedure for peptide toxins from the cyanobacterium *Microcystis aeruginosa* involving high-performance thin-layer chromatography, *J. Chromatogr.* **387**:551–555.

Porter, E. M., 1887, Investigation on supposed vegetation in the waters of some lakes in Minnesota, in: *Report of the Department of Agriculture of the University of Minnesota (for the period 1883 to 1886). Supplement 1 to the 4th Biennial Report of the Board of Regents, University of Minnesota*, Pioneer Press Company, St. Paul, Minnesota, pp. 95–96.

Prescott, G. W., 1948, Objectionable algae with reference to the killing of fish and other animals, *Hydrobiologia* **1**:1–13.

Ransom, R. E., Nerad, T. E., and Meier, P. G., 1978, Acute toxicity of some blue-green algae to the protozoan *Paramecium caudatum*, *J. Phycol.* **14**:114–116.

Rose, E. T., 1953, Toxic algae in Iowa lakes, *Proc. Iowa Acad Sci.* **60**:738–745.

Runnegar, M. T. C., and Falconer, I. R., 1981, Isolation, characterization, and pathology of the toxin from the blue-green alga *Microcystis aeruginosa*, in: *The Water Environment: Algal Toxins and Health* (W. W. Carmichael, ed.), Plenum Press, New York, pp. 325–342.

Sargunar, H. T. P., and Sargunar, A. A. A., 1979, Phycotoxins from *Microcystis aeruginosa*:

Their implication in human and animal health, in: 4th International IUPAC Symposium on Mycotoxins and Phycotoxins, Lausanne, *Chem. Rundsch.* **1979**:Abstract 610.

Sasner, J. J., Jr., Ikawa, M., Foxall, T. L., and Watson, W. H., 1981, Studies on aphantoxin from *Aphanizomenon flos-aquae* in New Hampshire, in: *The Water Environment: Algal Toxins and Health* (W. W. Carmichael, ed.), Plenum Press, New York, pp. 389–403.

Sasner, J. J., Jr., Ikawa, M., and Foxall, T. L., 1984, Studies on *Aphanizomenon* and *Microcystis* toxins, in: *Seafood Toxins* (E. P. Ragiles, ed.), American Chemical Society Symposium Series Number 262, Washington, D.C., pp. 391–406.

Sawyer, P. J., Gentile, J. H., and Sasner, J. J., Jr., 1968, Demonstration of a toxin from *Aphanizomenon flos-aquae* (L) Ralfs, *Can. J. Microbiol.* **14**:1199–1204.

Schwimmer, D., and Schwimmer, M., 1964, Algae and medicine, in: *Algae and Man* (D. Jackson, ed.), Plenum Press, New York, pp. 368–412.

Schwimmer, M., and Schwimmer, D., 1968, Medical aspects of phycology, in: *Algae, Man and the Environment* (D. F. Jackson, ed.), Syracuse University Press, Syracuse, New York, pp. 278–358.

Shelubski, M., 1951, Observations of the properties of a toxin produced by *Microcystis*, *Proc. Int. Assoc. Theoret. Appl. Limnol.* **11**:362–366.

Siegelman, H. W., Adams, N. H., Stoner, R. D., and Slakin, D. W., 1984, Toxins of *Microcystis aeruginosa* and their hematological and histological effects, in: *Seafood Toxins* (E. P. Ragiles, ed.), American Chemical Society Symposium Series Number 262, Washington, D. C., pp. 407–413.

Sivonen, K., Kononen, K., Carmichael, W. W., Dahlem, A. M., Rinehart, K. L., Kirivanta, J., and Niemlea, S. I., 1989a, Occurrence of the hepatotoxic cyanobacterium *Nodularia spumigena* in the Baltic Sea and structure of the toxin, *Appl. Environ. Microbiol.* **55**:1990–1995.

Sivonen, K., Himberg, K., Luukkainen, R., Niemila, S. I., Poon, G. K., and Codd, G. A., 1989b, Preliminary characterization of neurotoxic cyanobacteria blooms and strains from Finland, *Toxicity Assessment* **4**:339–352.

Skulberg, O. M., Codd, G. A., and Carmichael, W. W., 1984, Toxic blue-green algal blooms in Europe: A growing problem, *Ambio* 13:244–247

Skulberg, O. M., Carmichael, W. W., Andersen, R., Matsunaga, S., Moore, R. E., and Skulberg, R., 1992, Investigation of a new neurotoxic *Oscillatoria* (Cyanophyceae) and it toxin. Isolation and characterization of a homoanatoxin-a, *Environ. Toxicol. Chem.* (in press).

Soll, M. D., and Williams, M. C., 1985, Mortality of a white rhinoceros (*Ceratotherium simum*) suspected to be associated with blue-green algae *Microcystis aeruginosa, J. S. Afr. Vet. Assoc.* **1985**(March):49–51.

Solomatina, U. D., and Matchinskaya, S. F., 1972, Alteration of the amino acid content of the carpe induced by blue-green algae, *Hydraulic J.* **8**:81–87.

Spivak, C. E., Witkop, B., Albuguergue, E. X., 1980, Anatoxin-a: A novel, potent agonist at the nicotinic acid receptor, *Mol. Pharmacol.* **18**:384–394.

Tisdale, E. S., 1931, Epidemic of intestinal disorders in Charlestown, W. Va. occurring simultaneously with unprecedented water supply conditions, *Am. J. Public Health* **21**:198.

Utkilen, H. C., Aase, B., Bryn, K., and Jantzen, E., 1991, 3-Hydroxy-fatty acids as potential determinants for cyanobacterial endotoxins, a preliminary report, in: *Toxin Producing Algae–Research on Advance*, Norwegian Institute for Water Research, pg. 55.

Van der Westhuizen, A. J., and Eloff, J. N., 1985, Effect of temperature and light on the toxicity and growth of the blue-green alga *Microcystis aeruginosa* (UV-006), *Planta* **163**:55–59.

Watanabe, M. F., and Oishi, S., 1982, Toxic substance from a natural bloom of *Microcystis aeruginosa, Appl. Environ. Microbiol.* **43**:819–822.

Watanabe, M. F., and Oishi, S., 1985, Effects of environmental factors on toxicity of a cyanobacterium (*Microcystis aeruginosa*) under culture conditions, *Appl. Environ. Microbiol.* **49**:1342–1344.

Watanabe, M. F., Oishi, S., and Nakao, T., 1981, Toxic characteristics of *Microcystis aeruginosa*, *Verh. Int. Ver. Limnol.* **21:**1441–1443.

Watanabe, M. F., Oishi, S., Watanabe, Y., and Watanabe, M., 1986, Strong probability of lethal toxicity in the blue-green alga *Microcystis viridis* Lemmermann, *J. Phycol.* **22:**552–556.

Wheeler, R. E., Lackey, J. B., and Schott, S. A., 1942, Contribution of the toxicity of algae, *Public Health Rep.* **57:**1695–1701.

Wilkinson, S. G., 1988, in: *Microbiol. Lipids* (C. Ratledge and S. G. Wilkinson, eds.), Volume I, Academic Press, New York, pp. 436–440.

Yamagiski, H., and Aoyama, K., 1972, Ecological studies on dissolved oxygen and bloom of *Microcystis* in Lake Suwa. I. Horizontal distribution of dissolved oxygen in relation to drifting of *Microcystis* by wind, *Bull. Jpn. Soc. Sci. Fish.* **38:**9–16.

Potential and Commercial Applications for Photosynthetic Prokaryotes

8

NIGEL W. KERBY and PETER ROWELL

1. INTRODUCTION

The cyanobacteria, together with phototrophic green and purple bacteria and prochlorophytes, share a basic prokaryotic cellular organization and together constitute the photosynthetic prokaryotes (see Stanier *et al.*, 1981). A major distinction between the photosynthetic bacteria and cyanobacteria is the presence of oxygenic photosynthesis, with two photosystems acting in series, in cyanobacteria and of anoxygenic photosynthesis, using only one photosystem, in photosynthetic bacteria. In 1952 the first Algal Mass Culture Symposium was held to consider potential applications of micro-algae (Burlew, 1953) and there has since been an increasing interest in this field. Oxygenic photosynthesis is a unique means of utilizing cheap substrates (CO_2, H_2O, and solar energy) for the primary production of organic compounds and many potential applications of cyanobacteria rely on this process. Since photosynthetic bacteria carry out anoxygenic photosynthesis, their use requires the provision of organic or inorganic electron donors; for example, organic wastes. Certain species of cyanobacteria and photosynthetic bacteria have the ability to fix atmospheric dinitrogen, catalyzed by the enzyme nitrogenase. The agronomic potential of nitrogen fixation by cyanobacteria, particularly in the cultivation of rice, is well documented, as is the production of H_2, catalyzed by nitrogenase, in both photosynthetic bacteria and cyanobacteria.

This chapter attempts to cover the potential applications of cyanobac-

NIGEL W. KERBY and PETER ROWELL ● Agricultural and Food Research Council Research Group on Cyanobacteria, and Department of Biological Sciences, University of Dundee, Dundee DD1 4HN, United Kingdom.

Photosynthetic Prokaryotes, edited by Nicholas H. Mann and Noel G. Carr. Plenum Press, New York, 1992.

teria and photosynthetic bacteria and is therefore broad in scope. An exhaustive review of the literature is thus not possible and we refer readers to a number of recent books which comprehensively cover various aspects of the potential applications of microalgae, including cyanobacteria (Richmond, 1986; Borowitzka and Borowitzka, 1988; Stadler et al., 1988; Cresswell et al., 1989). We are not aware of any equivalent texts for the photosynthetic bacteria at the time of writing (late 1989/early 1990) and make no apologies for the apparent bias toward our own research interests, the cyanobacteria.

2. HYDROGEN PRODUCTION

Over the past two decades, and particularly following the oil crisis, alternative energy sources have received considerable attention. One such alternative is the conversion of solar energy into chemical energy. A variety of microorganisms can evolve H_2 according to the following equation:

$$2H^+ + 2e^- \rightleftharpoons H_2$$

using a reversible hydrogenase. However, in diazotrophs, the production of H_2 is also catalyzed by nitrogenase:

$$N_2 + 8H^+ + 8e^- + 16MgATP \rightarrow 2NH_3 + H_2 + 16MgADP + 16Pi$$

Under appropriate conditions cultures of N_2-fixing cyanobacteria evolve H_2, catalyzed by nitrogenase in vivo (Haystead et al., 1970; Benemann and Weare, 1974). Cyanobacterial H_2 metabolism is complex and three enzymes with hydrogenase activity are involved (Lambert and Smith, 1981; Houchins, 1984; Rao and Hall, 1988): (1) nitrogenase, which catalyzes H_2 evolution; (2) a membrane-bound "uptake" hydrogenase, which provides reductant for photosynthetic and respiratory electron transport; and (3) a soluble hydrogenase, which catalyzes either H_2 evolution or H_2 uptake when provided with suitable electron donors or acceptors. H_2 production by nitrogenase is irreversible, insensitive to CO, dependent on a supply of electrons and ATP, and always occurs during N_2 fixation. A minimum of 25% of the electron flux is diverted to the reduction of H^+ even at high N_2 partial pressures (Rivera-Ortiz and Burris, 1975) and optimal levels of ATP and reductant (Hageman and Burris, 1980). Therefore, such H_2 evolution consumes a substantial amount of reductant and ATP and is most likely a consequence of the nitrogenase mechanism (Mortensen, 1978). Rates of nitrogenase-catalyzed H_2 evolution are influenced by treatments which affect the provision of reductant and ATP and by factors that alter the stability and expression of nitrogenase. Other factors, including competing substrates such as N_2 and C_2H_2 or inhibitors of nitrogenase, such as CO, which inhibits all nitrogenase activities except H_2 evolution, will affect rates of H_2 production. In heterocystous cyanobacteria the source of reductant for H_2 evolution is ultimately generated in vegetative cells by the photosynthetic oxidation of H_2O. Therefore, these organisms constitute a biophotolytic system whereby H_2 and O_2 are generated from water using light energy, with the evolution of O_2 and H_2 occurring in different cells (vegetative cells and heterocysts, respectively). Since H_2-consuming enzymes occur in cyanobacteria, rates of H_2 evolution are not solely dependent on the activity of nitrogenase. H_2 evolved by nitrogenase is normally recycled, providing ATP or electrons for biosynthesis and removing O_2 from the site of nitrogenase. One such H_2-consuming enzyme is the membrane-associated "uptake" hydrogenase found in all heterocystous cyanobacteria so far examined (Houchins, 1984) and also in the non-N_2-fixing cyanobacterium *Anacystis nidulans* (Peschek, 1979a,b). However, certain nonheterocystous cyanobacteria apparently lack an "uptake" hydrogenase (Weisshaar and Böger, 1983; Van der Oost *et al.*, 1987), as do Ni-depleted cultures of *Anabaena cylindrica* (Daday *et al.*, 1985), thus increasing rates of H_2 production. "Uptake" hydrogenase has a high affinity for H_2 and is insensitive to O_2. The existence in cyanobacteria of another, "reversible" hydrogenase has been a matter of some controversy (Bothe, 1982; Houchins, 1984). It is now generally recognized to be present (Houchins and Burris, 1981a,b), but its function has not yet been fully elucidated. This hydrogenase is soluble or loosely associated with membranes, is capable of both H_2 evolution and consumption, and is inhibited by O_2.

H$_2$ evolution is widespread among photosynthetic bacteria, which in general utilize a wide spectrum of organic substances, such as carbohydrates, lipids, and fatty acids, as well as some inorganic sulfur compounds (Zürrer, 1982). Photosynthetic bacteria also have a Ni-requiring membrane-bound uptake hydrogenase which may recycle the H_2 evolved by nitrogenase (Willison *et al.*, 1983). Nitrogenase-deficient mutants of *Rhodopseudomonas capsulata* do not evolve H_2, confirming that nitrogenase is responsible for H_2 evolution (Wall *et al.*, 1975). Since nitrogenase is not protected in specialized cells, such as heterocysts, H_2 photoproduction must be performed under anaerobic conditions. Photosynthetic bacteria degrade a number of carbon sources to CO_2 and H_2 in the light and in the absence of N_2 and combined nitrogen, and the use of photosynthetic bacteria for H_2 production may be economical when waste substrates are utilized (Bollinger *et al.*, 1985). These include wastes from the milk industry, sugar refineries, and domestic water. Natural isolates of photosynthetic bacteria from sugar refinery waste ponds have been shown to utilize the untreated waste with higher yields of H_2 than laboratory strains of *Rhodospirillum rubrum* (Bollinger *et al.*, 1985). Additionally, natural isolates of photosynthetic bacteria have been screened for maximal H_2 production using organic acids as substrates (Mao *et al.*, 1986). It has been suggested that the coculture of photosynthetic bacteria with anaerobic bacteria could provide a more efficient system for H_2 production (Miyake *et al.*, 1984). In such a system an anaerobic bacterium, such as *Clostridium butyricum*, which itself shows high rates of H_2 production, would provide organic acids for conversion to H_2 by the photosynthetic bacteria. Other systems involving the coculture of microorganisms with photosynthetic bacteria have been reported and include the anaerobic coculture, at the expense of cellulose, of a *Cellulomonas* strain with *Rhodopseudomonas capsulata* (Odom and Wall, 1983) and coculture of the eukaryotic alga *Chlamydomonas reinhardtii* with *Rhodospirillum rubrum* (Miyamoto *et al.*, 1987).

Semicontinuous outdoor cultures of *Rhodopseudomonas sphaeroides* have been shown to produce H_2 for longer periods than batch cultures with equivalent H_2 evolution rates (Kim *et al.*, 1987a). Laboratory studies using continuous cultures of *Rhodospirillum rubrum* (Zürrer and Bachofen, 1982) and *R. rubrum* strain B6 (Kim *et al.*, 1987b) showed variable rates of H_2 production which were similar under continuous or periodic illumination and may therefore be suitable for outdoor production of H_2 (Kim *et al.*, 1987b). H_2 production rates reported for photosynthetic bacteria vary between 5 and 150 μl H_2 (mg dry weight)$^{-1}$ hr^{-1}, depending on the species and conditions employed, and higher rates have been reported for immobilized cells (Von Felten *et al.*, 1985; Planchard *et al.*, 1989) than for free-living cultures. Bacterial hydrogenases, including that from *Thiocapsa roseopersicina*, are being used to develop artificial photocatalytic systems (Nikandrov *et al.*, 1988) (see Section 5).

In general, the reported rates of H_2 production are lower in cyanobacteria than in photosynthetic bacteria, being 2–40 μl H_2 (mg dry weight)$^{-1}$ hr^{-1} for species of *Anabaena* (Daday *et al.*, 1977; Asada *et al.*, 1985; Xiankong *et al.*, 1983), *Oscillatoria* (Belkin and Padan, 1978), *Lyngbya* (Kuwada and Ohata 1987), *Synechococcus* (Mitsui *et al.*, 1983), and *Microcystis* (Asada and Kawamura, 1985). In species such as those of *Microcystis*, which lack the ability to fix N_2, dark H_2 production is probably via a reversible hydrogenase (Houchins, 1984; Asada and Kawamura, 1985). In a study of marine cyanobacteria, it was shown that nonheterocystous filamentous strains and unicellular, aerobic, N_2-fixing strains produced H_2 for longer periods and at higher rates than heterocystous, filamentous strains (Mitsui *et al.*, 1983) and this was attributed to higher nitrogenase activities and lower H_2 consumption.

The question of whether biological H_2 production will ever be economical remains unanswered. The prospects are not promising since, although H_2 is a low-value product with a large potential market (Benemann, 1989), present productivities are far from meeting this. The suggestion that photosynthetic bacteria may one day be used as a source of renewable fuel in the Shuttle Transportation System (Strayer *et al.*, 1985) seems somewhat optimistic! Outdoor systems utilizing solar radiation are being tested and screening programs are being employed to isolate and select strains with high rates of H_2 production. Further laboratory-based experiments are required to elucidate the optimal physiological conditions and maximal rates of production. To maximize H_2 production, strains with high growth, photosynthetic, and nitrogen-fixation rates, together with low rates of H_2 recycling, are required. Mutant strains of *Rhodopseudomonas capsulata* deficient in H_2 uptake have been shown to have higher rates of H_2 evolution (Takakuwa *et al.*, 1983) and strains with derepressed nitrogenase may be useful since they evolve H_2 in the presence of combined nitrogen (Wall and Gest, 1979).

3. WASTE WATER TREATMENT

Waste waters resulting from human activities include those of domestic, agricultural, and industrial origin. The nature of these wastes and the use of microalgae, including certain cyanobacteria, in waste water treatment have been reviewed (Goldman, 1979; Fallowfield and Garret, 1985; Abeliovich, 1986; Oswald, 1988a). The use of microalgae in waste water treatment inevitably involves the use of ponds, the most common being facultative and high-rate oxidation ponds. Facultative ponds are usually more than 1 m deep and are anoxic at the bottom, permitting surface algal growth, whereas high-rate ponds are usually shallower (<1 m deep), mixed, and aerobic (Oswald, 1988a). Algal systems are versatile, providing O_2 for BOD removal and also incorporating nutrients such as nitrogen and phosphorus into biomass, thereby reducing the eutrophication potential of receiving waters. However, the roles of microalgae in enhancing sedimentation, disinfection, and removal of heavy metals, nutrients, and toxic organic compounds are not as well researched. Algae grown on domestic sewage may contain toxic organic and inorganic compounds. The affinity of microalgae and cyanobacteria for metal ions is an advantage in waste water treatment, but a disadvantage when the biomass is used an animal feed, although it could be used for methane production (Matsunaga and Izumida, 1984; Oswald, 1988a). Species dominance depends primarily on organic load, and blooms of cyanobacteria are generally indicative of lower organic load (Palmer, 1969).

Spirulina has been used for secondary treatment of pig wastes (Chung *et al.*, 1978) and the biomass gave a protein efficiency ratio higher than *Chlorella* in rat feeding experiments, with no observed toxic effects. Raw sewage alone did not support the growth of *Spirulina* in outdoor shallow ponds, but additions of 1% (w/v) bicarbonate and 0.1% (w/v) nitrate were required for long-term stabilized biomass production (Saxena *et al.*, 1983). The dried cyanobacterial biomass was used as a substitute for groundnut cake in chicken feeding experiments, without apparent adverse effect, and produced yolks with deeper color than conventional oxycarotenoid sources (e.g., yellow maize and berseem seed). An integrated approach to village-scale waste water treatment and biomass production, for use as feed supplements, in developing countries has been studied by Fox (1988). Human wastes are fermented in an anaerobic digestor to produce methane and CO_2. *Spirulina* is grown in sand-bed-filtered, solar-sterilized liquid effluent from the digestor, which is supplemented with sea salt, sodium carbonate, and the CO_2 produced by fermentation. A large proportion of the human pathogens present in the excrement is removed during fermentation and sand-bed filtration. It is claimed that, in addition to providing a rich source of protein and vitamins, the integrated system will eventually, through improved sanitation, eliminate intestinal parasites, thereby increasing the effectiveness of local agriculture. Additionally, the methane produced can be used as a source of cooking fuel and the digestor sludge used as a compost. Fox (1988) states that the costs of such a project can be recovered during the first year, but are very variable. High rates of nitrogen and phosphorus removal from secondary effluent, supplemented with 1% sodium bicarbonate, by *Oscillatoria* have been reported (Hashimoto and Furukawa, 1989). Industrial-grade chitosan flakes, colonized by *Phormidium*, depleted inorganic nitrogen and orthophosphate from secondary urban effluent (De la Noue and Proux, 1988). However, uncolonized flakes were shown to remove orthophosphate, which was explained by Ca^{2+} release from chitosan resulting in orthophosphate precipitation.

Potential applications for photosynthetic bacteria in waste treatment include: the production of single-cell protein (SCP) utilizing agricultural or food wastes; H_2 production using wastes containing suitable sugars (e.g., sugar refinery waste water); the degradation of aromatic compounds; removal of reduced sulfur compounds in gas streams; and the removal of chlorinated hydrocarbons from waste waters.

Rhodopseudomonas sphaeroides strain P47 was selected for SCP production from pineapple peel waste, due to its ability to reduce the chemical oxygen demand (from 110 to 16 g liter^{-1} within 60 hr), its relatively high growth rate (0.3 hr^{-1}), and its usefulness as a feed supplement (Noparatnaraporn et al., 1986a). Pineapple waste medium, which is a rich source of sugars (40 g sucrose liter^{-1}, 24 g glucose liter^{-1}, and 14 g fructose liter^{-1}), was supplemented with $(NH_4)_2SO_4$, $(NH_4)_2HPO_4$, and vitamins (nicotinic acid, thiamine-HCl, and biotin) prior to inoculation. A growth yield of 0.45 g cells (g sugar consumed)$^{-1}$ was obtained and it was suggested that up to 2.5 tons dry cells day^{-1} could be recovered from a medium-sized cannery producing 150 tons of peel waste day^{-1}. Other wastes used for photosynthetic bacterial SCP production include soybean (Sasaki et al., 1981) and cassava (Noparatnaraporn et al., 1987); mixed cultures of *Rhodocyclus gelatinosus* and *Rhodopseudomonas sphaeroides* had higher growth yields (Noparatnaraporn et al., 1987).

Recent concern about the environmental fate of industrial compounds has stimulated interest in the anaerobic degradation of aromatic compounds. Harwood and Gibson (1988), using auxanography, have shown *Rhodopseudomonas palustris* to be more versatile in aromatic degradation than had been previously demonstrated (Dutton and Evans, 1978). Many phenolic, dihydroxylated, and methoxylated aromatic acids as well as aromatic aldehydes and hydroaromatic acids supported growth in the presence and absence of oxygen. The removal of chlorinated hydrocarbons from water and waste water by *Rhodopseudomonas sphaeroides* immobilized on magnetite particles has been studied (MacRae, 1985, 1986). Five insecticides [γ-hexachlorocyclohexane (HCH), dieldrin, heptachlor, aldrin, and p,p'-DDT] and α-HCH, which occurs as a major contaminant in the preparation of γ-HCH, were removed from both water and waste water (MacRae, 1986). The levels of p,p'-DDT, aldrin, and heptachlor were rapidly reduced to below the limits of detection. The sorption onto the immobilized bacteria showed an inverse relationship to the solubilities of the chlorinated hydrocarbons. However, comparable experiments with magnetite alone were not performed to enable the effectiveness of bacterial sorption to be estimated. A process for the removal of sulfur compounds formed during fossil fuel hydrogasification or hydroliquefaction processes from acid gas streams (Institute of Gas Technology, 1987) consists of contacting the gas stream with a photosynthetic bacterium, preferably *Chlorobium limicola forma thiosulphatum*. Elemental sulfur was formed by *C. thiosulphatum* in stoichiometric amounts when light energy and H_2S supply were optimally adjusted in the presence of nonlimiting CO_2 (Cork *et al.*, 1985). Rhodanese, isolated from *Rhodopseudomonas palustris* and immobilized, may have applications in the detoxification of cyanide (Vázquez *et al.*, 1988).

The microbial accumulation of heavy metals has been reviewed (Gadd, 1988; Reed and Gadd, 1990) and is a potential application for photosynthetic prokayotes. Cyanobacteria, together with eukaryotic filamentous algae, growing in tailing ponds and artificial streams, reduced the levels of Pb, Zn, Cu, and Mn from a lead mine to permissible discharge levels (Gale and Wixson, 1979). The cell walls of cyanobacteria are similar in structure to those of Gram-negative eubacteria, where most deposition occurs. Passive accumulation of Mn, Co, Zn, Sn, Ag, Cs, Hg, Np, Pu, and Am by *Synechococcus* sp. has been demonstrated (Fisher, 1985) with concentration factors ranging from 0 (Cs and Np) to approximately 10^6 (Sn, Hg, and Pu). Certain heavy metals are transported into cells and are deposited in cellular inclusions. *Synechococcus elongatus* accumulated uranium from seawater, leading to the formation of dense internal deposits (Horikoshi *et al.*, 1979), and *Anabaena cylindrica* and *Plectonema boryanum* sequestered aluminum and cadmium in polyphosphate bodies (Jensen *et al.*, 1982; Vymazal, 1987). A filamentous cyanobacterium, strain OL3, mobilized up to 18% and 51% uranium from coal and carbonate rock, respectively, after 80 days (Lorenz and Krumbein, 1985). Before the suitability of cyanobacteria for leaching neutral and alkaline low-grade uranium ores can be fully assessed, the process requires optimization and elucidation of the mechanisms of leaching.

4. CYANOBACTERIA AS BIOFERTILIZERS

4.1. Nitrogen Fixation

N_2-fixing cyanobacteria can utilize sunlight as the sole source of energy for the fixation of carbon and nitrogen, and therefore have potential as biofertilizer. The ability to fix N_2, which is widespread among cyanobacteria, has been considered in several reviews (W. D. P. Stewart, 1980a,b; Bothe et al., 1984; Gallon and Chaplin, 1987; Hallenbeck, 1987; Rowell and Kerby, 1991) and is not discussed in detail here. The nitrogenase of cyanobacteria is similar to that of other diazotrophs, being an FeMo-containing enzyme, and there is recent evidence for an alternative nitrogenase (Kentemich et al., 1988). Nitrogenase is extremely O_2-labile and has to be protected from irreversible inactivation. Aerobic N_2 fixation is invariably a property of heterocystous cyanobacteria and most of the nonheterocystous species known to fix N_2 do so only under anaerobic or microaerobic conditions. Spatial separation of photosynthetic O_2 evolution and N_2 fixation is achieved in filamentous cyanobacteria by the localization of the nitrogenase in specialized cells, the heterocysts. Additionally, certain nonheterocystous species can show a temporal separation of the photosynthetic and N_2-fixing activities. N_2 fixation is usually inhibited in the presence of combined nitrogen sources, such as NH_4^+ and NO_3^-, and the regulatory mechanisms controlling nitrogenase activity and biosynthesis are still largely unknown. Since ammonia or its metabolites repress the formation of heterocysts, it has been difficult to establish whether the inhibition of N_2 fixation is due to an inhibition of heterocyst differentiation or to a repression of nitrogenase synthesis. Furthermore, ammonia may inhibit nitrogenase activity in vivo by interrupting ATP or reductant supply.

4.2. Field Applications

The agronomic potential of cyanobacteria, either free-living or in symbiotic association with the water fern Azolla, has long been recognized (De, 1939; R. N. Singh, 1961; W. D. P. Stewart et al., 1979; Roger and Kulasooriya, 1980; Lumpkin and Plucknett, 1982; Watanabe and Roger, 1984; Achtnich et al., 1986; L. V. Venkataraman, 1986; Whitton and Roger, 1989) and the natural fertility of tropical paddy fields has often been attributed to N_2-fixing cyanobacteria. To improve crop productivity, there has been a tendency to breed and select high-yielding varieties which require higher inputs of fertilizer to realize their full potential. However, the production and distribution of nitrogen fertilizers is expensive and in developing countries their availability is a major factor limiting crop production. There is, therefore, renewed interest in the role of biological N_2 fixation as a source of nitrogen fertilizer, to reduce the dependency of agriculture on fossil fuel, which, in turn, would reduce the emission of CO_2 into the atmosphere.

The major environments in rice paddies are floodwater, oxidized soil surface layer (2–20 mm), anaerobic reduced soil, rice plant and rhizosphere, and subsoil. The floodwater is the photic zone where aquatic communities, including bacteria, cyanobacteria, algae, and aquatic weeds, provide organic matter to the soil. The productivity of aquatic photosynthetic biomass rarely exceeds 1000 kg dry weight ha^{-1} (Watanabe and Roger, 1984). In floodwater, aquatic plants and the basal portion of rice shoots are colonized by epiphytic bacteria, algae, and cyanobacteria. Epiphytic biological N_2 fixation becomes more significant in deepwater rice, where the submerged biomass is high (Watanabe *et al.*, 1982; Rother *et al.*, 1988). Cyanobacteria and photosynthetic bacteria colonize the flood water and the surface soil layer, but it is generally thought that photosynthetic bacteria make no significant nitrogen contribution to paddy fields, despite their abundance (Watanabe and Roger, 1984). Approximately 50% of the cyanobacterial genera found in rice paddies are heterocystous and include species of *Anabaena, Nostoc, Calothrix, Aulosira, Scytonema, Gloeotrichia, Tolypothrix, Fischerella,* and *Wollea* (Whitton and Roger, 1989). Certain countries encourage the use of cyanobacteria as biofertilizers. A soil-based starter culture containing *Aulosira, Tolypothrix, Scytonema, Nostoc, Anabaena,* and *Plectonema* is grown to provide a large inoculum for field application (L. V. Venkataraman, 1986). In many villages in India the cost of production has been estimated to be about $0.5 (U.S.) for 10 kg algal material. The dried cyanobacterial inoculum is broadcast over the floodwater at the rate of 8– 10 kg (soil-based material from production units) ha^{-1}, 1 week after transplanting rice seedlings. The cyanobacterial inoculation should be repeated for at least three consecutive seasons (L. V. Venkataraman, 1986). Data conclusively demonstrating the beneficial effects of algalization, which is apparently not as widely employed as has sometimes been implied, are lacking (Whitton and Roger, 1989). It is likely that indigenous populations would outgrow the introduced strains used for algalization.

Azolla is also grown in nurseries and may either be grown in rotation with rice and ploughed into fields prior to rice planting or grown together with rice in alternative rows. *Azolla* requires 5–10 cm of water in the field and the addition of phosphate and is temperature sensitive, which inhibits growth during the summer months in hot countries. Since *Azolla* is a weed, heavy growth may hamper rice growth and tiller production and its uncontrolled growth in irrigation channels and tanks may carry negative effects.

Various factors affect the growth of cyanobacteria in paddy fields and these include physical factors (light and temperature), biological factors (cyanophages, bacteria, fungi, and invertebrate grazers), soil factors (pH, micronutrients, and macronutrients), and agronomic practices (tillage and applications of fertilizers, herbicides, and pesticides). Plant size, nitrogen content, number of tillers, ears, and spikelets, and filled grains per panicle may be affected when rice paddies are inoculated with either free-living cyanobacteria or *Azolla* (Roger and Kulasooriya, 1980). Field experiments suggest an average yield increase of about 14% above the control (no added combined nitrogen), due to cyanobacterial N_2 fixation, which corresponds to about 450 kg ha^{-1} (Watanabe and Roger, 1984; L. V. Venkataraman, 1986). Grain and nitrogen yields of rice treated with cyanobacteria are often quoted as being comparable to those obtained following an application of about 30 kg nitrogen fertilizer ha^{-1} (Roger and Kulasooriya, 1980; L. V. Venkataraman, 1986; A. L. Singh and Singh, 1986, 1987). Similar results have been obtained using *Azolla* (Lumpkin and Plucknett, 1982; Achtnich *et al.*, 1986). In ^{15}N-tracer experiments, *Azolla* derived 50–60% of its nitrogen from fixation (equivalent to 11–14 kg N ha^{-1} in 14 days), rice plants incorporated *Azolla* N more efficently than urea N, and N uptake was better when the *Azolla* was incorporated at tillering than at transplanting (Kulasooriya *et al.*, 1988). Additionally, up to 40% of the cyanobacterial nitrogen can be utilized by rice plants within 60 days, as was demonstrated using ^{15}N (Mian and Stewart, 1985). Cyanobacterial nitrogen, because of its organic nature, is less susceptible to nitrogen loss than chemical fertilizer (Gunnison and Alexander, 1975).

Since the transfer of biologically fixed nitrogen to plants is not fully understood, and since only limited data are available on field experiments, the relative contribution of cyanobacterial nitrogen fixation versus other possible beneficial factors cannot be fully assessed. Other beneficial effects of cyanobacteria may include the mobilization of bound phosphorus (Bose *et al.*, 1971), the aggregation of soil particles, increasing aeration (Roychoudhury *et al.*, 1980), and the release of putative plant growth regulators. Filamentous cyanobacteria, which have been shown to produce compounds that stimulate the growth of plants, include: *Cylindrospermum muscicola* (G. S. Venkataraman and Neelakantan, 1967); *Calothrix anomola* (Dadhich *et al.*, 1969); and *Anabaena* sp., *Nostoc* sp., *Oscillatoria* sp., *Plectonema* sp., and *Nodularia* sp. (Rodgers *et al.*, 1979). To date, the characterization of plant growth stimulants extracted from cyanobacteria is lacking and it is hard to ascertain whether stimulation was caused by a particular plant growth regulator or solely by the nutritive value of such extracts.

The success of rhizobial strain selection in increasing the N_2-fixing capacity of legumes highlights the possibility of selecting superior strains of other N_2-fixing prokaryotes. W. D. P. Stewart *et al.* (1979) discuss the important traits that should be considered in the selection of cyanobacteria for use in the field, and these include: N_2-fixing strains capable of rapid growth; strains that can fix N_2 under different O_2 tensions and under photoautotrophic, photoheterotropic, and chemoheterotrophic conditions; strains with little or no H_2 evolution; strains with a derepressed nitrogenase; and strains that liberate nitrogenous compounds in significant amounts. Since 1979, to our knowledge, only a limited number of studies have dealt with these issues.

4.3. Photoproduction of Ammonia

We have been interested in the development of strains with one of the traits discussed by W. D. P. Stewart *et al.* (1979), the photoproduction of nitrogenous compounds [see for a review Kerby *et al.* (1989)]. Ammonia liberation by N_2-fixing cyanobacteria is not a normal physiological function, other than in symbiotic associations (W. D. P. Stewart *et al.*, 1983: Rowell *et al.*, 1985), but can be achieved by the use of certain enzyme inhibitors, or by the selection of ammonia-liberating strains. W. D. P. Stewart and Rowell (1975) demonstrated that over 90% of the newly-fixed nitrogen was released extracellularly as ammonia when glutamine synthetase (GS) was inhibited by L-methionine-D,L-sulfoximine (MSX). Other GS inhibitors which induce ammonia liberation by cyanobacteria are 5-hydroxylysine (Ladha *et al.*, 1978) and the herbicide phosphinothricin (Lea *et al.*, 1984; N. W. Kerby and P. Rowell, unpublished data). Musgrave *et al.* (1982) was the first to demonstrate the technical feasibility of sustained ammonia production using MSX-treated, immobilized N_2-fixing cyanobacteria (*Anabaena* sp. ATCC 27893). MSX was applied either continuously or intermittently to a variety of continuous-flow bioreactors and specific rates of production varied from 4 to 40 μmol ammonia mg Chl a^{-1} hr^{-1}, depending on the conditions employed and with durations of over 800 hr (Musgrave *et al.*, 1982, 1983; Kerby *et al.*, 1983). Similar results were obtained for the photoproduction of ammonia from nitrate by free-living *Anacystis nidulans* (Ramos *et al.*, 1982a,b) and free-living *Anabaena* sp. ATCC 33047 (Ramos *et al.*, 1984). Subsequently, other groups have used MSX in similar studies using a variety of free-living or immobilized cyanobacteria (D. O. Hall *et al.*, 1985; Brouers and Hall, 1986; Jeanfils and Loudeche, 1986; Vincenzini *et al.*, 1986; Zimmerman and Boussiba, 1987; Martinez *et al.*, 1989). MSX is of value only in demonstrating the technical feasibility of ammonia production by cyanobacteria.

Much lower rates of ammonia liberation have been reported for a free-living paddy field strain of *Anabaena* (approximately 2.5 nmol ammonia mg Chl a^{-1} hr^{-1}) (Subramanian and Shanmugasundaram, 1986) and for *Anabaena azollae* immobilized in polyvinyl foams (0.12 μmol ammonia mg Chl a^{-1} hr^{-1}) (Shi *et al.*, 1987), in the absence of metabolic inhibitors. High rates of ammonia production (10–56 μmol ammonia mg Chl $^{-1}$ hr^{-1} (Kerby *et al.*, 1986) have been obtained using mutant strains of N$_2$-fixing cyanobacteria with a reduced ammonia-assimilating capacity. Strains have been obtained by selecting for resistance to ethylenediamine (EDA) following chemical mutagenesis (Polukhina *et al.*, 1982; Kerby *et al.*, 1986). EDA-resistant strains of *Anabaena variabilis* (ED81 and ED92) have low growth rates under N$_2$-fixing conditions or in the presence of ammonia, but show enhanced growth rates in the presence of glutamine (Sakhurieva *et al.*, 1982). Nitrogenase activity was higher in strains ED81 and ED92 than the parent strain and was depressed with respect to ammonia. The GS activity of strain ED92 was much reduced (approximately 25%), as measured by both the biosynthetic and transferase activities, and this reduction in GS activity corresponded to a similar reduction in GS protein (Kerby *et al.*, 1986) and GS mRNA (Hien *et al.*, 1988). Strain ED81 had normal levels of GS protein, GS mRNA, and GS transferase activity, but had a much reduced GS biosynthetic activity (approximately 10%) (Kerby *et al.*, 1986; Hien *et al.*, 1988). There were no apparent differences in structure and kinetic properties of GS from ED81, but V_{max} was reduced (Kerby *et al.*, 1990). These data imply that ED92 may be a regulatory mutant synthesizing reduced levels of GS, whereas ED81 appears to have a catalytically deficient GS.

MSX-resistant strains liberate ammonia and have derepressed nitrogenase (Spiller *et al.*, 1986). A MSX-resistant strain of *Anabaena variabilis* (strain SA1) had a low level of GS biosynthetic activity which was not inhibited by MSX. An ammonium-releasing, MSX-resistant mutant of *Anabaena siamensis* has also been described (Thomas *et al.*, 1990) which appears to have properties similar to ED81 (see above). Strain SA1 enhanced the growth of rice plants grown in nitrogen-free medium; both dry weight and nitrogen content of rice plants were increased by this strain, but not by the parent strain (Latorre *et al.*, 1986).

The nitrogenase of cyanobacteria with reduced GS biosynthetic activity (either mutant strains or cells treated with GS inhibitors) is derepressed with respect to ammonia. Stewart *et al.* (1979) commented on the desirability, for use in rice paddies, of strains of N$_2$-fixing cyanobacteria which would continue to fix N$_2$ in the presence of combined nitrogen. Indeed, in some paddy soils ammonium ion concentrations may increase to 300 ppm as organic matter decays on waterlogging (Ponnamperuma, 1976) and such levels would inhibit nitrogenase activity and synthesis. We have characterized a mutant strain of *Anabena variabilis*, resistant to the ammonium analogue methylammonium, which has a deficient active ammonium transport system (Reglinski *et al.*, 1989). The nitrogenase activity of this strain was less sensitive to exogenous ammonia than that of the parent strain at pH 7.0, but not at pH 9.0 where ammonia is probably transported by passive diffusion (Kerby *et al.*, 1990).

Mutant strains with the ability to liberate ammonia would probably be unable to compete under field conditions due to their lower growth rates than indigenous strains. Future work may benefit from the selection of mutant strains derived from natural isolates, particularly those with high growth rates.

The creation of novel associations between plants and N_2-fixing cyanobacteria has great environmental and economic potential. Two approaches have been employed in laboratory studies. The first relies on the regeneration of plants from callus inoculated with cyanobacteria (Gusev and Korzhenevskaya, 1990) whereas the second involves the co-cultivation of wheat seedlings with cyanobacteria (Gantar *et al.*, 1991a,b).

5. PHOTOCHEMICAL AND PHOTOELECTROCHEMICAL SYSTEMS

Photosynthetic microorganisms can be used in fuel cells with suitable mediators to convert light energy into electrical energy. Intact *Phormidium* sp. and *Mastigocladus laminosus* immobilized on SnO_2 semiconductor electrodes are capable of light-dependent electron transfer to the electrode (Ochiai *et al.*, 1980, 1983). Other fuel cells using cyanobacteria have been developed and it has been shown that the electrical current is dependent on endogenous glycogen reserves and light (Tanaka *et al.*, 1985, 1988).

A need for the development of rapid, simple, and low-cost toxicity screening procedures for continuous monitoring and control of pollution has been identified (Gaisford and Rawson, 1989). Whole-cell biosensors utilizing cyanobacteria as biocatalysts are being developed (Rawson *et al.*, 1987, 1989). The photosynthetic activity of cyanobacteria is monitored by coupling electron flow, via an electrochemical mediator, $(FeCN)_6^{3-/4-}$, to an electrode. Perturbation of electron flow, induced, for instance, by a herbicide, is readily monitored. The reduction of ferricyanide by intact cells is thought to be due to the action of NADPH dehydrogenase located in the plasma membrane. Therefore, for nonpermeating mediators, such as ferricyanide, the coupling to photosynthetic electron transport is indirect. Results obtained using *Synechococcus* sp. show herbicide detection levels as low as 20 μg liter^{-1} and that biosensors possess a working life of 7 days (Rawson *et al.*, 1989).

6. PRODUCTS FROM PHOTOSYNTHETIC PROKARYOTES

6.1. Biomass

Cyanobacteria have served as a source of food in various parts of the world; in the past, the main focus has been on the production of SCP, although other applications that require mass cultivation are apparent. In the 1940s several potential applications for microalgae were recognized (Burlew, 1953) and some of these have become commercial realities. At present, several microalgae are being produced on a large scale for their pigments, as vitamin supplements (e.g., β-carotene), and as health foods. The cyanobacterium *Spirulina* is grown commercially and the biomass used as a health food or for its pigments. It has also been reported that large-scale commercial facilities are being developed to grow N_2-fixing cyanobacteria, for use as biofertilizers (Karuna-Karan, 1987). The current cost of producing microalgal biomass, using outdoor commercial facilities, is estimated by Benemann (1989) to be between approximately $500 and $3000 tonne^{-1} for large (approximately 1000 ha) and small (approximately 10 ha) outdoor facilities, respectively. Therefore, the production of low-value commodities such as foods, feeds, fuels, and fertilizers using air, sunlight, and water as feedstocks appears just as distant as it did in the 1940s.

Cyanobacteria can be cultured either in enclosed bioreactors or in open pond systems utilizing sunlight. Both culture systems have been the subject of recent reviews (Benemann *et al.*, 1979; Goldman, 1979; Terry and Raymond, 1985; Gudin and Thepenier, 1986; Lee, 1986; Oswald, 1988b; Borowitzka and Borowitzka, 1989), as has the economics of production (Tsur and Hochman, 1986; Tapie and Bernard, 1988), and are discussed in full in Chapter 6, as is the nutritional value of cyanobacterial biomass.

Photosynthetic bacteria have also been identified as a source of single-cell protein when grown on wastes or low-cost raw materials (Litchfield, 1983) (see Section 3). Generally these organisms grow in mixed culture with other bacteria and culture densities are of the order of 1–2 g dry weight liter^{-1} (Kobayashi and Kurata, 1978). Protein and nucleic acid contents account for 61% and 6%, respectively, of *Rhodopseudomonas capsulata* biomass. The amino acid composition of cell hydrolysates is broadly similar to that of *Chlorella vulgaris* except for the higher methionine content of *R. capsulata* (Kobayashi and Kurata, 1978). In addition to high protein contents, the biomass contains vitamins and carotenoids which may be useful if it is used as a food supplement. For instance, when *Rhodocyclus gelatinosa* and *Rhodobacter sphaeroides* were grown in mixed culture on cassava waste the vitamin B_{12} and carotenoid contents reached 44 and 230 μg (g dry weight)$^{-1}$, respectively (Norparatnaraporn *et al.*, 1987). Supplementing feed for laying hens with photosynthetic bacteria at a very low rate (0.01%) was reported to increase the rate of egg laying by 10% and to improve the quality of eggs (Kobayashi and Kurata, 1978). This stimulatory effect was apparently lost on heat treating the cells, but the component(s) responsible for the effect are not yet identified. In general, little data appear to be available on the nutritional aspects of photosynthetic bacteria as a food source.

6.2. Restriction Endonucleases

Cyanobacteria, and to a lesser extent photosynthetic bacteria, are a rich source of type II restriction endonucleases, and some of these are currently marketed. Restriction endonucleases are part of the restriction modification system that protects prokaryotic organisms against foreign genetic information. This modification system consists of an enzyme complex which modifies and protects the host-cell DNA, usually by methylation, against restriction endonucleases which cut and destroy unprotected DNA. Methylation occurs at a limited number of sequences within DNA, which are the recognition sites for restriction endonucleases. Cyanobacteria contain different endonucleases in different combinations and a detailed list is given by Ciferri *et al.* (1989). Some are isoschizomers, even in unrelated strains, and this is apparently not unusual, as the same restriction enzymes have been found in different bacterial genera (R. J. Roberts, 1987). The difficulties encountered in obtaining reliable gene transfer systems for cyanobacteria are probably due, in part at least, to the presence of restriction endonucleases. Several endonucleases have been identified in photosynthetic bacteria and one of these, *Rsr*I, which has been purified from *Rhodobacter sphaeroides* (Greene *et al.*, 1988; Aiken and Gumport, 1988), is an isoschizomer of *Eco*RI, the most extensively characterized restriction enzyme.

6.3. Amino Acids

Cyanobacteria liberate small quantities of amino acids, polypeptides, and proteins into their growth medium (Kerby *et al.*, 1989, and references therein). The release of amino acids in the natural environment is normally associated with stationary-phase cells, with small amounts of a broad spectrum of amino acids being liberated. The amount liberated is dependent on the stage of growth, is highest during the lag and stationary phases (Fogg, 1952; W. D. P. Stewart, 1963), and may be due, in part, to cell lysis. Fogg (1952) reported traces of free amino acids in culture filtrates of *Anabaena cylindrica*, but concluded that there were not of quantitative significance. More recently, *Anabaena siamensis,* isolated from rice paddies, was shown to liberate small amounts (170 nmol total amino acids (g protein)$^{-1}$), mainly phenylalanine, threonine, glutamate, and glycine at late logarithmic phase of growth (Antarikanonda, 1984). Out of 200 strains of marine cyanobacteria isolated from coastal areas of Japan, several strains liberated glutamate (0.6–6.3 µg (mg dry weight)$^{-1}$ day^{-1}) (Matsunaga *et al.*, 1988). A *Synechococcus* sp. NKBG 040607 excreted glutamate (83% of the total amino acids liberated) at the highest rate. Continuous photoproduction of glutamate was also demonstrated in Ca-alginate-immobilized cells with a duration of over 7 days.

The overproduction of amino acids can be achieved, in strains that do not normally liberate significant quantities of amino acids, by the selection of mutant strains resistant to amino acid analogues (Yamada *et al.* 1972). The photoproduction of amino acids by cyanobacteria is of interest since they utilize inexpensive feedstocks (CO_2 and either N_2 or combined nitrogen, usually as NO_3^-). Spontaneous mutants of *Synechocystis* sp. ATCC 29108, *Synechococcus* sp. 602, *Anacystis nidulans*, and *Anabaena* sp. ATCC 29151 resistant to analogues of methionine, proline, tryptophan, and phenylalanine liberated amino acids which supported the growth of auxotrophic strains of *Bacillus subtilis* (G. Hall *et al.*, 1980). Mutant strains of *Spirulina platensis* resistant to analogues of proline, methionine, and aromatic amino acids have also been selected (Riccardi *et al.*, 1981a) and individual amino acids liberated were quantified (up to 30 mg liter^{-1}) by measuring the growth of auxotrophic bacteria. This method does not take into account the effects of other extracellular cyanobacterial products on the growth of bacteria, nor does it show the range of amino acids liberated. These mutant strains were cross-resistant to other amino acid analogues and some may have acquired resistance through changes in transport systems or in the mechanisms of incorporation of amino acids into proteins (Riccardi *et al.*, 1981b). A spontaneous mutant of *Synechocystis* sp. PCC 6803 resistant to *p*-fluorophenylalanine liberated considerable quantities of aromatic amino acids (180 mg liter^{-1}), principally phenylalanine (Labarre *et al.*, 1987).

We have demonstrated the continuous photoproduction of amino acids by free-living and Ca-alginate-immobilized mutant strains of *Anabaena variabilis* (Kerby *et al.*, 1987) and *Synechocystis* sp. 6803 (Kerby *et al.*, 1988) resistant to tryptophan and methionine analogues. *A. variabilis* strains resistant to 6-fluorotryptophan (FT) liberated either a broad range of amino acids, with alanine invariably being the major amino acid liberated, or aromatic amino acids, with phenylalanine predominating (Kerby *et al.*, 1989). Up to 200 mg amino acids liter^{-1} at stationary phase in batch culture and 50 mg amino acids (liter medium)$^{-1}$ in continuous culture were obtained. Tryptophan liberation by FT-resistant strains of *A. variabilis* was only detected when cells were grown in the presence of the detergent MYRJ 45 (Niven *et al.*, 1988a). This detergent probably enhances tryptophan release by causing changes in the composition of the plasma membrane, resulting in the uncoupling of specific amino acid transport systems, as has been shown for glutamate production by *Corynebacterium glutamicum* (Clément and Lanéelle, 1986). A key point of control of aromatic amino acid biosynthesis in cyanobacteria is 3-deoxy-D-arabinoheptulosonate 7-phosphate (DAHP) synthase (Jensen and Hall, 1982). We have isolated two (iso)enzymes of DAHP synthase from *A. variabilis*, one inhibited by tyrosine and the other by phenylalanine. Only the tyrosine-inhibited form was detected in certain strains, whereas in others, having both forms of the enzyme, the phenylalanine-inhibited form was deregulated with respect to phenylalanine inhibition (Niven *et al.*, 1988b).

Low-molecular-mass metabolites, including amino acids, can be released from free-living and immobilized cyanobacteria following osmotic shock (Reed *et al.*, 1986). Transfer of unicellular cyanobacteria from a medium of high salt to one of low salt resulted in a transient loss of plasma membrane integrity and a loss of organic compounds normally retained intracellularly. This method could be repeated for several cycles without loss of cell viability and may be a useful method of recovering intracellular metabolites without having to lyse cells.

At present, commercial production of amino acids from photosynthetic microorganisms is restricted to the production of labeled amino acids by *Anacystis nidulans* (see below) and the current status of unlabeled amino acid production is at the research stage.

6.4. Labeled Organic Compounds

Phototrophic organisms are currently used in the manufacture of uniformly labeled organic compounds. Amino acids and nucleotides uniformly labeled with ^{14}C have been obtained from the eukaryotic microalga *Chlorella pyrenoidosa* grown on $^{14}CO_2$ (Nejedly *et al.*, 1968) or from yeast grown on uniformly ^{14}C-labeled invert sugar (Laufer *et al.*, 1964). The use of phototrophic organisms is preferred since $^{14}CO_2$ is a simple and convenient starting material and phototrophs can incorporate about 90% of the substrate supplied into cellular material (Tovey *et al.*, 1974). This contrasts with the much lower incorporation of substrate into cellular components (approximately 30%) by bacteria (R. B. Roberts *et al.*, 1955). The cyanobacterium *Anacystis nidulans* is employed for the production of uniformly labeled organic compounds (United Kingdom Atomic Energy Authority, 1973; Tovey *et al.*, 1974) and has several advantages over *Chlorella;* shorter doubling times, even in the presence of high levels of radiation; greater resistance to radiation damage; higher DNA content for the production of uniformly labeled nucleotides; and higher protein yield for the production of uniformly labeled amino acids (Tovey *et al.*, 1974). The process involves growing *A. nidulans* in a sealed vessel in the presence of up to 6 Ci $^{14}CO_2$ (60 Ci mol^{-1}), harvesting cells, purifying RNA and DNA, and digesting the nucleic acids to yield nucleotides, which are subsequently separated. The residual fraction, which is mainly protein with some carbohydrate, is hydrolyzed to yield amino acids, which are purified (Tovey *et al.*, 1974). This process involves harvesting and disruption of cells for the production of labeled organic compounds. The use of mutant strains, which liberate amino acids into their growth media (see Section 5.3), may be a means of enhancing the production of a desired amino acid (University of Dundee, 1989). Other labeled compounds could be extracted from the biomass. We have demonstrated the extracellular production of uniformly ^{14}C-labeled phenylalanine by a mutant strain of *Anabaena variabilis* grown in the presence of $^{14}CO_2$ (N. W. Kerby and P. Rowell, unpublished data). Such strains may also be applicable to the production of amino acids uniformly labeled with stable isotopes (e.g., ^{15}N and ^{13}C) (Cox *et al.*, 1988).

6.5. Pigments

Cyanobacterial pigments are currently being evaluated for use as food-colorants and as diagnostic probes. The pigments of commercial interest are the phycobiliproteins phycocyanin and phycoerythrin, which are photosynthetic accessory pigments. The Dai Nippon Inks and Chemical Co. is producing phycobiliproteins ("Linablue") from *Spirulina* in California and these are sold as a food colorant in Japan. There is currently a medium-sized market for these products, which have a value of >$100 kg^{-1} (Benemann, 1989). Phycobiliproteins may also be used as fluorescent probes ("phycofluors") (Glazer and Stryer, 1984) and, although the market size is small, the product has a high value (>$10,000 kg^{-1}) (Benemann, 1989). Phycobiliproteins, especially phycoerythrin, are stable, can be stored for long periods of time, and show high fluorescence which is not quenched by a number of biomolecules. Fluorescence is stable over a broad pH range and does not vary markedly with temperature. Certain species of cyanobacteria such as the picoplanktonic cyanobacterium *Synechococcus* sp. WH7803, which uses phycoerythrin as a nitrogen reserve (Wyman *et al.*, 1985), contain considerable quantities of phycoerythrin, facilitating rapid purification with high yields (N. W. Kerby and P. Rowell, unpublished data). Such strains may, therefore, be applicable to commercial production of phycobiliproteins.

6.6. Other Biochemicals

ATP can be synthesized using chromatophores from photosynthetic bacteria or cyanobacteria. Chromatophores are usually immobilized, but do not have long-term stability. ATP can apparently be generated photosynthetically using intact cells of the thermophilic cyanobacterium *Mastigocladus* sp. in the presence of ADP, Pi, and phenazine methosulfate (PMS) or 1-methoxy-PMS, which was required for maximal rates (Sawa *et al.*, 1982). This unique system has been used for photosynthetic glutathione production by intact cells of *Phormidium lapideum* supplied with glutamate, cysteine, and glycine (Sawa *et al.*, 1986). The biotransformation of steroids has been reported for certain species of cyanobacteria (Abul-Hajj and Qian, 1986). For example, several strains transformed 4-androstenedione to testosterone with overall yields of up to 80% after 7 days of incubation. Testosterone proved to be the major metabolite in the majority of cultures, which were unable to perform the reverse reaction, the oxidation of the

17β-hydroxyl group. Cyanobacteria contain a variety of saturated and unsaturated sterols (Kohlhase and Pohl, 1988).

A series of patents (Kyowa-Hakko, 1985, 1986; Mitsubishi Gas and Chemicals, 1986; Ergo-Forsch., 1985) relating to the production of coenzyme Q_{10} from *Rhodopseudomonas sphaeroides*, *Rhodopseudomonas sulfidophila*, and *Rhodospirillum rubrum* have been filed. Coenzyme Q_{10} is available as a health-food supplement (Bliznakov and Hunt, 1987) and apparently has therapeutic properties in the treatment of heart disease (Lenaz, 1985).

The extracellular production of 5-aminolevulinate (ALA), an inhibitor of ALA dehydratase, from levulinic acid by dimethylsulfoxide-permeabilized *Anabaena variabilis* (Avissar, 1983) and by *Rhodobacter sphaeroides* grown under anaerobic conditions in the light (Sasaki *et al.*, 1987, 1990) has been reported. ALA is claimed to be a new selective herbicide but is at present chemically synthesized and is therefore expensive.

A two-phase process for the enrichment of vitamin B_{12} in *Rhodopseudomonas gelatinosa* SCP, grown on cassava starch, has been demonstrated (Noparatnaraporn *et al.*, 1986b). Initially, cells were cultivated aerobically in the dark for 19 hr, then transferred to microaerobic conditions, which increased concentrations of carotenoid, bacteriochlorophyll, and vitamin B_{12} from 230, 0, and 25 μg (g cells)$^{-1}$ to 310, 960, and 38 μg (g cells)$^{-1}$, respectively. Vitamin B_{12} formation by *Rhodospirillum rubrum* was stimulated by changing the concentrations of Co^{2+}, Fe^{2+}, Mg^{2+}, and glycine in the culture medium (Hirayama and Katsuta, 1988). When the Co^{2+} concentration was increased in the presence of $FeSO_4$ the rate of intracellular accumulation of vitamin B_{12} increased while maintaining adequate growth. The commercial potential of vitamins is considerable and some are expensive in pure form (approximately \$4600 kg^{-1} vitamin B_{12}). Although they appear to be excreted by most microalgae and cyanobacteria, there are limited data available. Vitamin contents of cyanobacteria and reports of vitamin excretion are summarized by Borowitzka (1988). The vitamin contents of certain cyanobacteria and photosynthetic bacteria may be of value when biomass is used as a food supplement.

6.7. Animal Cell Growth Stimulants

Cyanobacterial extracts have been reported to support the growth of cultured animal cells and to decrease or eliminate the requirement for animal sera, such as calf fetal serum. Hot water extracts of *Spirulina* decrease the requirement for calf fetal serum from 10% to 1% (v/v), give accelerated growth rates, and allow normal successive cultivation of animal cells (Chlorella, 1983). A dialyzed aqueous extract of *Synechococcus elongatus*, when added to basal medium, enhanced the growth of human cell lines to a greater extent than did a mixture of insulin, transferrin, ethanolamine, and selenite (Shinohara *et al.*, 1986). Fractionation of the dialyzate revealed that one of the growth-promoting compounds may be a phycobiliprotein (Shinohara *et al.*, 1988). The growth-promoting activity of allophycocyanin was higher than that of phycocyanin. However, the specific activity did not increase during purification of phycobiliproteins, and other compounds present in the dialyzate may be responsible for the observed stimulation of growth of cultured animal cells.

6.8. Other Biologically Active Compounds

Various biological activities have been demonstrated for cyanobacterial extracts, but often the structures of the compounds reponsible have not been elucidated (see, for example, Flores and Wolk, 1986; Cannell *et al.*, 1988; Kellam *et al.*, 1988; Bloor and England, 1991). Here we will only detail biologically active compounds with known structures [for recent reviews see Moore (1982), Faulkner (1984), Metting and Pyne (1986), Moore *et al.* (1988) and Glombitzka and Koch (1989)].

An algicide, cyanobacterin, a chlorine-containing γ-lactone, is produced by the cyanobacterium *Scytonema hofmanni* (Mason *et al.*, 1982; Pignatello *et al.*, 1983). It is effective against cyanobacteria, eukaryotic algae, and higher plants (Mason *et al.*, 1982; Gleason and Baxa, 1986; Gleason and Case, 1986). Cyanobacterin is an inhibitor of photosynthetic electron transport, probably photosystem II (Gleason and Paulson, 1984; Gleason and Case, 1986). Ethanolic extracts of *Lyngbya aestuarii* showed herbicidal activity against *Lemna* sp. and the active constituent was identified as the fatty acid 2,5-dimethyldodecanoic acid (Entzeroth *et al.*, 1985).

Various compounds have been reported to have antifungal activities and include: the indole alkaloid hapalindole A isolated from *Hapalosiphon fontinalis* (Moore *et al.*, 1984), which, with 18 other indoles isolated from *H. fontinalis*, has been patented (Moore and Patterson, 1986); the depsipeptide majusculamide C from *Lyngbya majuscula* (Carter *et al.*, 1984); the macrolides scytophycins A, B, C, D, and E from *Scytonema pseudohofmanni* (Ishibashi *et al.*, 1986); a nucleoside tubercidin from *Tolypothrix byssoidea* (Entzeroth *et al.*, 1986); and the 5′-α-D-glucopyranose derivatives of tubercidin and toyocamycin from *Plectonema radiosum* and *Tolypothrix tenuis* (J. B. Stewart *et al.*, 1988).

A number of compounds isolated from cyanobacteria are reported to have cytotoxic and antineoplastic activities (see Patterson *et al.*, 1991) and include the following compounds: the macrolides acutiphycin and 20,21-didehydroacutiphycin from *Oscillatoria acutissima* (Barchi *et al.*, 1984); scytophycins from *Scytonema pseudohofmanni* (Moore *et al.*, 1986); the γ-pyrone hormothamnione from *Hormothamnion enteromorphoides*, an inhibitor of RNA synthesis (Gerwick *et al.*, 1986); malyngamide D from *Lyngbya majuscula* (Gerwick *et al.*, 1987); tubercidin from *Tolypothrix byssoidea* (Barchi *et al.*, 1983); a phenolic compound, debromoaplysiatoxin (Mynderse *et al.*, 1977), and an indole alkaloid, lyngbyatoxin A (which is identical to telocidin A) (Cardellina *et al.*, 1979a), both isolated from *L. majuscula;* the alkaloids, tantazoles, from *Scytonema mirabile* (Carmeli *et al.*, 1991a); the polysaccharide spirulinan from *Spirulina subsalsa* (Seiko-Epson, 1986); tudercidin, toyocamycin, and their 5'-α-D-glucopyranose derivatives (J. B. Stewart *et al.*, 1988); paracyclophanes from *Cylindrospermum licheniforme* and *Nostoc linckia* (B. S. Moore *et al.*, 1990); and mirabilene isonitrites from *S. mirabile*, which also have mild anti-microbial activity (Carmeli *et al.*, 1990b).

Certain cyanobacteria of the family Oscillatoriaceae can cause dermatitis and also may promote tumors in laboratory animals. These compounds are not themselves carcinogenic, but can act as cocarcinogens or tumor promoters that accelerate the development of tumors (Moore 1982; Fujiki *et al.*, 1984). These compounds include lyngbyatoxin A (telocidin A), aplysiatoxin, and debromoaplysiatoxin.

Antibacterial activity has been demonstrated for several compounds: the δ-lactone malyngolide isolated from *L. majuscula* (Cardellina *et al.*, 1979b); the *O*-methyl acid malyngamide from *L. majuscula* (Gerwick *et al.*, 1987); the hapalindoles from *Hapalosiphon fontinalis* (Moore and Patterson, 1986); hormothamnins which are peptides from *Hormothamnion enteromorphoides* (Gerwick *et al.*, 1989) and which were also found to be toxic to goldfish but not to brine shrimps.

Six biindoles isolated from *Rivularia firma* (Norton and Wells, 1982) have antiinflammatory activities. Indolenines from *Fischerella* sp. inhibit ^{3}H-arginine-vasopressin binding to kidney tissue (Schwartz *et al.*, 1987).

Recently, extracts of unialgal, but nonaxenic, *Lyngbya lagerheimii* and *Phormidium tenue* were shown to contain compounds that were very active against human immunodeficiency virus (HIV-1) (Gustafson *et al.*, 1989). The compounds were identified as sulfonic acid-containing glycolipids and are presently being evaluated for their anti-AIDS potential, particularly to complement the action of nucleosides such as azidothymidine (AZT).

Other toxins, including the peptide microcystin isolated from *Microcystis aeruginosa* and anatoxins, are discussed by Utkilen (Chapter 7, this volume).

Toxins active against mosquito larvae are produced by spore-forming bacteria (e.g., *Bacillus thuringiensis* and *Bacillus sphaericus*) and genes coding for the toxin have been expressed in cyanobacteria (Tandeau de Marsac *et al.*, 1987; Chungjatupornchai, 1990). This may provide an efficient method for mosquito control.

In the foregoing discussion it is clear that the structures of several biologically active compounds derived from cyanobacteria have now been elucidated, although in many cases there is a dearth of published data on their activities and modes of action. Many of these reports are for natural isolates of cyanobacteria and some of these are now being maintained in laboratory culture (Moore *et al.*, 1988). However, activities detected in field collections may not be apparent when isolates are grown in laboratory cultures. Axenic cyanobacterial isolates are obviously required to unequivocally attribute these activities to cyanobacteria rather than contaminating microorganisms. To our knowledge, none of these biologically active compounds is produced commercially, but since they may have a high value, the cost of cultivation in enclosed bioreactors may be economically viable. Cultivation of microalgae is more expensive than that of bacteria and fungi, and growth and product yields are lower, adding to the costs. To maximize productivities, further studies are clearly required on the biosynthesis of these secondary metabolites and on the conditions under which they are synthesized.

REFERENCES

Abeliovich, A., 1986, Algae in wastewater oxidation ponds, in: *CRC Handbook of Microalgal Mass Culture* (A. Richmond, ed.), CRC Press, Boca Raton, Florida, pp. 331–338.

Abul-Hajj, Y. J., and Qian, X., 1986, Transformation of steroids by algae, *J. Nat. Prod.* **49:**244–248.

Achtnich, W., Moawad, A. M., and Johal, A. M., 1986, *Azolla*, a biofertilizer for rice, *Int. J. Trop. Agric.* **4:**188–211.

Aiken, C., and Gumport, R. I., 1988, Restriction endonuclease *Rsr*I from *Rhodobacter sphaeroides*, an isoschizomer of *Eco*RI: Purification and properties, *Nucleic Acids Res.* **16:**7901–7916.

Antarikanonda, P., 1984, Production of extracellular free amino acids by cyanobacterium *Anabaena siamensis* Antarikanonda, *Curr. Microbiol.* **11:**191–196.

Asada, Y., and Kawamura, S., 1985, Hydrogen evolving activity among the genus, *Microcystis*, under dark and anaerobic conditions, *Rep. Ferment. Res. Inst. Japan* **63:**39–54.

Asada, Y., Tomizuka, N., and Kawamura, S., 1985, Prolonged hydrogen production by a cyanobacterium (blue-green alga), *Anabaena* sp., *J. Ferment. Technol.* **63:**85–90.

Avissar, Y. J., 1983, 5-Aminolevulinate synthesis is permeabilized filaments of the blue-green alga *Anabaena variabilis*, *Plant Physiol.* **72:**200–203.

Barchi, J. J., Norton, T. R., Furusawa, E., Patterson, G. M. L., and Moore, R. E., 1983, Identification of a cytotoxin from *Tolypothrix byssoidea* as tubercidin, *Phytochemistry* **22:**2851–2852.

Barchi, J. J., Moore, R. E., and Patterson, G. M. L., 1984, Acutiphycin and 20,21-dihydroacutiphycin, new antineoplastic agents from the cyanophyte *Oscillatoria acutissima*, *J. Am. Chem. Soc.* **106:**8193–8197.

Belkin, S., and Padan, E., 1978, Hydrogen metabolism in the faculative anoxygenic cyanobacteria (blue-green algae) *Oscillatoria limnetica* and *Aphanothece halophytica*, *Arch. Microbiol.* **116:**109–111.

Benemann, J. R., 1989, The future of microalgal biotechnology, in: *Algal and Cyanobacterial Biotechnology* (R. C. Cresswell, T. A. V. Rees, and N. Shah, eds.), Longman, Harlow, pp. 317–337.

Benemann, J. R., and Weare, N. M., 1974, Nitrogen fixation by *Anabaena cylindrica*. III. Hydrogen-supported nitrogenase activity, *Arch. Microbiol.* **101:**401–408.

Benemann, J. R., Weismann, J. C., and Oswald, W. J., 1979, Algal biomass, in: *Microbial Biomass, Ecomomic Microbiology*, (Volume 4 (A. H. Rose, ed.)), Academic Press, London, pp. 177–206.

Bliznakov, E. G., and Hunt, G. L., 1987, *The Miracle Nutrient Coenzyme Q_{10}*, Bantam Books, New York.

Bloor, S., and England, R. R., 1991, Elucidation and optimization of the medium constituents controlling antibiotic production by the cyanobacterium *Nostoc muscorum, Enzyme Microb. Technol.* **13**:76–81.

Bollinger, R., Zürrer, H., and Bachofen, R., 1985, Production of molecular hydrogen from waste water of a sugar refinery by photosynthetic bacteria, *Appl. Microbiol. Biotechnol.* **23**:147–151.

Borowitzka, M. A., 1988, Vitamins and fine chemicals from microalgae, in: *Micro-algal Biotechnology* (M. A. Borowitzka and L. J. Borowitzka, eds.), Cambridge University Press, Cambridge, pp. 153–196.

Borowitzka, M. A., and Borowitzka, L. J. (eds.), 1988, *Micro-algal Biotechnology*, Cambridge University Press, Cambridge.

Borowitzka, L. J., and Borowitzka, M. A., 1989, Industrial production: Methods and economics, in: *Algal and Cyanobacterial Biotechnology* (R. C. Cresswell, T. A. V. Rees, and N. Shah, eds.), Longman, Harlow, pp. 294–316.

Bose, P., Nagpal, U. S., Venkataraman, G. S., and Goyal, S. K., 1971, Solubilization of tricalcium phosphate by blue green algae, *Curr. Sci.* **7**:165–166.

Bothe, H., 1982, Hydrogen production by algae, *Experientia* **38**:59–64.

Bothe, H., Nelles, H., Hager, K.-P., Papen, H., and Neuer, G., 1984, Physiology and biochemistry of N_2-fixation by cyanobacteria, in: *Advances in Nitrogen Fixation Research* (C. Veeger and W. E. Newton, eds.), Martinus Nijhoff/Dr. W. Junk, The Hague, pp. 199–210.

Brouers, M., and Hall, D. O., 1986, Ammonium and hydrogen production by immobilized cyanobacteria, *J. Biotechnol.* **3**:307–321.

Burlew, J. S. (ed.), 1953, *Algal Culture from Laboratory to Pilot Plant*, Carnegie Institute, Washington, D. C.

Cannell, R. J. P., Owsianka, A. M., and Walker, J. M., 1988, Results of a large-scale screening programme to detect antibacterial activity from freshwater algae, *Br. Phycol. J.* **23**:41–44.

Cardellina, J. H., Marner, F.-J., and Moore, R. E., 1979a, Seaweed dermatitis: Structure of lyngbyatoxin A, *Science* **204**:193–195.

Cardellina, J. H., Moore, R. E., Arnold, E. V., and Clardy, J., 1979b, Structure and absolute configuration of malyngolide, an antibiotic from the marine blue-green alga *Lyngbya majuscula* Gomont, *J. Org. Chem.* **44**:4039–4042.

Carmeli, S., Moore, R. E., Patterson, G. M. L., Corbett, T. H., and Valeriate, F. A., 1990a, Tantazoles: Unusual cytotoxic alkaloids from the blue-green alga *Scytonema mirabile, J. Am. Chem. Soc.* **112**:8195–8197.

Carmeli, S., Moore, R. E., Patterson, G. M. L., Mori, Y., and Suzuki, M., 1990b, Isonitriles from the blue-green alga *Scytonema mirabile, J. Org. Chem.* **55**:4431–4438.

Carter, D. C., Moore, R. E., Mynderse, J. S., Niemczura, W. P., and Todd, J. S., 1984, Structure of majusculamide C, a cyclic depsipeptide from *Lyngbya majuscula, J. Org. Chem.* **49**:236–241.

Chlorella, 1983, Method of human cell culture, U. S. Patent 4468–460.

Chung, P., Pond, W. C., Kingsburg, J. M., Walker, E. F. and Krook, L., 1978, Production and nutritive values of *Arthospira platensis*, a spiral blue-green alga grown on swine wastes, *J. Anim. Sci.* **47**:319–330.

Chungjatupornchai, W., 1990, Expression of the mosquitocidal-protein genes of *Bacillus thuringiensis* subsp. *israelensis* and the herbicide-resistance gene *bar* in Synechocystis PCC 6803, *Curr. Microbiol.* **21**:283–288.

Ciferri, O., Tiboni, O., and Sanagelantoni, A. M., 1989, The genetic manipulation of cyanobacteria and its potential uses, in: *Algal and Cyanobacterial Biotechnology* (R. C. Cresswell, T. A. V. Rees, and N. Shah, eds.), Longman, Harlow, pp. 239–271.

Clément, Y., and Lanéelle, G., 1986, Glutamate excretion mechanism in *Corynebacterium glu-*

tamicum: Triggering by biotin starvation or by surfactant addition, *J. Gen. Microbiol.* **132:**925–929.

Cork, D., Mathers, J., Maka, A., and Srnak, A., 1985, Control of oxidative sulfur metabolism of *Chlorobium limicola forma thiosulfatophilum*—effect of light energy and molar flow rate, *Appl. Environ. Microbiol.* **49:**269–272.

Cox, J., Kyle, D., Radmer, R., and Delente, J., 1988, Stable-isotope-labeled biochemicals from microalgae, *Trends Biotechnol.* **6:**279–282.

Cresswell, R. C., Rees, T. A. V., and Shah, N. (eds.), 1989, *Algal and Cyanobacterial Biotechnology,* Longman, Harlow.

Daday, A., Platz, R. A., and Smith, G. D., 1977, Anaerobic and aerobic hydrogen gas formation by the blue-green alga *Anabaena variabilis, Appl. Environ. Microbiol.* **34:**478–483.

Daday, A., Mackerras, A. H., and Smith, G. D., 1985, The effect of nickel on hydrogen metabolism and nitrogen fixation in the cyanobacterium *Anabaena cylindrica, J. Gen. Microbiol.* **131:**231–238.

Dadhich, K. S., Varma, A. K., and Venkataraman, G. S., 1969, The effect of *Calothrix* inoculation on vegetable crops, *Plant Soil* **31:**377–379.

De, P. K., 1939, The role of blue-green algae in nitrogen fixation in rice fields, *Proc. R. Soc. Lond. B* **127:**121–139.

De la Noue, J., and Proulx, D., 1988, Tertiary treatment of urban wastewaters by chitosan-immobilized *Phormidium* sp., in: *Algal Biotechnology* (T. Stadler, J. Mollion, M.-C., Verdus, Y. Karamanos, H. Morvan, and D. Christiaen, eds.), Elsevier, London, pp. 159–168.

Dutton, P. L., and Evans, W. C., 1978, Metabolism of aromatic compounds by Rhodospirillaceae, in: *The Photosynthetic Bacteria* (R. K. Clayton and W. R. Sistrom, eds.), Plenum Press, New York, pp. 719–726.

Entzeroth, M., Mead, D. J., Patterson, G. M. L., and Moore, R. E., 1985, A herbicidal fatty acid produced by *Lyngbya aestuarii, Phytochemistry* **24:**2875–2876.

Entzeroth, M., Moore, R. E., Niemczura, W. P. and Patterson, G. M. L., 1986, *O*-Acetyl-*O*-butyl-*O*-carbamoyl-*O,O*-dimethyl-α-cyclodextrins from the cyanophyte *Tolypothrix byssoidea, J. Org. Chem.* **51:**5307–5310.

Ergo-Forsch., 1985, Coenzyme Q$_{10}$ production (ubiquinone), West German Patent 3416–854.

Fallowfield, H. J., and Garrett, M. K., 1985, The treatment of wastes by algal culture, *J. Appl. Bacteriol. Symp. Suppl.* **1985:**187s–205s.

Faulkner, D. J., 1984, Marine natural products: Metabolites of marine algae and herbivorous marine molluscs, *Nat. Prod. Rep.* **1:**251–280.

Fisher, N. S., 1985, Accumulation of metals by marine picoplankton, *Mar. Biol.* **87:**137–142.

Flores, E., and Wolk, C. P., 1986, Production, by filamentous, nitrogen-fixing cyanobacteria, of a bacteriocin and other antibiotics that kill related strains, *Arch. Microbiol.* **145:**215–219.

Fogg, G. E., 1952, The production of extracellular nitrogenous substances by a blue-green alga, *Proc. R. Soc. Lond. B* **139:**372–397.

Fox, R. D., 1988, Nutrient preparation and low cost basin construction for village production of *Spirulina,* in: *Algal Biotechnology* (T. Stadler, J. Mollion, M.-C. Verdus, Y. Karamanos, H. Morvan, and D. Christiaen, eds.), Elsevier, London, pp. 355–364.

Fujiki, H., Suganuma, M., Tahira, T., Yoshioka, A., Nakayasu, M., Endo, Y., Shudo, K., Takayama, S., Moore, R. E., and Sugimura, T., 1984, New class of tumour promoters: Teleocidin, aplysiatoxin, and palytoxin, in: *Cellular Interactions by Environmental Tumour Promoters* (H. Fujiki, E. Hecker, R. E. Moore, T. Sugimura, and I. B. Weinstein, eds.), Japan Scientific Society Press, Tokyo/VNU Science Press, Utrecht, pp. 37–45

Gadd, G. M., 1988, Accumulation of metals by microorganisms and algae, in: *Biotechnology—A Comprehensive Treatise,* Volume 6b (H.-J. Rehm, ed.), VCH Verlagsgesellschaft, Weinheim, pp. 401–433.

Gaisford, W. C., and Rawson, D. M., 1989, Biosensors for environmental monitoring, *Measurement Control* **22:**183–186.

Gale, N. L., and Wixson, B. G., 1979, Removal of heavy metals from industrial effluents by algae, *Dev. Ind. Microbiol.* **20:**259–273

Gallon, J. R., and Chaplin, A. E., 1987, *An Introduction to Nitrogen Fixation*, Cassell, London.

Gantar, M., Kerby, N. W., Rowell, P., and Obreht, Z., 1991a, Colonization of wheat (*Triticum vulgare* L.) by N_2-fixing cyanobacteria: I. A survey of soil cyanobacterial isolates forming associations with roots, *New Phytol.* **118:**477–483.

Gantar, M., Kerby, N. W., and Rowell, P., 1991b, Colonization of wheat (*Triticum vulgare* L.) by N_2-fixing cyanobacteria: II. An ultrastructural study, *New Phytol.* **118:**485–492.

Gerwick, W. H., Lopez, A., Van Duyne, G. D., Clardy, J., Ortiz, W., and Baez, A., 1986, Hormothamnione, a novel cytotoxic styrylchromone from the marine cyanophyte *Hormothamnion enteromorphoides* Grunow, *Tetradedron Lett.* **27:**1979–1982.

Gerwick, W. H., Reyes, S., and Alvardo, B., 1987, Two malyngamides from the Caribbean cyanobacterium *Lyngbya majuscula*, *Phytochemistry* **26:**1701–1704.

Gerwick, W. H., Mrozek, C., Moghaddam, M. F., and Agarwal, S. K., 1989, Novel cytotoxic peptides from the tropical marine cyanobacterium, *Hormothamnion enteromorphoides*, 1. Discovery, isolation, and initial chemical and biological characterization of the hormothamnins from wild and cultured material, *Experientia* **45:**115–121.

Glazer, A., and Stryer, L., 1984, Phycofluor probes, *Trends Biochem. Sci.* **8:**423–427.

Gleason, F. K., and Baxa, C. A., 1986, Activity of the natural algicide, cyanobacterin, on eukaryotic microorganisms, *FEMS Microbiol. Lett.* **33:**85–88.

Gleason, F. K., and Case, D. E., 1986, Activity of the natural algicide, cyanobacterin, on angiosperms, *Plant Physiol.* **80:**834–837.

Gleason, F. K., and Paulson, J. L., 1984, Site of action of the natural algicide, cyanobacterin, in the blue-green alga, *Synechococcus* sp., *Arch. Microbiol.* **138:**273–277.

Glombitza, K.-W. and Koch, M., 1989, Secondary metabolites of pharmaceutical potential, in: *Algal and Cyanobacterial Biotechnology* (R. C. Cresswell, T. A. V. Rees, and N. Shah, eds.), Longman, Harlow, pp. 161–238.

Goldman, J. C., 1979, Outdoor algal mass cultures. I. Applications, *Water Res.* **13:**1–19.

Greene, P. J., Ballard, B. T., Stephenson, F., Kohr, W. J., Rodriguez, H., Rosenberg, J. M., and Boyer, H. W., 1988, Purification and characterization of the restriction endonuclease *Rsr*I, an isoschizomer of *Eco*RI, *Gene* **68:**43–52.

Gudin, C., and Thepenier, C., 1986, Bioconversion of solar energy into organic chemicals by microalgae, *Adv. Biotechnol. Processes* **6:**73–110.

Gunnison, D., and Alexander, M., 1975, Resistance and susceptibility of algae to decomposition by various microbial communities, *Limnol. Oceanogr.* **20:**64–70.

Gusev, M. V. and Korhenevskaya, G., 1990, Artificial Associations, in: *CRC Handbook of Symbiotic Cyanobacteria* (A. N. Raj, ed.), CRC Press, Boca Raton, Florida, pp. 173–230.

Gustafson, K. R., Cardellina, J. H., Fuller, R. W., Weislow, S., Kiser, R. F., Snader, K. M., Patterson, G. M. L., and Boyd, M. R., 1989, AIDS-antiviral sulfolipids from cyanobacteria (blue-green algae), *J. Natl. Cancer Inst.* **81:**1255–1258.

Hageman, R. V., and Burris, R. H., 1980, Electron allocation to alternative substrates of *Azotobacter* nitrogenase is controlled by the electron flux through dinitrogenase, *Biochim. Biophys. Acta* **591:**63–75.

Hall, D. O., Affolter, D. A., Brouers, M., Shi, D. J., Wang, L. W., and Rao, K. K., 1985, Photobiological production of fuels and chemicals by immobilized algae, *Proc. Phytochem. Soc. Eur.* **26:**161–185.

Hall, G., Flick, M. B., and Jensen, R. A., 1980, Approach to the recognition of regulatory mutants of cyanobacteria, *J. Bacteriol.* **143:**981–988.

Hallenbeck, P. C., 1987, Molecular aspects of nitrogen fixation by photosynthetic prokaryotes, *CRC Critical Rev. Microbiol.* **14:**1–48.

Harwood, C. S., and Gibson, J., 1988, Anaerobic and aerobic metabolism of diverse aromatic compounds by the photosynthetic bacterium *Rhodopseudomonas palustris*, *Appl. Environm. Microbiol.* **54:**712–717.

Hashimoto, S., and Furukawa, K., 1989, Nutrient removal from secondary effluent by filamentous algae, *J. Ferment. Bioeng.* **67**:62–69.

Haystead, A., Robinson, R., and Stewart, W. D. P., 1970, Nitrogenase activity in extracts of heterocystous and non-heterocystous blue-green algae, *Arch. Mikrobiol.* **74**:235–243.

Hien, N. T., Kerby, N. W., Machray, G. C., Rowell, P., and Stewart, W. D. P., 1988, Expression of glutamine synthetase in mutant strains of the cyanobacterium *Anabaena variabilis* which liberate ammonia, *FEMS Microbiol. Lett.* **56**:337–342.

Hirayama, O., and Katsuta, Y., 1988, Stimulation of vitamin B_{12} in *Rhodospirillum rubrum* G-9 BM, *Agric. Biol. Chem.* **52**:2949–2951.

Horikoshi, T., Nakajima, A., and Sakaguchi, T., 1979, Uptake of uranium from sea water by *Synechococcus elongatus*, *J. Ferment. Technol.* **57**:191–194.

Houchins, J. P., 1984, The physiology and biochemistry of hydrogen metabolism in cyanobacteria, *Biochim. Biophys. Acta* **768**:227–255.

Houchins, J. P., and Burris, R. H., 1981a, Occurrence and localization of two distinct hydrogenases in the heterocystous cyanobacterium *Anabaena* sp. strain 7120, *J. Bacteriol.* **146**:209–214

Houchins, J. P., and Burris, R. H., 1981b, Comparative characterization of two distinct hydrogenases from *Anabaena* sp. strain 7120, *J. Bacteriol.* **146**:215–221.

Institute of Gas Technology, 1987, Removing sulfur compounds and carbon oxides from gas streams, U. S. Patent 4666–852.

Ishibashi, M., Moore, R. E., and Patterson, G. M. L., 1986, Sctophycins, cytotoxic and antimycotic agents from a cyanophyte *Scytonema pseudohofmanni*, *J. Org. Chem.* **51**:5300–5306

Jeanfils, J., and Loudeche, R., 1986, Photoproduction of ammonia by immobilized heterocystic cyanobacteria. Effect of nitrite and anaerobiosis, *Biotechnol. Lett.* **8**:265–270.

Jensen, R. A., and Hall, G. C., 1982, Endo-oriented control of pyramidally arranged metabolic branch points, *Trends Biochem. Sci.* **7**:177–185.

Jensen, T. E., Baxter, M., Rachlin, J. W., and Jani, V., 1982, Uptake of heavy metals by *Plectonema boryanum* (Cyanophyceae) into cellular components, especially polphosphate bodies: An X-ray energy dispersive study, *Environ. Pollut. A* **27**:119–127.

Karuna-Karan, A., 1987, Product formulations from commercial scale culture of microalgae, in: *World Biotech Report*, Volume 1, Part 4, Online, London, pp. 37–44.

Kellam, S. J., Cannell, R. J. P., Owsianka, A. M., and Walker, J. M., 1988, Results of large-scale screening programme to detect antifungal activity from marine and freshwater microalgae in laboratory culture, *Br. Phycol. J.* **23**:45–47.

Kentemich, T., Danneberg, G., Hundeshagen, B., and Bothe, H., 1988, Evidence for the occurrence of the alternative vanadium-containing nitrogenase in the cyanobacterium *Anabaena variabilis*, *FEMS Microbiol. Lett.* **51**:19–24.

Kerby, N. W., Musgrave, S. C., Codd, G. A., Rowell, P., and Stewart, W. D. P., 1983, Photoproduction of ammonia by immobilized cyanobacteria, in: *Biotech '83 Proceedings of the International Conference on the Commercial Applications and Implications of Biotechnology*, Online, Northwood, pp. 1029–1036.

Kerby, N. W., Musgrave, S. C., Shestakov, S. V., Rowell, P., and Stewart, W. D. P., 1986, Photoproduction of ammonium by immobilized mutant strains of *Anabaena variabilis*, *Appl. Microbiol. Biotechnol.* **24**:42–46.

Kerby, N. W., Niven, G. W., Rowell, P., and Stewart, W. D. P., 1987, Photoproduction of amino acids by mutant strains of N_2-fixing cyanobacteria, *Appl. Microbiol. Biotechnol.* **25**:547–552.

Kerby, N. W., Niven, G. W., Rowell, P., and Stewart, W. D. P., 1988, Ammonia and amino acid production by cyanobacteria, in: *Algal Biotechnology* (T. Stadler, J. Mollion, M.-C. Verdus, Y. Karamanos, H. Morvan, and D. Christiaen, eds.), Elsevier, London, pp. 277–286.

Kerby, N. W., Rowell, P., and Stewart, W. D. P., 1989, The transport, assimilation and production of nitrogenous compounds by cyanobacteria and microalgae, in: *Algal and Cyanobacterial Biotechnology* (R. C. Cresswell, T. A. V. Rees, and N. Shah, eds.), Longman, Harlow, pp. 50–90.

Kerby, N. W., Rowell, P., and Reglinski, A., 1990, Characterization of ammonia analogue resistant mutants of the cyanobacterium *Anabaena variabilis*, in: *Inorganic Nitrogen Uptake and Metabolism in Plants and Microorganisms* (W. R. Ullrich, C. Rigano, A. Fuggi, and P. J. Aparicio, eds.), Springer-Verlag, Berlin, pp. 106–112.

Kim, J. S., Ito, K., Izaki, K., and Takahashi, H., 1987a, Production of molecular hydrogen by a semi-continuous outdoor culture of *Rhodopseudomonas sphaeroides*, *Agric. Biol. Chem.* **51:**1173–1174.

Kim, J. S., Ito, K., Izaki, K., and Takahashi, H., 1987b, Production of molecular hydrogen by a continuous culture under laboratory conditions, *Agric. Biol. Chem.* **51:**2591–2593.

Kobayashi, M., and Kurata, S., 1978, The mass culture and cell utilization of photosynthetic bacteria, *Process Biochem.* **13:**27–30.

Kohlhase, M., and Pohl, P., 1988, Saturated and unsaturated sterols of nitrogen-fixing blue-green algae (cyanobacteria), *Phytochemistry* **27:**1735–1740.

Kulasooriya, S. A., Seneviratne, P. R. G., De Silva, W. S. A. G., Abeysekera, S. W., Wijesundra, C., and De Silva, A. P., 1988, Isotopic studies on N_2-fixation and the availability of its nitrogen to rice, *Symbiosis* **6:**151–166.

Kuwada, Y., and Ohata, Y., 1987, Hydrogen production by an immobilized cyanobacterium, *Lyngbya* sp., *J. Ferment. Technol.* **65:**597–602.

Kyowa-Hakko, 1985, Process for producing coenzyme Q_{10}, Japanese Patent J6 0075-293.

Kyowa-Hakko, 1986, Process for producing coenzyme Q_{10}, Japanese Patent J6 0256-390.

Labarre J., Thuriaux, P., and Chauvat, F., 1978, Genetic analysis of amino acid transport in the facultatively heterotrophic cyanobacterium *Synechocystis* sp. strain 6803, *J. Bacteriol.* **169:**4668–4673.

Ladha, J. K., Rowell, P., and Stewart W. D. P., 1978, Effects of 5-hydroxylysine on acetylene reduction and NH_4^+ assimilation in the cyanobacterium *Anabaena cylindrica*, *Biochem. Biophys. Res. Commun.* **83:**688–696.

Lambert, G. R., and Smith, G. D., 1981, The hydrogen metabolism of cyanobacteria, *Biol. Rev.* **56:**589–660.

Latorre, C., Lee, J. H., Spiller, H., and Shanmugam, K. T., 1986, Ammonium ion excreting cyanobacterial mutant as a source of nitrogen for the growth of rice: A feasibility study, *Biotechnol. Lett.* **8:**507–512.

Laufer, L., Gutcho, S., Castro, T., and Grennen, R., 1964, Preparation of radioactive biochemicals by use of yeast, *Biotechnol. Bioeng.* **6:**127–146.

Lea, P. J., Joy, K. W., Ramos, J. L., and Guerrero, M. G., 1984, The action of 2-amino-4-(methylphosphinyl)-butanoic acid (phosphinothricin) and its 2-oxo-derivative on the metabolism of cyanobacteria and higher plants, *Phytochemistry* **23:**1–6.

Lee, Y.-K., 1986, Enclosed bioreactors for the mass cultivation of phytosynthetic microorganisms: The future trend, *Trends Biotechnol.* **4:**186–189.

Lenaz, G., 1985, *Coenzyme Q: Biochemistry, Bioenergetics and Clinical Applications of Ubiquinone,* Wiley, New York.

Litchfield, J. H., 1983, Single cell proteins, *Science* **219:**740–746.

Lorenz, M. G., and Krumbein, W. E., 1985, Uranium mobilization from low-grade ore by cyanobacteria, *Appl. Microbiol. Biotechnol.* **21:**374–377.

Lumpkin, T. A., and Plucknett, D. L., 1982, *Azolla as a Green Manure. Use and Management in Crop Production,* Westview Press, Bowker Publishing Co., Epping.

MacRae, I. C., 1985, Removal of pesticides in water by microbial cells adsorbed to magnetite, *Water Res.* **19:**825–830.

MacRae, I. C., 1986, Removal of chlorinated hydrocarbons from water and wastewater by bacterial cells adsorbed to magnetite, *Water Res.* **20:**1149–1152.

Mao, X.-Y., Miyake, J., and Kawamura, S., 1986, Screening photosynthetic bacteria for hydrogen production from organic acids, *J. Ferment. Technol.* **64:**245–249.

Martinez, A., Llama, M. J., Alana, A., and Serra, J. L., 1989, Sustained photoproduction of ammonia from nitrate or nitrite by permeabilized cells of the cyanobacterium *Phormidium laminosum, J. Photochem. Photobiol. B* **3:**269–279.

Mason, C. P., Edwards, K. R., Carlson, R. E., Pignatello, J., Gleason, F. K., and Wood, J. M., 1982, Isolation of chlorine-containing antibiotic from the freshwater cyanobacterium *Scytonema hofmanni*, *Science* **215**:400–402.

Matsunaga, T., and Izumida, H., 1984, Seawater-based methane production from blue-green algae biomass by marine bacteria coculture, *Biotechnol. Bioeng. Symp.* **14**:407–418.

Matsunaga, T., Nakamura, N., Tsuzaki, N., and Takeda, H., 1988, Selective production of glutamate by an immobilized marine blue-green alga, *Synechococcus* sp., *Appl. Microbiol. Biotechnol.* **28**:373–376.

Metting, B., and Pyne, J. W., 1986, Biologically active compounds from microalgae, *Enzyme Microb. Technol.* **8**:386–394.

Mian, M. H., and Stewart, W. D. P., 1985, Fate of nitrogen applied as *Azolla* and blue-green-algae (cyanobacteria) in waterlogged rice soils—A ^{15}N tracer study, *Plant Soil* **83**:363–370.

Mitsubishi Gas and Chemicals, 1986, Production of coenzyme Q_{10}, Japanese Patent J6 1192–294.

Mitsui, A., Phlips, E. J., Kumazawa, S., Reddy, K. J., Ramachandran, S., Matsunaga, T., Haynes, L., and Ikemoto, H., 1983, Progress in research toward outdoor biological hydrogen production using solar energy, sea water, and marine photosynthetic microorganisms, *Ann. N. Y. Acad. Sci.* **413**:515–530.

Miyake, J., Mao, X.-Y., and Kawamura, S., 1984, Photoproduction of hydrogen from glucose by a co-culture of a photosynthetic bacterium and *Clostridium butyricum*, *J. Ferment. Technol.* **62**:531–535.

Miyamoto, K., Ohata, S., Nawa, Y., Mori, Y., and Miura, Y., 1987, Hydrogen production by a mixed culture of a green alga, *Chlamydomonas reinhardtii* and *a photosynthetic bacterium, Rhodospirillum rubrum*, *Agric. Biol. Chem.* **51**:1319–1324.

Moore, B. S., Chen, J. L., Patterson, M. L., and Moore, R. E., 1991, Paracyclophanes from blue-green algae, *J. Am. Chem. Soc.*, **112**:4061–4063.

Moore, R. E., 1982, Toxins, anticancer agents, and tumour promoters from marine prokaryotes, *Pure Appl. Chem.* **54**:1919–1934.

Moore, R. E., and Patterson, G. M. L., 1986, Hapalindoles, European Patent Application EP 171.283.

Moore, R. E., Cheuk, C., and Patterson, G. M. L., 1984, Hapalindoles: New alkaloids from the blue-green alga *Hapalosiphon fontinalis*, *J. Am. Chem. Soc.* **106**:6456–6457.

Moore, R. E., Patterson, G. M. L., Mynderse, J. S., Barchi, J. J., Norton, T. R., Furusawa, E., and Furusawa, S., 1986, Toxins from cyanophytes belonging to the Scytonemataceae, *Pure Appl. Chem.* **58**:263–271.

Moore, R. E., Patterson, G. M. L., and Carmichael, W. W., 1988, New pharmaceuticals from cultured blue-green algae, *Mem. Calif. Acad. Sci.* **1988**:143–150.

Mortensen, L. E., 1978, The role of dihydrogen and hydrogenase in nitrogen fixation, *Biochimie* **60**:219–223.

Musgrave, S. C., Kerby, N. W., Codd, G. A., and Stewart, W. D. P., 1982, Sustained ammonia production by immobilized filaments of the nitrogen-fixing cyanobacterium *Anabaena* 27893, *Biotechnol. Lett.* **4**:647–652.

Musgrave, S. C., Kerby, N. W., Codd, G. A., Rowell, P., and Stewart, W. D. P., 1983, Reactor types for the utilization of immobilized photosynthetic microorganisms, *Process Biochem.* (Suppl.) **1983**:184–190.

Mynderse, J. S., Moore, R. E., Kashiwagi, M., and Norton, T. R., 1977, Antileukemia activity in the Oscillatoriaceae: Isolation of debromoaplysiatoxin from *Lyngbya*, *Science* **196**:538–540.

Nejedly, Z., Filip, J., and Grunberger, D., 1968, Preparation of [14]Clabelled nucleic acid components of high specific activity from *Chlorella pyrenoidosa*, in: *Proceedings of the Second International Conference on Methods of Preparing and Storing Labelled Compounds* (J. Sirchia, ed.), European Atomic Energy Commission, Brussels, pp. 527–536.

Nikandrov, V. V., Shlyk, M. A., Gogotov, I. N., and Krasnovsky, A. A., 1988, Efficient photoinduced electron transfer from inorganic semiconductor TiO_2 to bacterial hydrogenase, *FEBS Lett.* **234:**111–114.

Niven, G. W., Kerby, N. W., Rowell, P., Foster, C. A., and Stewart, W. D. P., 1988a, The effect of detergents on amino acid liberation by the N_2-fixing cyanobacterium *Anabaena variabilis, J. Gen. Microbiol.* **134:**689–695.

Niven, G. W., Kerby, N. W., Rowell, P., and Stewart, W. D. P., 1988b, The regulation of aromatic amino acid biosynthesis in amino acid liberating mutant strains of *Anabaena variabilis, Arch. Microbiol.* **150:**272–277.

Noparatnaraporn, N., Wongkornchawalit, W., Kantachote, D., and Nagai, S., 1986a, SCP production of *Rhodopseudomonas sphaeroides* on pineapple wastes, *J. Ferment. Technol.* **64:**137–143.

Noparatnaraporn, N., Sasaki, K., Nishizawa, Y., and Nagai, S., 1986b, Stimulation of vitamin B_{12} formation in aerobically-grown *Rhodopseudomonas gelatinosa* under microaerobic condition, *Biotechnol. Lett.* **8:**491–496.

Noparatnaraporn, N., Trakulnaleumsai, S., Silveira, R. G., Nishizawa, Y., and Nagai, S., 1987, SCP production by a mixed culture of *Rhodocyclus gelatinosus* and *Rhodobacter sphaeroides* from cassava waste, *J. Ferment. Technol.* **65:**11–16.

Norton, R. S., and Wells, R. J., 1982, A series of chiral polybrominated biindoles from the marine blue-green alga *Rivularia firma.* Application of ^{13}C NMR spin–lattice relaxation data and $^{13}C–^{1}H$ coupling constants to structure elucidation, *J. Am. Chem. Soc.* **104:**3628–3635.

Ochiai, H., Shibata, H., Sawa, Y., and Katoh, T., 1980, "Living electrode" as a long-lived photoconverter for biophotolysis of water, *Proc. Natl. Acad. Sci. USA* **77:**2442–2444.

Ochiai, H., Shibata, H., Sawa, Y., Shoga, M., and Ohta, S., 1983, Properties of semiconductor electrodes coated with living films of cyanobacteria, *Appl. Biochem. Biotechnol.* **8:**289–303.

Odom, J. M., and Wall, J. D., 1983, Photoproduction of H_2 from cellulose by an anaerobic bacterial coculture, *Appl. Environ. Microbiol.* **45:**1300–1305.

Oswald, W. J., 1988a, Micro-algae and waste-water treatment, in: *Micro-algal Biotechnology* (M. A. Borowitzka and L. J. Borowitzka, eds.), Cambridge University Press, Cambridge, pp. 305–328.

Oswald, W. J., 1988b, Large-scale algal culture systems (engineering aspects), in: *Micro-algal Biotechnology* (M. A. Brotowitzka and L. J. Borowitzka, eds.), Cambridge University Press, Cambridge, pp. 357–394.

Palmer, C. M., 1969, A composite rating of algae tolerating organic loading, *J. Phycol.* **5:**78–82.

Patterson, M. L., Baldwin, C. L., Bolis, C. M., Caplan, F. R., Karuso, H., Larsen, L. K., Levine, I. A., Moore, R. E., Nelson, C. S., Tschappat, D., and Tuang, G. D., 1991, Antineoplastic activity of cultured blue-green algae (Cyanophyta), *J. Phycol.* **27:**530–536.

Peschek, G. A., 1979a, Aerobic hydrogenase activity in *Anacystic nidulans* the oxyhydrogen reaction, *Biochim. Biophys. Acta* **548:**203–215.

Peschek, G. A., 1979b, Evidence for two functionally distinct hydrogenases in *Anacystis nidulans, Arch. Microbiol.* **123:**81–92.

Pignatello, J. J., Porwoll, J., Carlson, R. E., Xavier, A., Gleason, F. K., and Wood, J. M., 1983, Structure of the antibiotic cyanobacterin, a chlorine-containing τ-lactone from the freshwater cyanobacterium *Scytonema hofmanni, J. Org. Chem.* **48:**4035–4037.

Planchard, A., Mignot, L., Jouenne, T., and Junter, G.-A., 1989, Photoproduction of molecular hydrogen by *Rhodospirillum rubrum* immobilized in composite agar layer/microporous membrane structures, *Appl. Microbiol. Biotechnol.* **31:**49–54.

Polukhina, L. E., Sakhurieva, G. N., and Shestakov, S. V., 1982, Ethylenediamine-resistant *Anabaena variabilis* mutants with derepressed nitrogen-fixing system, *Microbiology* **51:**90–95.

Ponnamperuma, F. N., 1976, Physiochemical properties of submerged soils in relation to fertility, in: *The Fertility of Paddy Soils and Fertilizer Applications.* Compiled by Food and Fertilizer Technology Centre for the Asian and Pacific Regions, Taiwan, pp. 1–27.

Ramos, J. L., Guerrero, M. G., and Losada, M., 1982a, Photoproduction of ammonia from nitrate by *Anacystis nidulans* cells, *Biochim. Biophys. Acta* **679**:323–330.

Ramos, J. L., Guerrero, M. G., and Losada, M., 1982b, Sustained photoproduction of ammonia from nitrate by *Anacystis nidulans, Appl. Environ, Microbiol.* **44**:1020–1025

Ramos, J. L., Guerrero, M. G., and Losada, M., 1984, Sustained photoproduction of ammonia from dinitrogen and water by the nitrogen-fixing cyanobacterium *Anabaena* sp. strain ATCC 33047, *Biotechnol. Bioeng.* **24**:566–571.

Rao, K. K., and Hall, D. O., 1988, Hydrogenases: Isolation and assay, *Meth. Enzymol.* **167**:501–509.

Rawson, D. M., Willmer, A. J., and Cardosi, M. F., 1987, The development of whole cell biosensors for on-line screening of herbicide pollution of surface waters, *Toxicity Assessment* **2**:325–340.

Rawson, D. M., Willmer, A. J., and Turner, A. P. F., 1989, Whole-cell biosensors for environmental monitoring, *Biosensors* **4**:299–311.

Reed, R. H., and Gadd, G. M., 1990, Metal tolerance in eukaryotic and prokaryotic algae, in: *Heavy Metal Tolerance in Plants—Evolutionary Aspects* (J. Shaw, ed.), CRC Press, Boca Raton, Florida, pp. 105–118.

Reed, R. H., Warr, S. R. C., Kerby, N. W., and Stewart, W. D. P., 1986, Osmotic shock-induced release of low molecular weight metabolites from free-living and immobilised cyanobacteria, *Enzyme Microb. Technol.* **8**:101–104.

Reglinski, A., Rowell, P., Kerby, N. W., and Stewart, W. D. P., 1989, Characterization of methylammonium/ammonium transport in mutant strains of *Anabaena variabilis* resistant to ammonium analogues, *J. Gen. Microbiol.* **135**:1441–1451.

Riccardi, G., Sora, S., and Ciferri, O., 1981a, Production of amino acids by analog-resistant mutants of the cyanobacterium *Spirulina platensis, J. Bacteriol.* **147**:1002–1007.

Riccardi, G., Sanangelatoni, D., Carboera, D., Savi, A., and Ciferri, O., 1981b, Characterization of mutants of *Spirulina platensis* resistant to amino acid analogues, *FEMS Microbiol. Lett.* **12**:333–336.

Richmond, A. (ed.), 1986, *CRC Handbook of Microalgal Mass Culture*, CRC Press, Boca Raton, Florida.

Rivera-Ortiz, J. M., and Burris, R. H., 1975, Interactions among substrates and inhibitors of nitrogenase, *J. Bacteriol.* **123**:537–545.

Roberts, R. B., Cowie, D. B., Abelson, P. H., Bolton, E. T., and Britten, R. J., 1955, *Studies of Biosynthesis in Escherichia coli*, Carnegie Institute of Washington, Washington, D.C.

Roberts, R. J., 1987, Restriction and modification enzymes and their isoschizomers, *Nucleic Acid Res.* **15**:r189–r218.

Rodgers, G. A., Bergman, B., Henriksson, E., and Urdis, M., 1979, Utilization of blue green algae as biofertilizers, *Plant Soil* **52**:99–107.

Roger, P. A., and Kulasooriya, S. A., 1980, *Blue-Green Algae and Rice*, IRRI, Manila.

Rother, J. A., Aziz, A., Karim, N. H., and Whitton, B. A., 1988, Ecology of deepwater rice-fields in Bangladesh 4. Nitrogen fixation by blue-green algal communities, *Hydrobiologia* **169**:43–56.

Rowell, P., and Kerby, N. W., 1991, Cyanobacteria and their symbionts in: *Biology and Biochemistry of Nitrogen Fixation* (M. J. Dilworth and A. R. Glenn, eds.), Elsevier, New York, pp. 373–407.

Rowell, P., Rai, A. N., and Stewart, W. D. P., 1985, Studies on the nitrogen metabolism of the lichens *Peltigera aphthosa* and *Peltigera canina*, in: *Lichen Physiology and Cell Biology* (D. H. Brown, ed.), Plenum Press, New York, pp. 145–160.

Roychoudhury, P., Krishnamurti, G. S. R., and Venkataraman, G. S., 1980, Effect of algal inoculation on soil aggregation in rice soils, *Phykos* **19**:224–227.

Sakhurieva, G. N., Polukhina, L. E., and Shestakov, S. V., 1982, Glutamine synthetase in *Anabaena variabilis* mutants with derepressed nitrogenase, *Microbiology* **51**:308–312.

Sasaki, K., Noparatnaraporn, N., Hayashi, M., Nishizawa, Y., and Nagai, S., 1981, Single-cell protein production by treatment of soybean wastes with *Rhodopseudomonas gelatinosa, J. Ferment. Technol.* **59**:471–477.

Sasaki, K., Ikeda, S., Nishizawa, Y., and Hayashi, M., 1987, Production of 5-aminolevulinic acid by photosynthetic bacteria, *J. Ferment. Technol.* **65**:511–515.

Sasaki, K., Tanaka, T., Nishizawa, Y., and Hayashi, M., 1990, Production of a herbicide, 5–aminolevulinic acid, by *Rhodobacter spaeraides* using the effluent of swine waste from an anaerobic digester, *Appl. Microbiol. Biotechnol.* **32**:727–731.

Sawa, Y., Kanayama, K., and Ochiai, H., 1982, Photosynthetic regeneration of ATP using a strain of thermophilic blue-green algae, *Biotechnol. Bioeng.* **24**:305–315.

Sawa, Y., Shindo, H., Nishimura, S., and Ochiai, H., 1986, Photosynthetic glutathione production using intact cyanobacterial cells, *Agric. Biol. Chem.* **50**:1361–1363.

Saxena, P. N., Ahmad, M. R., Shyam, R., and Amla, D. V., 1983, Cultivation of *Spirulina* in sewage for poultry feed, *Experientia* **39**:1077–1083.

Schwartz, R. E., Hirsch, C. F., Springer, J. P., Pettibone, D. J., and Zink, D. L., 1987, Unusual cyclopropane-containing hapalindolinones from a cultured cyanobacterium, *J. Org. Chem.* **52**:3704–3706.

Seiko-Epson, 1986, Cytostatic drug containing polysaccharides, Japanese Patent J6 1158–926.

Shi, D.-J., Brouers, M., Hall, D. O., and Robins, R. J., 1987, The effects of immobilization on the biochemical, physiological and morphological features of *Anabaena azollae, Planta* **172**:298–308.

Shinohara, K., Okura, Y., Koyano, T., Murakami, H., Kim, E.-H., and Omura, H., 1986, Growth-promoting effects of an extract of a thermophillic blue-green alga, *Synechococcus elongatus* var. on human cell lines, *Agric. Biol. Chem.* **50**:2225–2230.

Shinohara, K., Okura, Y., Koyano, T., Murakami, H., and Omura, H., 1988, Algal phyco-cyanins promote growth of human cells in culture, *In Vitro Cell. Dev. Biol.* **24**:1057–1060.

Singh, A. L., and Singh P. K., 1986, Comparative effects of *Azolla* and blue-green algae in combination with chemical N fertilizer on rice crop, *Proc. Indian Acad. Sci. Plant Sci.* **96**:147–152.

Singh, A. L., and Singh, P. K., 1987, Comparative study on *Azolla* and blue-green algae dual culture with rice, *Isr. J. Bot.* **36**:53–61.

Singh, R. N., 1961, *The Role of Blue-Green Algae in Nitrogen Economy of Indian Agriculture*, Indian Council of Agricultural Research, New Delhi.

Spiller, H., Latorre, C., Hassan, M. E., and Shanmugam, K. T., 1986, Isolation and charac-terisation of nitrogenase-derepressed mutant strains of the cyanobacterium *Anabaena variabilis, J. Bacteriol.* **132**:596–603.

Stadler, T., Mollion, J., Verdus, M.-C., Karamanos, Y., Morvan, H., and Christiaen, D. (eds.), 1988, *Algal Biotechnology*, Elsevier, London.

Stanier, R. Y., Pfennig, N., and Trüper, H. G., 1981, Introduction to the phototrophic pro-karyotes, in: *The Prokaryotes*, Vol. 1 (M. P. Starr, H. Stolp, H. G. Trüper, A. Balowa, and H. G. Schlegel, eds.), Springer-Verlag, Berlin, pp. 197–211.

Stewart, J. B., Bornemann, V., Chen, J. L., Moore, R. E., Caplan, F. R., Karuso, H., Larsen, L. K., and Patterson, G. M. L., 1988, Cytotoxic, fungicidal nucleosides from blue green algae belonging to the Scytonemataceae, *J. Antibiot.* **41**:1048–1056.

Stewart, W. D. P., 1963, Liberation of extracellular nitrogen by two nitrogen-fixing blue-green algae, *Nature* **200**:1020–1021.

Stewart, W. D. P., 1980a, Some aspects of structure and function in N_2-fixing cyanobacteria, *Annu. Rev. Microbiol.* **34**:497–536.

Stewart, W. D. P., 1980b, Systems involving blue-green algae (cyanobacteria), in: *Methods for Evaluating Biological Nitrogen Fixation* (F. J. Bergersen, ed.), Wiley, London, pp. 583–635.

Stewart, W. D. P., and Rowell, P., 1975, Effects of L-methionine-D,L-sulphoximine on the assimilation of newly fixed NH_3, acetylene reduction and heterocyst production in *Anabaena cylindrica*, *Biochem. Biophys. Res. Commun.* **65**:846–857.

Stewart, W. D. P., Rowell, P., Ladha, J. K., and Sampaio, M. J. A. M., 1979, Blue-green algae (cyanobacteria)—Some aspects related to their role as sources of fixed nitrogen in paddy soils, in: *Proceedings of Nitrogen and Rice Symposium*, IRRI, Manila, pp. 263–283.

Stewart, W. D. P., Rowell, P., and Rai, A. N., 1983, Cyanobacteria–eukaryotic plant symbioses, *Ann. Microbiol. (Inst. Pasteur)* **134B**:205–228.

Strayer, R. F., Baska, D. F., and Knott, W. M., 1985, Biological hydrogen production as a potential renewable fuel source for the Shuttle Transportation System, *Abstr. Ann. Meet. Am. Soc. Microbiol.* **1985**:792.

Subramanian, G., and Shanmugasundaram, S., 1986, Uninduced ammonia release by the nitrogen-fixing cyanobacterium *Anabaena*, *FEMS Microbiol. Lett.* **37**:151–154.

Takakuwa, S., Odom, J. M., and Wall, J. D., 1983, Hydrogen uptake deficient mutants of *Rhodopseudomonas capsulata*, *Arch. Microbiol.* **136**:20–25.

Tanaka, K., Tamamushi, R., and Ogawa, T., 1985, Bioelectrochemical fuel-cells operated by the cyanobacterium *Anabaena variabilis*, *J. Chem. Technol. Biotechnol.* **35B**:191–197.

Tanaka, K., Kashiwagi, N., and Ogawa, T., 1988, Effects of light on the electrical output of bioelectrochemical fuel cells containing *Anabaena variabilis* M-2: mechanisms of the post-illumination burst, *J. Chem. Technol. Biotechnol.* **42**:235–240.

Tandeau de Marsac, N., de la Torre, F., and Szulmajster, J., 1987, Expression of the larvicidal gene of *Bacillus sphaericus* 1593M in the cyanobacterium *Anacystic nidulans* R2, *Mol. Gen. Genet.* **209**:396–398.

Tapie, P., and Bernard, A., 1988, Microalgae production: Technical and economic evaluations, *Biotechnol. Bioeng.* **32**:873–885.

Terry, K. L., and Raymond, L. P., 1985, System design for the autotrophic production of microalgae, *Enzyme Microb. Technol.* **7**:474–487.

Thomas, S. P., Zaritsky, A., and Boussiba, S., 1990, Ammonium excretion by an L-methionine-DL-sulfoximine-resistant mutant of the rice field cyanobacterium *Anabaena siamensis*, *Appl. Environ. Microbiol.* **56**:3499–3504.

Tovey, K. C., Spiller, G. H., Oldham, K. G., Lucas, N., and Carr, N. G., 1974, A new method for the preparation of uniformly ^{14}C-labelled compounds by using *Anacystis nidulans*, *Biochem. J.* **142**:47–56.

Tsur, Y., and Hochman, E., 1986, Economic aspects of the management of algal production, in: *CRC Handbook of Microalgal Mass Culture* (A. Richmond, ed.), CRC Press, Boca Raton, Florida, pp. 473–483.

United Kingdom Atomic Energy Authority, 1973, Improvements in or relating to ^{14}C-labelled compounds, UK Patent GB 1342098.

University of Dundee, 1989, Production of organic compounds, International Patent Application PCT/GB88/00510.

Van der Oost, J., Kannevorff, W. A., Krab, K., and Kraayenhof, R., 1987, Hydrogen metabolism of three unicellular nitrogen-fixing cyanobacteria, *FEMS Microbiol. Lett.* **48**:41–45.

Vázquez, E., Buzaleh, A. M., Wider, E., and Batlle, A. M. C., 1988, Soluble and immobilized *Rhodopseudomonas palustris* rhodanese: Optimal conditions, *Biotechnol. Appl. Biochem.* **10**:131–136.

Venkataraman, G. S., and Neelakantan, S., 1967, Effect of cellular constituents of the nitrogen-fixing blue-green alga *Cylindrospermum muscicola* on the growth of rice seedlings, *J. Gen. Microbiol.* **13**:53–58.

Venkataraman, L. V., 1986, Blue-green algae as biofertilizer, in: *CRC Handbook of Microalgal Mass Culture* (A. Richmond, ed.), CRC Press, Boca Raton, Florida, pp. 455–471.

Vincenzini, M., De Philippis, R., Ena, A., and Florenzano, G., 1986, Ammonia photoproduction by *Cyanospira rippkae* cells "entrapped" in dialysis tube, *Experientia* **42**:1040–1043.

Von Felten, P., Zürrer, H., and Bachofen, R., 1985, Production of molecular hydrogen with immobilised cells of *Rhodospirillum rubrum, Appl. Microbiol. Biotechnol.* **23**:15–20.

Vymazal, J., 1987, Toxicity and accumulation of cadmium with respect to algae and cyanobacteria: A review, *Toxicity Assessment* **2**:387–415.

Wall, J. D., and Gest, H., 1979, Derepression of nitrogenase activity in glutamine auxotrophs of *Rhodopseudomonas capsulata, J. Bacteriol.* **137**:1459–1463.

Wall, J. D., Weaver, P. F., and Gest, H., 1975, Genetic transfer of nitrogenase-hydrogenase activity in *Rhodopseudomonas capsulata, Nature* **258**:630–631.

Watanabe, I., and Roger, P. A., 1984, Nitrogen fixation in wetland rice, in: *Current Developments in Biological Nitrogen Fixation* (N. S. Subba Rao, ed.), Edward Arnold, London, pp. 237–276.

Watanabe, I., Ventura, W., Cholitkul, W., Roger, P. A., and Kulasooriya, S. A., 1982, Potential of biological nitrogen fixation in deep water rice, in: *Proceedings of the 1981 International Deepwater Rice Workshop,* IRRI, Manila, pp. 191–200.

Weisshaar, H., and Böger, P., 1983, Nitrogenase activity of the non-heterocystous cyanobacterium *Phormidium foveolarum, Arch. Microbiol.* **136**:270–274.

Whitton, B. A., and Roger, P. A., 1989, Use of blue-green algae and *Azolla* in rice culture, in: *Microbial Inoculation of Crop Plants* (R. Campbell and R. M. Macdonald, eds.), IRL Press, Oxford, pp. 89–100.

Willison, J. C., Jouanneau, Y., Colbeau, A., and Vignais, P. M., 1983, H_2 metabolism in photosynthetic bacteria and relationship to N_2 fixation, *Ann. Microbiol. (Inst. Pasteur)* **134B**:115–135.

Wyman, M., Gregory, R. P. F., and Carr, N. G., 1985, Novel role for phycoerythrin in a marine cyanobacterium, *Synechococcus* strain DC2, *Science* **230**:818–820.

Xiankong, Z., Haskell, J. B., Tabita, R., and Van Baalen, C., 1983, Aerobic hydrogen production by the heterocystous cyanobacteria *Anabaena* spp. strains CA and 1F, *J. Bacteriol.* **156**:1118–1122.

Yamada, K., Kinoshita, S., Tsunoda, T., and Aida, K., 1972, *The Microbiol Production of Amino Acids,* Halstead Press, New York.

Zimmerman, W. J., and Boussiba, S., 1987, Ammonia assimilation and excretion in an asymbiotic strain of *Anabaena azollae* from *Azolla filiculoides* Lam, *J. Plant Physiol.* **127**:443–450.

Zürrer, H., 1982, Hydrogen production by photosynthetic bacteria, *Experientia* **38**:64–67.

Zürrer, H., and Bachofen, R., 1982, Aspects of growth and hydrogen production of the photosynthetic bacterium *Rhodospirillum rubrum* in continuous culture. *Biomass* **2**:165–174.

Index

Organisms are entered in the index by the names used by the authors of each chapter. Where an organism is referred to more than ten times in a particular chapter, only the chapter is indicated rather than individual page numbers. The following abbreviations have been used to refer to particular culture collections: ATCC, American Type Culture Collection; PCC, Pasteur Culture Collection; UTEX, Culture Collection of Algae at the University of Texas; WH, Woods Hole Oceanographic Institution.